MW00845279

LOW-NOISE ELECTRONIC SYSTEM DESIGN

LOW-NOISE ELECTRONIC SYSTEM DESIGN

C. D. MOTCHENBACHER
Consultant
Honeywell Inc., Engineering Fellow

J. A. CONNELLY
Professor and Associate Director
School of Electrical Engineering
Georgia Institute of Technology

A Wiley-Interscience Publication
JOHN WILEY & SONS, INC.
New York / Chichester / Brisbane / Toronto / Singapore

This text is printed on acid-free paper.

Library of Congress Cataloging in Publication Data:

Motchenbacher, C. D., 1931-
Low noise eletronic system design / C. D. Motchenbacher, J. A. Connelly.
p. cm.
Includes index.
ISBN 0-471-57742-1
1. Electronic circuit design. 2. Electronic circuits–Noise.
3. Electronic circut design--Data processing. I. Connelly, J. A. (Joseph Alvin), 1942- . II. Title.
TK7867.M692 1993
621.3815–dc20 92-39598

Printed in the United States of America

10 9 8 7 6 5 4 3 2 1

To Our Fathers,
Chris Motchenbacher
and
Joe Connelly

CONTENTS

PREFACE

This is a new book to replace *Low-Noise Electronic Design* (© 1973). All the relevant topics from the original book are included and updated to today's technology. Since the emphasis has been expanded to address the total system design from sensor to simulation to design, the title has been changed to reflect the new scope.

A significant improvement is the change of technological emphasis. The first book emphasized discrete component technology with extensions to ICs. The new book focuses on IC design concepts with added support for discrete design where necessary. Additionally, considerable theoretical expansion has been included for many of the practical concepts discussed. This makes the new book serve very well as a textbook.

Six completely new chapters have been added to support the current direction of technology. These new chapters cover the use of SPICE and PSpice for low-noise analysis and design, noise in feedback amplifiers which are extensively used in IC designs, noise mechanisms in analog/digital and digital/analog converters, noise models for many popular sensors, power supplies and voltage references, and useful low-noise amplifier designs.

This book is intended for use by practicing engineers and by students of electronics. It can be used for self-study or in an organized classroom situation as a quarter or semester course or short course. A knowledge of electric circuit analysis, the principles of electronic circuits, and mathematics through basic calculus is assumed.

The approach used in this text is practice or design oriented. The material is not a study of noise theory, but rather of noise sources, models, and methods to deal with the every-present noise in electronic systems.

This new book serves the following users:

1. The academic community, where the new book can be used (and has been at Georgia Tech) as a textbook for a one-quarter or one-semester electrical engineering course in low-noise electronic design. Derivations are added for clarity and completeness. Many original problems developed in over 15 years of teaching this course are included at the end of each chapter. Key points are summarized in every chapter.

2. Electronic design engineers who design with integrated circuits or discrete components. The new book describes the fundamental noise mechanisms and introduces useful noise models. It shows the details of how feedback can be employed to meet system and circuit specifications. Derivations and modeling approaches are given which are useful for determining the noise in active and passive components as well as in power supplies. This approach is directed toward supporting the total system design concept. Typical, low-noise design examples are provided, analyzed, and discussed. Laboratory techniques for noise measurement and instrumentation complete the coverage.

3. Project engineers who design and build low-noise integrated circuits. The new book contains complete descriptions of the noise models of all important active devices: MOSFETs, JFETs, GaAs FETs, and BJTs. Modeling is done in terms of the SPICE, Gummel-Poon, Curtice, and hybrid-π models. Chapters on low-noise design methodology and single-stage and multistage amplifier design approaches will aid and direct the project design engineers. Furthermore, the chapter on noise measurement will permit them to test and characterize their new devices and new circuit designs.

4. Digital designers who convert very low level analog signals into suitable digital logic levels. The chapters on noise mechanisms, amplifier noise modeling, and sensor noise models enable them to define fundamental noise limits and to specify design requirements. The chapters on low-noise design methodology plus the included sample circuits will enable them to produce functional preamplifier and amplifier interface stages which bridge the analog and digital technologies. Finally, the material on the noise mechanisms in A/D and D/A converters will enable them to evaluate and solve the critical analog–digital interface and partitioning problems.

There is no other book in the present market that directly competes with *Low-Noise Electronic System Design*. Usually, noise in ICs is addressed in textbooks as a section or maybe as a complete chapter. Industrial IC manuals treat noise mostly from a specification and test point of view. The PSpice manual produced by MicroSim Corporation for use with their simulation programs only covers noise modeling. One can find trade journal articles and application notes on low noise design methodology. Digital noise is now being addressed more in the literature. Noise measurement is covered in some textbooks and application notes. But *all* these subjects are covered in *Low-Noise Electronic System Design*. In addition, this book pulls together the whole subjects of noise combined with design. It provides descriptions of noise sources and noise models, addresses the practical problems of circuit design for low-noise employing negative feedback, filtering, component noise, measurement techniques and instrumentation, and finally gives many examples of practical amplifier designs.

In summary, *Low-Noise Electronic System Design* is a textbook for the study of low-noise design, an IC design textbook, and a design manual covering the complete subject of noise from theory, to modeling, to design, to final application.

This book is an outgrowth of our many years of research, development, design, and teaching experience. Special recognition is given to Honeywell, Inc., and the Georgia Institute of Technology for their cooperation and support.

ACKNOWLEDGMENTS

It would be impossible for us to enumerate all those whose assistance contributed to completing this book. However, there are some who deserve special acknowledgment. We wish to thank our many colleagues at Honeywell for their helpful suggestions and critiques especially Steve Baier, Jeff Haviland, Jim Ravis, Dr. Mike Liu, Dr. Paul Kruse, Scott Crist, Dave Erdmann, and Don Benz.

We wish to give special thanks to the many students over the years in the course "Low-Noise Electronic Design" at Georgia Tech who posed key questions and provided alternative solutions to problems and design specifications. Special thanks go to Tim Holman, Katherine Taylor, Admad Dowlatabadi, Jose Perez, Jeff Hall, Dan Shamanski, Ben Blalock, and Walter Thain.

The encouragement and support of Dr. Roger P. Webb, Electrical Engineering School Director, was essential to the success of this book. Appreciation is expressed to faculty members at Georgia Tech, especially Drs. Phillip E. Allen, Martin M. Brooke, Georgio Casinovi, Thomas E. Brewer, W. Marshall Leach, David R. Hertling, Robert K. Feeney, William E. Sayle, Steve DeWeerth, and Miran Milkovic. Ms. Paula Brooks provided very valuable assistance with the manuscript preparation.

Suggestions from our reviewers were very much appreciated, especially Drs. Albin J. Gasiewski of Georgia Tech, E. J. Kennedy of the University of Tennessee, and Eugene Chenette of the University of Florida. Finally, the support of instrumentation for laboratory measurements and computers for integrated circuit layout and simulation is very gratefully acknowledged from AT&T Bell Laboratories and Hewlett-Packard.

LOW-NOISE ELECTRONIC SYSTEM DESIGN

INTRODUCTION

Sensors, detectors, and transducers are basic to the instrumentation and control fields. They are the "fingers," "eyes," and "ears" that reach out and measure. They must translate the characteristics of the physical world into electrical signals. We process and measure these signals and interpret them to be the reactions taking place in a chemical plant, the environment of an orbiting satellite, or the odor of an onion. An engineering problem often associated with sensing systems is the level of electrical noise generated in the sensor and in the electronic system.

In recent years, new high-resolution sensors and high-performance systems have been developed. All sensors have a basic or limiting noise level. The system designer must interface the sensor with electronic circuitry that contributes a minimum of additional noise. To raise the signal level an amplifier must be designed to complement the sensor. Achieving optimum system performance is the primary consideration of this book.

The following chapters answer several important questions. When given a sensor with specific impedance, signal and noise characteristics, how do you design or select an amplifier for minimum noise contribution, and concurrently maximize the signal-to-noise ratio of the system? If we have a sensor–amplifier system complete with known signal and noise, we must determine the major noise source: Is it sensor, amplifier, or pickup noise? Are we maximizing the signal? Is improvement possible? The methods for design analysis, and solution are provided.

Low Noise Electronic System Design, as the title says, is a study of low-noise design and not a treatise on noise as a physical phenomenon. It is divided into four parts to improve organization and facilitate study.

Fundamental Concepts, Part I, of the book describes the four fundamental sources of noise including excess noise. Useful noise voltage and noise current models for IC's and other amplifiers then follow. This part is concluded with methods for utilizing the noise model with feedback around the amplifier to determine the effects of the feedback components upon the total equivalent input and output noises. These first three chapters provide a complete overview of noise, enabling the designer to understand the subject and to proceed with specifying system noise and design objectives.

Noise Modeling, Part II, covers noise modeling in considerable detail. Noise models of all types of sensors and active devices such as FETs and BJTs are derived. Methods of using SPICE and PSpice for circuit analysis are illustrated, and new techniques for incorporating the preceding noise models into practical circuits are shown. A methodology is developed for the selection of an active device and operating point to provide an optimum noise match for maximum signal-to-noise ratio for any sensor type, sensor impedance, and operating frequency range.

Designing for Low Noise, Part III, addresses the practical ways to design a low-noise circuit. Expressions for equivalent input and output noise voltages and currents are derived, as well as gains of discrete and cascaded stages. Design examples further illustrate the critical elements of the amplifiers. A detailed derivation of the noise model of the popular differential amplifier circuit and others are included. The fundamental noise limits in analog/digital and digital/analog converters are established. Usually limited by digital noise pickup, the final limit of converter resolution is determined by fundamental noise mechanisms. This limit is derived for several popular converter circuits and methods of reducing fundamental noises are shown.

Low-Noise Design Applications, Part IV, focuses on applications of low-noise approaches. It discusses noise mechanisms in passive components, power supplies, and voltage references for low-noise circuits, noise measurement methods using modern instrumentation, and cites several practical design examples.

The appendixes contain much useful noise data on many commercial operational amplifiers, preamplifiers, and discrete devices. Answers are given to many of the problems posed at the end of each chapter.

PART I

FUNDAMENTAL CONCEPTS

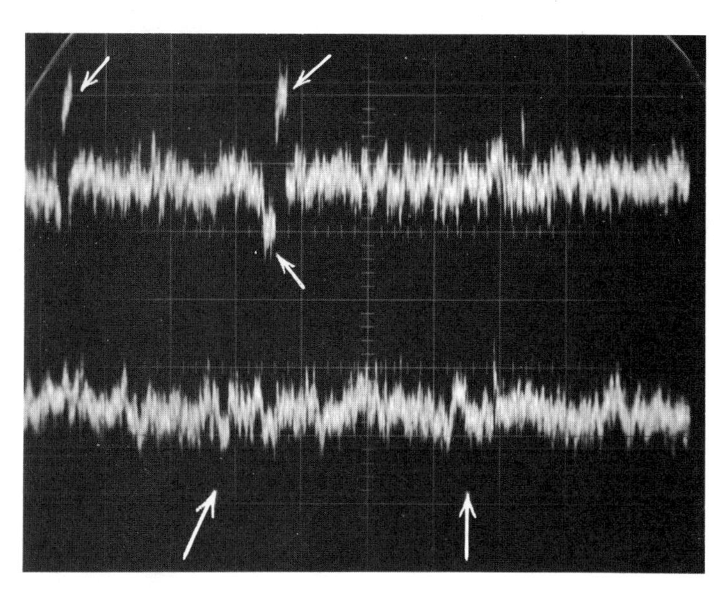

"Popcorn noise," discussed in Chap. 5, is shown in the traces. The top trace is considered to represent a moderate level of this noise. The bottom trace is a low level. Some devices exhibit popcorn noise with five times the amplitude shown in the top trace. Horizontal sensitivity is 2 ms/cm.

CHAPTER 1

FUNDAMENTAL NOISE MECHANISMS

The problems caused by electrical noise are apparent in the output device of an electrical system, but the sources of noise are unique to the low-signal-level portions of the system. The "snow" that may be observed on a television receiver display is the result of internally generated noise in the first stages of signal amplification.

This chapter defines the fundamental types of noise present in electronic systems and discusses methods of representing these sources for the purpose of noise circuit analysis. In addition, concepts such as noise bandwidth and spectral density are introduced.

1-1 NOISE DEFINITION

Noise, in the broadest sense, can be defined as *any unwanted disturbance that obscures or interferes with a desired signal.* Disturbances often come from sources external to the system being studied and may result from electrostatic or electromagnetic coupling between the circuit and the ac power lines, radio transmitters, or fluorescent lights. Cross-talk between adjacent circuits, hum from dc power supplies, or microphonics caused by the mechanical vibration of components are all examples of unwanted disturbances. With the exception of noise from electrical storms and galactic radiation, most of these types of disturbances are caused by radiation from electrical equipment; they can be eliminated by adequate shielding, filtering, or by changing the layout of circuit components. In extreme cases, changing the physical location of the test system may be warranted.

We use the word "noise" to represent basic random-noise generators or spontaneous fluctuations that result from the physics of the devices and materials that make up the electrical system. Thus the thermal noise apparent in all electrical conductors at temperatures above absolute zero is an example of noise as discussed in this book. This fundamental or true noise cannot be predicted exactly, nor can it be totally eliminated, but it can be manipulated and its effects minimized.

Noise is important. The limit of resolution of a sensor is often determined by noise. The dynamic range of a system is determined by noise. The highest signal level that can be processed is limited by the characteristics of the circuit, but the smallest detectable level is set by noise.

In addition to the familiar effects of noise in communication systems, noise is a problem in digital, control, and computing systems. For example, the presence of spikes of random noise makes it difficult to design a circuit that triggers (switches) at a specific signal amplitude. When noise of varying amplitude is mixed with the signal, noise peaks can cause a level detector to trigger falsely. To reduce the probability of false triggering, noise reduction is necessary.

Suppose that we have a system that is too noisy, but are uncertain whether the noisiness is caused by electrical equipment disturbances or by fundamental noise. We add shielding. A general rule for frequencies above 1000 Hz or impedance levels over 1000 Ω is to use conductive shielding (aluminum or copper). For low frequencies and lower impedances, we can use magnetic shielding (super-malloy, mu-metal) and twisted-wire pairs. We can also put the preamplifier on a separate battery supply. If these efforts help, we can try more shielding. The work may be moved to another location, or measurements can be made during the quieter evening hours. If these techniques do not reduce the disturbance, then look to fundamental noise mechanisms. Fundamental or true noise is the type considered almost exclusively in this book.

1-2 NOISE PROPERTIES

Noise is a totally random signal. It consists of frequency components that are random in both amplitude and phase. Although the long-term rms value can be measured, the exact amplitude at any instant of time cannot be predicted. If the instantaneous amplitude of noise could be predicted, noise would not be a problem.

It is possible to predict the randomness of noise. Much noise has a Gaussian or normal distribution of instantaneous amplitudes with time [1]. The common Gaussian curve is depicted in Fig. 1-1 along with a photograph of the associated electrical noise as obtained from an oscilloscope.

The Gaussian distribution predicts the probability of the measured noise signal having a specific value at a specific point in time. A noise signal with a

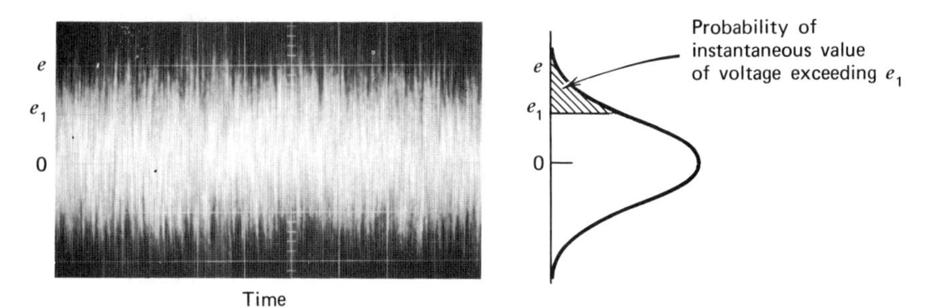

Figure 1-1 Noise waveform and Gaussian distribution of amplitudes.

zero mean Gaussian distribution has the highest probability of having a value of zero at any instant of time. The Gaussian curve is the limiting case produced by overlaying an imaginary coordinate grid structure on the noise waveform. If one could sample a large collection of data points and tally the number of occurrences when the noise voltage level is equal to or greater than a particular level, the Gaussian curve would result. Mathematically, the distribution can be described as

$$f(x) = \frac{1}{\sigma\sqrt{2\pi}} \exp - \left[\frac{(x-\mu)^2}{2\sigma^2} \right] \tag{1-1}$$

where μ is the mean or average value and σ is the standard deviation or root mean square (rms) value of the variable x. The function $f(x)$ is referred to as the probability density function, or pdf.

The area under the Gaussian curve represents the probability that a particular event will occur. Since probability can only take on values from 0 to 1, the total area must equal unity. The waveform is centered about a mean or average voltage level μ, corresponding to a probability of .5 that the instantaneous value of the noise waveform is either above or below μ. If we consider a value such as e_1, the probability of exceeding that level at any instant in time is shown by the cross-hatched area in Fig. 1-1. To a good engineering approximation, common electrical noise lies within $\pm 3\sigma$ of the mean μ. In other words, the peak-to-peak value of the noise wave is less than six times the rms value for 99.7% of the time.

The rms definition is based on the equivalent heating effect. True rms voltmeters measure the applied time-dependent voltage $v(t)$ according to

$$V_{\text{rms}} = \sqrt{\frac{1}{T_p} \int_0^T v^2(t)\, dt} \tag{1-2}$$

where T_p is the period of the voltage. Applying Eq. 1-2 to a sine wave of peak value V_m volts gives the familiar result $V_{rms} = 0.707V_m$.

The most common type of ac volmeter rectifies the wave to be measured, measures the average or dc value, and indicates the rms value on a scale calibrated by multiplying the average value by 1.11 to simulate the rms. This type of meter correctly indicates the rms values of a sine wave, *but noise is not sinusoidal*, and the reading of a noise waveform will be 11.5% low. Correction can be made by multiplying the reading by 1.13. Chapter 15 discusses specific noise measurement instrumentation and these correction factors in much greater detail.

1-3 THERMAL NOISE

Three main types of fundamental noise mechanisms are thermal noise, shot noise, and low-frequency ($1/f$) noise. Thermal noise is the most often encountered and is considered first. The other two types of noise are defined in later sections of this chapter. A special case of thermal noise limited by shunt capacitance called kT/C noise is also defined. Additional discussions of the effects of these types of noise in devices and circuits will be found throughout this book.

Thermal noise is caused by the random thermally excited vibration of the charge carriers in a conductor. This carrier motion is similar to the Brownian motion of particles. From studies of Brownian motion, thermal noise was predicted. It was first observed by J. B. Johnson of Bell Telephone Laboratories in 1927, and a theoretical analysis was provided by H. Nyquist in 1928. Because of their work thermal noise is called Johnson noise or Nyquist noise.

In every conductor or resistor at a temperature above absolute zero, the electrons are in random motion, and this vibration is dependent on temperature. Since each electron carries a charge of 1.602×10^{-19} C, there are many little current surges as electrons randomly move about in the material. Although the average current in the conductor resulting from these movements is zero, instantaneously there is a current fluctuation that gives rise to a voltage across the terminals of the conductor.

The *available noise power* in a conductor, N_t, is found to be proportional to the absolute temperature and to the bandwidth of the measuring system. In equation form this is

$$N_t = kT\Delta f \qquad (1\text{-}3)$$

where k is Boltzmann's constant (1.38×10^{-23} W-s/K), T is the temperature of the conductor in kelvins (K), and Δf is the *noise bandwidth* of the measuring system in hertz (Hz).

At room temperature (17°C or 290 K), for a 1.0-Hz bandwidth, evaluation of Eq. 1-3 gives $N_t = 4 \times 10^{-21}$ W. This is -204 dB when referenced to

1 W. In RF communications, 1 mW is often taken as the reference standard and dB_m is used to indicate this standard.

$$\text{Noise power in } dB_m = 10\log_{10}\left(\frac{4\times10^{-21}}{10^{-3}}\right)$$

$$= -174\ dB_m \tag{1-4}$$

This level of $-174\ dB_m$ is often referred to as the "noise floor" or minimum noise level that is practically achievable in a system operating at room temperature. It is not possible to achieve any lower noise unless the temperature is lowered.

The noise power predicted by Eq. 1-3 is that caused by thermal agitation of the carriers. Other noise mechanisms can exist in a conductor, but they are excluded from consideration here. Thus the thermal noise represents a minimum level of noise in a restrictive element.

In Eq. 1-3 the noise power is proportional to the noise bandwidth. There is equal noise power in each hertz of bandwidth; the power in the band from 1 to 2 Hz is equal to that from 1000 to 1001 Hz. This results in thermal noise being called "white" noise. "White" implies that the noise is made up of many frequency components just as white light is made up of many colors. A Fourier analysis gives a flat plot of noise versus frequency. The comparison to white light is not exact, for white light consists of equal energy per wavelength, not per hertz. Thermal noise ultimately limits the resolution of any measurement system. Even if an amplifier could be built perfectly noise-free, the resistance of the signal source would still contribute noise.

It is considerably easier to measure noise voltage than noise power. Consider the circuit shown in Fig. 1-2. *The available noise power is the power that can be supplied by a resistive source when it is feeding a noiseless resistive load equal to the source resistance.* Therefore, $R_S = R_L$ and $E_o = E_t/2$ represents the true rms noise voltage. The power supplied to R_L is N_t and is given by Eqs. 1-5 and 1-3:

$$N_t = \frac{E_o^2}{R_L} = \frac{E_t^2}{4R_L} = \frac{E_t^2}{4R_S} = kT\,\Delta f \tag{1-5}$$

Solving Eq. 1-5 for E_t, the rms thermal noise voltage E_t of a resistance

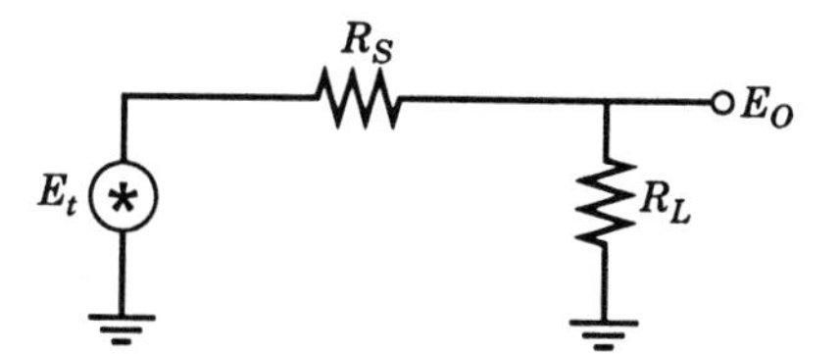

Figure 1-2 Circuit for determination of noise voltage.

$R = R_S$ is*

$$\boxed{E_t = \sqrt{4kTR\,\Delta f}} \qquad (1\text{-}6)$$

where R is the resistance or real part of the conductor's impedance, T is the temperature in kelvins, (room temperature = 17°C = 290 K), and k is Boltzmann's constant (1.38×10^{-23} W-s/K). Solving for $4kT$,

$$4kT = 1.61 \times 10^{-20} \qquad \text{(at 290 K)} \qquad (1\text{-}7)$$

Example 1-1 Using Eq. 1-6, the thermal noise of a 1-kΩ resistor produces a noise voltage of 4 nV rms in a noise bandwidth of 1 Hz. This is a good number to memorize for a reference level. Using 4 nV as a standard, we can easily scale up or down by the square root of the resistance and/or the bandwidth.

This discussion might lead one to consider using a large-valued resistor and a wide bandwidth (which can produce several volts) in series with a diode in an attempt to power a load such as a transistor radio. It should be obvious that this will not work, but can you explain the flaw in the reasoning?

Several important observations can be made from Eq. 1-6. Noise voltage is proportional to the square root of the bandwidth, no matter where the frequency band is centered. Reactive components do not generate thermal noise. The resistance used in the equation is not simply the dc resistance of the device or component, but is more exactly defined as the real part of the complex impedance. In the case of inductance, it may include eddy current losses. For a capacitor, it can be caused by dielectric losses. It is obvious that cooling a conductor decreases its thermal noise.

Equation 1-6 is very important in noise analysis. It provides the noise limit that must always be watched. Today, low-noise amplifying devices are so quiet that system performance is often limited by thermal noise. We shall see that the measure of an amplifier's performance, its signal-to-noise ratio (S/N) and its noise figure (NF) are only measures of the noise the amplifier adds to the thermal noise of the source resistance.

The effect of broadband thermal noise must be minimized. Equation 1-6 implies that there are several practical ways to do this. The sensor resistance must be kept as low as possible, and additional series resistance elements must be avoided. Also, it is desirable to keep the system bandwidth as narrow as possible, while maintaining enough bandwidth to pass the desired signal.

*A more complete expression for thermal noise is $E_t^2 = 4kTRp(f)\,df$, where $p(f)$ is referred to as the Planck factor: $p(f) = (hf/kT)(e^{hf/kt} - 1)^{-1}$. $h = 6.62 \times 10^{-34}$ J-s is Planck's constant. The term $p(f)$ is usually ignored since $hf/kT \ll 1$ at room temperature for frequencies into the microwave band. Therefore, $p(f) = 1$ for most purposes [2].

When designing a system, frequency limiting should be incorporated in one of the later stages. For laboratory applications, frequency limiting is usually obtained with spectrum analyzers or tuned filters. It is normally undesirable to do the frequency limiting at the sensor or the input coupling network. This tends to decrease both the signal and the sensor noise but it does not attenuate the amplifier noise that is generated following the coupling network.

Even though we have shown that there is a time-varying current and available power in every conductor, this is not a new power source! Recalling the previous teaser question, you cannot put a diode in series with a noisy resistor and use it to power a transistor radio. If the conductor were connected to a load (another conductor), the noise power of each would merely be transferred to the other. If a resistor at room temperature were connected in parallel with a resistor at absolute zero, 0 K, there would indeed be a power transfer from the higher temperature resistor to the lower. The warmer resistor would try to cool down and the other would try to warm up until they came into thermal equilibrium. At that point there would be no further power transfer.

Thermal noise has been extensively studied. Expressions are available for predicting the number of maxima per second present in thermal noise, and also the number of zero crossings expected per second present in thermal noise, and also the number of zero crossings expected per second in the noise waveform. These quantities are dependent on the width of the passband. Formulas are given in Prob. 1-17.

1-4 NOISE BANDWIDTH

Noise bandwidth is not the same as the commonly used -3-dB bandwidth. There is one definition of bandwidth for signals and another for noise. The bandwidth of an amplifier or a tuned circuit is classically defined as the frequency span between half-power points, the points on the frequency axis where the signal transmission has been reduced by 3 dB from the central or midrange reference value. A -3-dB reduction represents a loss of 50% in the power level and corresponds to a voltage level equal to 0.707 of the voltage at the center frequency reference.

The noise bandwidth, Δf, is the frequency span of a rectangularly shaped *power* gain curve equal in area to the area of the actual power gain versus frequency curve. Noise bandwidth is the area under the power curve, the integral of power gain versus frequency, divided by the peak amplitude of the curve. This can be stated in equation form as

$$\Delta f = \frac{1}{G_o} \int_0^\infty G(f)\, df \tag{1-8}$$

where $G(f)$ is the power gain as a function of frequency and G_o is the peak power gain. Generally, we only know the frequency behavior of the *voltage* gain of the system and since power gain is proportional to the network voltage gain squared, the equivalent noise bandwidth can also be written as

$$\Delta f = \frac{1}{A_{vo}^2} \int_0^\infty |A_v(f)|^2 \, df \tag{1-9}$$

where A_{vo} is the peak magnitude of the voltage gain and $|A_v(f)|^2$ is the square of the magnitude of the voltage gain over frequency—the square of the magnitude of a Bode plot. Equation 1-9 is a more useful expression for Δf.

The plot shown in Fig. 1-3*a* is typical of a broadband amplifier with maximum gain at dc. The shape of the curve may appear strange because it

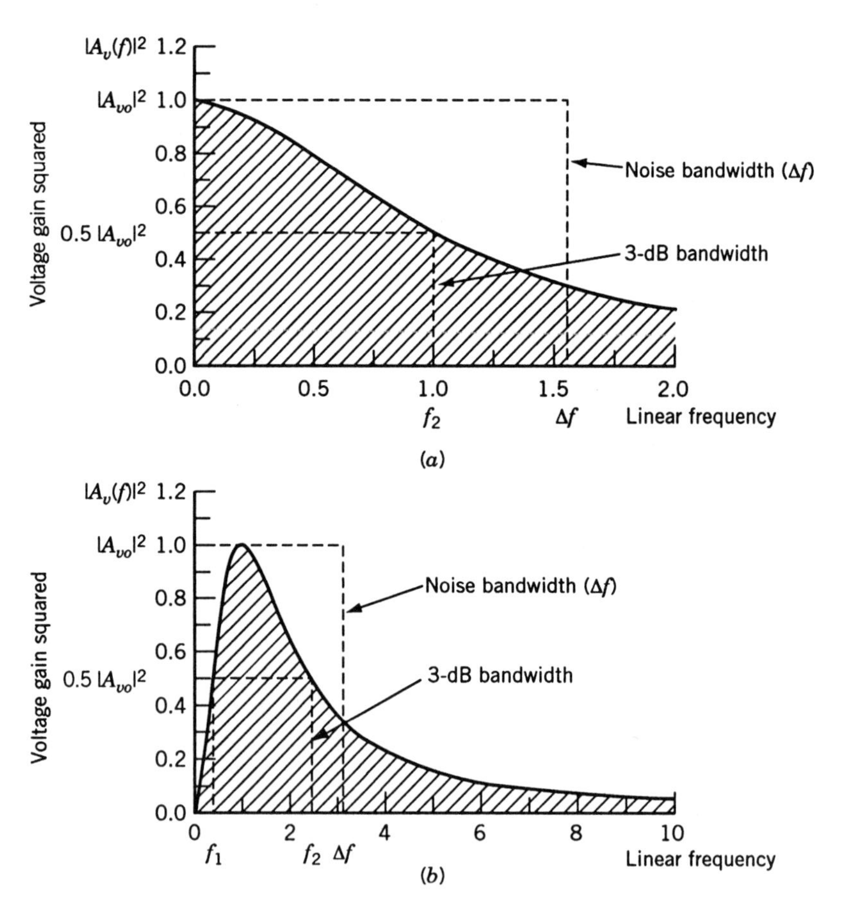

Figure 1-3 Definition of noise bandwidth.

has a linear frequency scale instead of the more common logarithmic scale. If the gain peak does not occur at dc such as with the bandpass amplifier shown in Fig. 1-3*b*, the maximum voltage gain must be found and used to normalize the noise bandwidth calculation. The area of the dashed rectangle is equal to the area of the integration in Eq. 1-9. Thus the noise bandwidth, Δf, is not equal to the half-power or -3-dB bandwidth, f_2. The noise bandwidth will always be greater than f_2.

As an example of Δf determination, consider a first-order low-pass filter whose signal transmission varies with frequency according to

$$A_v(f) = \frac{1}{1 + jf/f_2} \tag{1-10}$$

where f_2 is the conventional -3-dB cutoff frequency and the low-frequency and midband voltage gain has been normalized to unity. The magnitude of the voltage gain is

$$|A_v(f)| = \frac{1}{\sqrt{1 + (f/f_2)^2}} \tag{1-11}$$

Then from Eq. 1-9 the noise bandwidth is

$$\Delta f = \int_0^{\infty} \frac{df}{1 + (f/f_2)^2} \tag{1-12}$$

Now change variables so that

$$f = f_2 \tan\theta \qquad \text{and} \qquad df = f_2 \sec^2\theta \, d\theta$$

The new limits of integration become 0 to $\pi/2$ such that

$$\Delta f = \int_0^{\pi/2} \frac{f_2 \sec^2\theta \, d\theta}{1 + \tan^2\theta}$$

$$\Delta f = f_2 \int_0^{\pi/2} d\theta = \frac{\pi f_2}{2} = 1.571 f_2 \tag{1-13}$$

Note that the noise bandwidth is 57% larger than the conventional -3-dB bandwidth for the first-order low-pass filter.

As a second example consider two identical, first-order, low-pass filters cascaded with appropriate buffering to prevent loading as shown in Fig. 1-4.

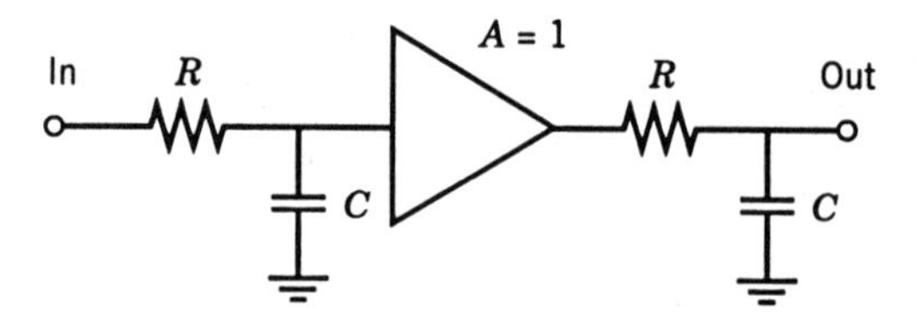

Figure 1-4 Cascaded low-pass filter.

The voltage transfer function for this circuit is

$$A_v(f) = \left| \frac{1}{1 + jf/f_2} \right|^2 \tag{1-14}$$

where f_2 is the -3-dB high-frequency corner of each RC time constant. The noise bandwidth is now

$$\Delta f = \int_0^\infty \left| \frac{1}{1 + (f/f_2)^2} \right|^2 df \tag{1-15}$$

Again making the same change of variable and change of limit substitution,

$$\Delta f = \int_0^{\pi/2} \frac{f_2 \, d\theta}{1 + \tan^2 \theta} = \frac{\pi f_2}{4} = 0.785 f_2 \tag{1-16}$$

However, it must be remembered here that f_2 is the conventional -3-dB cutoff frequency of each stage and not the system's -3-dB cutoff frequency which we will denote as f_a. The amplifier system cutoff frequency can be found from

$$\frac{1}{\sqrt{2}} = \frac{1}{1 + (f_a/f_2)^2} \tag{1-17}$$

Solving gives

$$f_a = 0.6436 f_2 \tag{1-18}$$

Therefore, the noise bandwidth of the amplifier system is

$$\Delta f = \frac{\pi f_2}{4} = \frac{\pi f_a}{4 \times 0.6436} = 1.222 f_a \tag{1-19}$$

Note that the noise bandwidth is 22% larger than the conventional -3-dB bandwidth for the system. As the high-frequency roll-off becomes sharper by

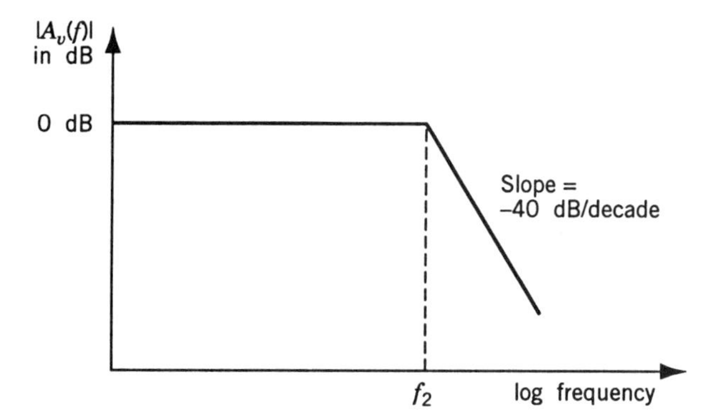

Figure 1-5 Frequency response of the cascaded low-pass filter.

using a large number of cascaded stages, Δf approaches the -3-dB bandwidth. For example, Prob. 1-8 shows that a cascade of three identical low-pass stages produces a Δf 1.155 times the system's -3-dB bandwidth.

Often an exact evaluation of the integration necessary to find the noise bandwidth in a convenient closed form is not possible and an approximation technique becomes necessary. To illustrate one such approximation, consider again the cascaded low-pass filter example of Fig. 1-4. The magnitude of its frequency response is as shown in Fig. 1-5.

The equation for the asymptotic frequency response of the magnitude of $A_v(f)$ in decibels is

$$|A_v(f)|_{\text{dB}} = \begin{cases} 0 \text{ dB} & \text{for} \quad 0 \le f \le f_2 \\ -40\log_{10}(f/f_2) & \text{for} \quad f_2 \le f \le \infty \end{cases} \tag{1-20}$$

The approximate noise bandwidth for the asymptotic response is

$$\Delta f = \int_0^{f_2} df + \int_{f_2}^{\infty} \left(\frac{f_2}{f}\right)^4 df = 1.333 f_2 \tag{1-21}$$

It was found previously in Eq. 1-19 that $\Delta f = 1.222 f_a = 0.785 f_2$ for a double time constant circuit, where f_2 is the time constant of each pole. In this example, the approximate analysis technique introduced an error into the Δf evaluation of $1.333/0.785 = 1.70$ or 70% error, which dramatically shows that this approximation method cannot be used!

Often the frequency response equations become very complicated and it is very difficult to determine the noise bandwidth using direct mathematical integration for large or complex electronic systems. In these cases, other approximation techniques may be successfully employed using numerical integration routines available in MathCAD, DERIVE, or similar computer programs. Alternatively, the noise bandwidth can be found using circuit simulators like SPICE and the analysis techniques explained in Chap. 4. Finally, a graphical approach may also be a viable method. Here $|A_v(f)|^2$ is plotted on linear graph paper and the number of squares underneath the curve counted. The noise bandwidth is found by dividing this total by $|A_{vo}|^2$.

Care must be exercised in measurements pertaining to noise. We must not allow the measuring equipment to change the bandwidth of the system. Also, we must bear in mind that to arrive at useful results, the network response must continue to fall off as we reach higher and higher frequencies.

The term *spectral density* is used to describe the noise content in a 1 Hz unit of bandwidth. It can be related to narrowband noise as will be presented later. Spectral density has units of volts^2 per hertz and is symbolized as $S(f)$ to show that in general it varies with frequency. For a thermal noise source the spectral density $S(f)$ is

$$S(f) = \frac{E_t^2}{\Delta f} = 4kTR \qquad \mathrm{V^2/Hz} \tag{1-22}$$

It is characteristic of white-noise sources that the plot of $S(f)$ versus frequency is a simple horizontal line.

When measuring noise we work with the rms value of a noise quantity. Thus we obtain the spectral density by dividing the mean square value of a noise voltage by the noise bandwidth. If we take the square root of this mathematical operation, it can be interpreted as simply the rms noise voltage in 1 Hz of bandwidth. Note that the square root of the spectral density (symbolized as $E/\sqrt{\Delta f}$) is a quantity that can be measured; the units are volts per hertz$^{1/2}$. Often this density function is symbolized by $E/\sqrt{\sim}$, or in the case of a current, $I/\sqrt{\sim}$, using the cycle symbol ($\sim$) to indicate frequency. This is not correct since frequency is in cycles per second not cycles, so volts per hertz$^{1/2}$ should be used. Since a bandwidth of 1 Hz is almost always used, the units for these functions are referred to as "volts per root hertz" and "amps per root hertz."

Spectral density is a narrowband noise and generally varies with frequency. In order to obtain the *total* wideband noise, the spectral density function must be integrated over the frequency band of interest. Consider the system shown in Fig. 1-6 where a resistor is used as a noise source which is amplified by an ideal bandpass amplifier. The total output noise measured by a true rms voltmeter is given by

$$E_{no}^2 = \int_0^\infty 4kTR|A_v(f)|^2\,df = 4kTR\int_0^\infty |A_v(f)|^2\,df$$

$$E_{no}^2 = 4kTRA_{vo}^2\left[\frac{1}{A_{vo}^2}\int_0^\infty |A_v(f)|^2\,df\right]$$

$$E_{no}^2 = 4kTRA_{vo}^2\,\Delta f \tag{1-23}$$

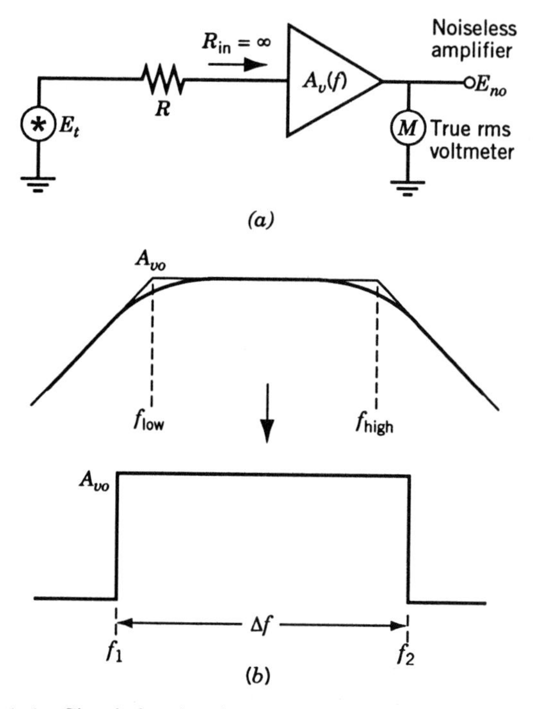

Figure 1-6 Circuit for showing narrowband and wideband noises.

Here E_{no}^2 represents the wideband mean square output noise voltage. The narrowband output noise voltage spectrum or spectral density is given by

$$E_{no}^2/\Delta f = S(f) = 4kTRA_{vo}^2 \tag{1-24}$$

1-5 THERMAL NOISE EQUIVALENT CIRCUITS

In order to perform a noise analysis of an electronic system, every element that generates thermal noise is represented by an equivalent circuit composed of a noise voltage generator in series with a noiseless resistance. Suppose, then, that we have a noisy resistance R connected between terminals a and b. For analysis, we substitute the equivalent shown in Fig. 1-7a, a noiseless resistance of the same ohmic value, and a series noise generator with rms value E_t equal to $(4kTR\,\Delta f)^{1/2}$. This generator is supplying the circuit with multifrequency noise; it is specified by the rms value of its total output. The $*$ symbols representing the voltage and current generators are used for noise sources exclusively.

According to Norton's theorem, the series arrangement shown in Fig. 1-7a can be replaced by an equivalent constant-current generator in parallel with a resistance. The noise current generator I_t will have a rms value equal to

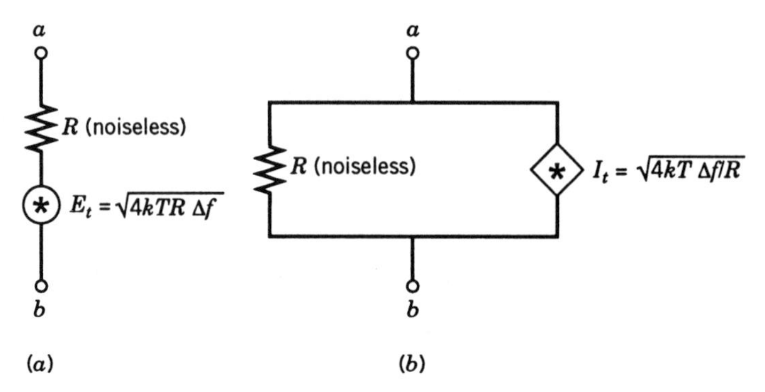

Figure 1-7 Equivalent circuits for thermal noise: (*a*) Thevenin equivalent circuit and (*b*) Norton equivalent circuit.

E_t/R, or in this instance

$$I_t = \sqrt{\frac{4kT\,\Delta f}{R}} = \sqrt{4kTG\,\Delta f} \tag{1-25}$$

where $G = 1/R$ is the conductance in siemens.

If a voltmeter with infinite input impedance and zero self-noise were connected between a and b, the thermal noise voltage could be measured. However, because most voltmeters also contribute noise, a direct reading will usually be too large.

The system of symbols that we employ in noise analysis uses the letters E and I to represent noise quantities. The letter V is reserved for signal voltage. Because noise generators do not have an instantaneous phase characteristic as is attributed to sine waves in the phasor method of representation, no specific polarity indication is included in the noise source symbols in Fig. 1-7. Polarity of noise sources (correlation) is discussed in Chap. 2.

1-6 ADDITION OF NOISE VOLTAGES

When two sinusoidal signal voltage sources of equal amplitude and the *same frequency and phase* are connected in series, the resultant voltage has twice the common amplitude, and combined they can deliver four times the power of one source. If, on the other hand, they differ in phase by 180°, the net voltage and power from the pair is zero. For other phase conditions they may be combined using the familiar rules of phasor algebra.

If two sinusoidal signal voltage sources of different non-harmonic frequencies with rms amplitudes V_1 and V_2 are connected in series, the resultant voltage has an rms amplitude equal to $(V_1^2 + V_2^2)^{1/2}$. The mean square value

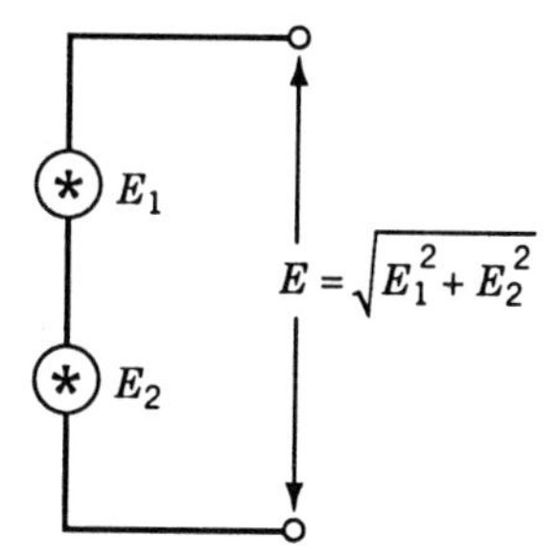

Figure 1-8 Addition of uncorrelated noise voltages.

of the resultant wave V_r^2 is the sum of the mean square values of the components ($V_r^2 = V_1^2 + V_2^2$).

Equivalent noise generators represent a very large number of component frequencies with a random distribution of amplitudes and phases. When independent noise generators are series connected, the separate sources neither help nor hinder one another. The output power is the sum of the separate output powers, and, consequently, it is valid to combine such sources so that the resultant mean square voltage is the sum of the mean square voltages of the individual generators. This statement can be extended to noise current sources in parallel.

The generators E_1 and E_2 shown in Fig. 1-8 represent uncorrelated noise sources. We form the sum of these voltages by adding their mean square values. Thus the mean square of the sum, E^2, is given by

$$E^2 = E_1^2 + E_2^2 \tag{1-26}$$

Taking the square root of the quantity such as E^2 represents the rms. It is not valid to linearly sum the rms voltages of series noise sources, they must be rms summed.

To a good engineering approximation, one can often neglect the smaller of the two noise signals when their rms values are in a 10:1 ratio. In this case, the smaller signal adds less than 1% to the overall voltage. A 3:1 ratio has only a 10% effect on the total. If two resistors are connected in parallel, the total thermal noise voltage is that of the equivalent resistance. Similarly, with two resistors in series, the total noise voltage is determined by the arithmetic sum of the resistances.

As an example, consider two simple circuits composed of resistive elements as shown in Fig. 1-9*a* and *b*. In both circuits we want to determine the output noise voltage, E_{no}, due to the thermal noise of the source resistor or resistors. For simplicity, we neglect the thermal noise of the load resistor. If we apply conventional linear circuit techniques, the 4-nV/Hz$^{1/2}$ thermal noise produced by R_S in Fig. 1-9*a* will be attenuated by a factor of 2 and produce an output noise voltage of

$$E_{no} = R_L E_t/(R_S + R_L) = 0.5(4\ \text{nV/Hz}^{1/2}) = 2\ \text{nV/Hz}^{1/2} \tag{1-27}$$

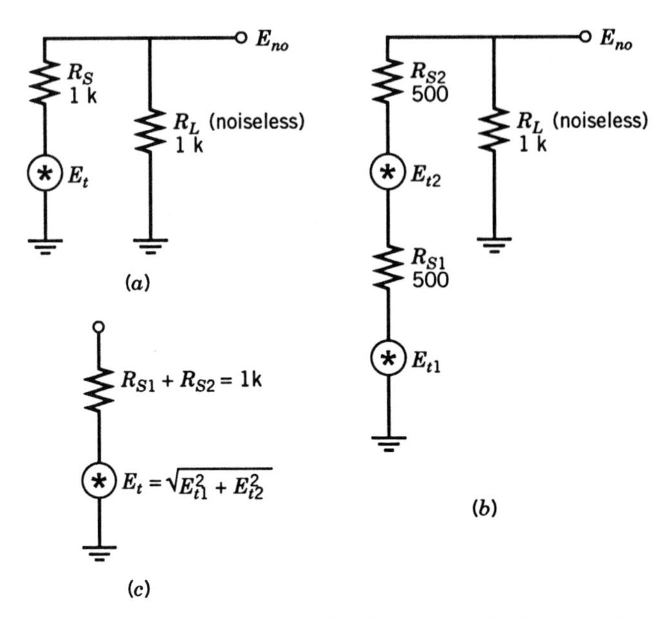

Figure 1-9 Circuits with noise voltages: (*a*) simple circuit, (*b*) equivalent circuit, and (*c*) correct resultant circuit.

If we apply the same linear analysis approach to the equivalent circuit of Fig. 1-9*b* which has two source resistors in series totaling 1 kΩ, we get an entirely different result

$$E_{no} = R_L E_{t1}/(R_{S1} + R_{S2} + R_L) + R_L E_{t2}/(R_{S1} + R_{S2} + R_L)$$

$$= 0.5(2.82 \text{ nV/Hz}^{1/2}) + 0.5(2.82 \text{ nV/Hz}^{1/2}) = 2.82 \text{ nV/Hz}^{1/2} \tag{1-28}$$

The difficulty with the second approach is that noise voltages do not combine in a linear manner and hence the principle of superposition does not apply here. The correct analysis approach for the circuit of Fig. 1-9*b* is

$$E_{no}^2 = R_L^2 E_{t1}^2/(R_{S1} + R_{S2} + R_L)^2 + R_L^2 E_{t2}^2/(R_{S1} + R_{S2} + R_L)^2$$

$$= (0.5)^2(2.82 \text{ nV/Hz}^{1/2})^2 + (0.5)^2(2.82 \text{ nV/Hz}^{1/2})^2$$

$$= (0.5)(2.82 \text{ nV/Hz}^{1/2})^2 = 4 \times 10^{-18} \text{ V}^2/\text{Hz}$$

$$E_{no} = 2 \text{ nV/Hz}^{1/2} \tag{1-29}$$

This example illustrates a common problem when combining noise sources. To avoid this problem, first combine any series or parallel elements into a single equivalent element. Then calculate the noise contribution due to the equivalent element. For example, the two resistors and their noise sources in Fig. 1-9*b* combine to the correct equivalent network as shown in Fig. 1-9*c*. Just remember, if you try adding noise sources such as

$$(E_{t1} + E_{t2})^2 = E_{t1}^2 + E_{t1}^2 + 2E_{t1}E_{t2} \tag{1-30}$$

you will get an extra cross-product term which defines the correlation coefficient which is presented in the next section.

1-7 CORRELATION

When noise voltages are produced independently and there is no relationship between the instantaneous values of the voltages, they are uncorrelated. Uncorrelated voltages are treated according to the discussion of the preceding section.

Two waveforms that are of identical shape are said to be 100% correlated even if their amplitudes differ. An example of correlated signals would be two sine waves of the same frequency and phase. The instantaneous or rms values of fully correlated waveforms can be added arithmetically.

A problem arises when we have noise voltages that are partially correlated. This can happen when each contains some noise that arises from a common phenomenon, as well as some independently generated noise. In order to sum partially correlated waves, the general expression is

$$E^2 = E_1^2 + E_2^2 + 2CE_1E_2 \tag{1-31}$$

The term C is called the correlation coefficient and can have any value between -1 and $+1$, including 0. When $C = 0$, the voltages are uncorrelated, and the equation is the same as given in Fig. 1-8. When $C = 1$, the signals are totally correlated. Then rms values E_1 and E_2 can be added linearly. A -1 value for C implies subtraction of correlated signals, for the waveforms are then 180° out of phase.

Very often one can assume the correlation to be zero with little error. The maximum error will occur when the two voltages are equal and fully correlated. Summing gives 2 times their separate rms values, whereas the uncorrelated summing is 1.4 times their separate rms values. Thus the maximum error caused by the assumption of statistical independence is 30%. If the signals are partially correlated or one is much larger than the other, the error is smaller. When one signal is 10 times the other, the error is 8.6% maximum which is pretty good accuracy for noise measurements.

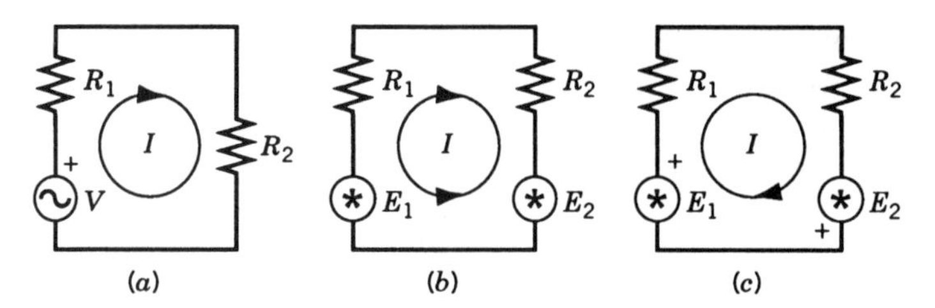

Figure 1-10 Circuits for analysis examples.

1-8 NOISE CIRCUIT ANALYSIS

An introduction to the circuit analysis of noisy networks was given in the preceding two sections. Here we expand on these discussions in order to clarify the theory and extend it to further applications.

Refer to Fig. 1-10*a*. A sinusoidal voltage source is feeding two noiseless resistances. Kirchhoff's voltage law allows us to write

$$V = IR_1 + IR_2 \tag{1-32}$$

Now suppose that we wanted to equate the mean square values of the three terms in Eq. 1-32. Let us square each term.

$$V^2 = (IR_1)^2 + (IR_2)^2 \tag{1-33}$$

This operation is not valid! Why? Because there is 100% correlation between IR_1 and IR_2, for they contain the same current I. Therefore, a correlation term must be present, and the correct expression is

$$V^2 = (IR_1)^2 + (IR_2)^2 + 2CIR_1IR_2 \tag{1-34}$$

Since C must equal unity here because only one current exists, the equation becomes

$$V^2 = I^2(R_1 + R_2)^2 \tag{1-35}$$

The rule for series circuit analysis is simply that *when resistances or impedances are series-connected, they should be summed first, and then, when dealing with mean square quantities, the sum should be squared.*

If V had been a noise source E, the same rule applies, for there is only one current in the circuit.

Now consider the circuit shown in Fig. 1-10*b*. Two uncorrelated noise voltage sources (or sinusoidal sources of different frequencies) are in series with two noiseless resistances. The current in this circuit must be expressed in mean square terms:

$$I^2 = \frac{E_1^2 + E_2^2}{(R_1 + R_2)^2} \tag{1-36}$$

No correlation term is present.

A convenient method for noise circuit analysis, when more than one source is present, employs superposition. The superposition principle states: *In a linear network the response for two or more sources acting simultaneously is the sum of the responses for each source acting alone with the other voltage sources short-circuited and the other current sources open-circuited*. Let us use superposition on the circuit of Fig. 1-10*b*. The loop currents caused by E_1 and E_2, each acting independently, are

$$I_1 = \frac{E_1}{R_1 + R_2} \qquad I_2 = \frac{E_2}{R_1 + R_2} \tag{1-37}$$

And, for *uncorrelated* quantities,

$$I^2 = I_1^2 + I_2^2 \tag{1-38}$$

Therefore,

$$I^2 = \frac{E_1^2}{(R_1 + R_2)^2} + \frac{E_2^2}{(R_1 + R_2)^2} = \frac{E_1^2 + E_2^2}{(R_1 + R_2)^2} \tag{1-39}$$

This agrees with Eq. 1-36.

In Fig. 1-10*c* sources E_1 and E_2 are correlated. Polarity symbols have been added to show that the generators are aiding. Then

$$I^2 = \frac{E_1^2 + E_2^2 + 2CE_1E_2}{(R_1 + R_2)^2} \tag{1-40}$$

where $0 < C \leq +1$ for aiding generators. If the polarity symbol on either generator were at its opposite terminal, C would take on values between 0 and -1. Note that when full correlation exists, it is valid to equate rms quantities ($E = IR_1 + IR_2$).

Suppose that we have a circuit such as shown in Fig. 1-11 in which there are several uncorrelated noise currents. We wish to determine the total current I_1 through R_1. For this example superposition is used; the contribution of E_1 to I_1 is termed I_{11}, and the contribution of E_2 to I_1 is I_{12}. It

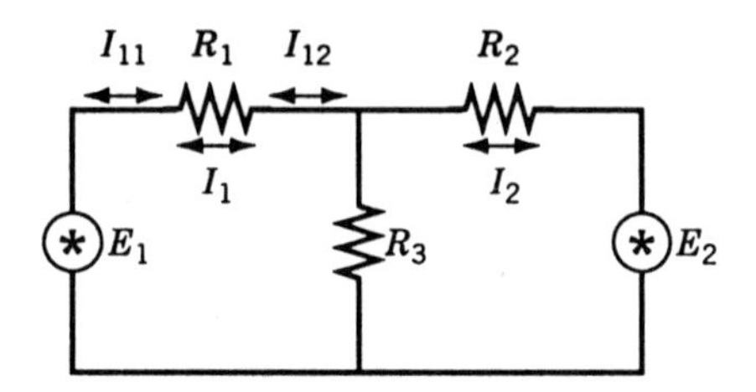

Figure 1-11 Two-loop circuit.

follows that

$$E_1^2 = I_{11}^2\left[R_1 + \frac{R_2R_3}{(R_2 + R_3)}\right]^2 \quad \text{and} \quad E_2^2 = I_2^2\left[R_2 + \frac{R_1R_3}{(R_1 + R_3)}\right]^2 \tag{1-41}$$

Next observe that

$$I_{12} = I_2R_3/(R_1 + R_3) \tag{1-42}$$

Therefore, we can write

$$I_{11}^2 = \frac{E_1^2(R_2 + R_3)^2}{(R_1R_2 + R_1R_3 + R_2R_3)^2}$$

and

$$I_{12}^2 = \frac{E_2^2R_3^2}{(R_1R_2 + R_1R_3 + R_2R_3)^2} \tag{1-43}$$

Hence

$$I_1^2 = \frac{E_1^2(R_2 + R_3)^2 + E_2^2R_3^2}{(R_1R_2 + R_1R_3 + R_2R_3)^2} \tag{1-44}$$

When finding the total current resulting from several uncorrelated noise sources, the contributions from each source must be added in such a way so that the magnitude of the total current is increased by each contribution. Therefore, neither the E_1 nor the E_2 terms in Eq. 1-44 could accept negative signs. An argument based on the heating effect of the currents, or one based on combining currents of different frequencies, can be used to justify this statement.

When performing a noise analysis of multisource networks, it is convenient to ascribe polarity symbols to uncorrelated sources in order that the proper addition (and no subtraction) of effects takes place.

1-9 EXCESS NOISE

We previously discussed the fundamental thermal noise in a resistor. Now it is time to point out that there can also be an additional excess noise source in a resistor or semiconductor, but *only when* a direct current is flowing [3]. Excess noise is so named because it is present in addition to the fundamental thermal noise of the resistance. Excess noise usually occurs whenever current flows in a discontinuous medium such as an imperfect semiconductor lattice. For example, in the base region of a BJT there are discontinuities or impurities that act as traps to the current flow and cause fluctuations in the base current.

As described in Chap. 12, many resistors also exhibit excess noise when a dc current is flowing. This noise contribution is greatest in composition carbon resistors and is usually not important in metal film resistors. A carbon resistor is made up of carbon granules squeezed together, and current tends to flow unevenly through the resistor. There are something like microarcs between the carbon granules. Excess noise in a resistor can be measured in terms of a noise index expressed in decibels. *The noise index is the number of microvolts of noise in the resistor per volt of dc drop across the resistor in each decade of frequency*. Thus, even though the noise is caused by current flow, it can be expressed in terms of the direct voltage drop rather than resistance or current. The noise index of some brands of resistors may be as high as 10 dB which corresponds to 3 μV/dc V/decade. This can be a significant contribution.

This excess noise exhibits a $1/f$ noise power spectrum. A $1/f$ spectrum means that the noise power varies inversely with frequency. Thus the noise voltage increases as the square root of the decreasing frequency. By decreasing the frequency by a factor of 10, the noise voltage increases by a factor of approximately 3.

Since excess noise has a $1/f$ power spectrum, most of the noise appears at low frequencies. This is why excess noise is often called low-frequency noise.

1-10 LOW-FREQUENCY NOISE

Low-frequency or $1/f$ noise has several unique properties. If it were not such a problem it would be very interesting. The spectral density of this noise increases without limit as frequency decreases. Firle and Winston [4] have measured $1/f$ noise as low as 6×10^{-5} Hz. This frequency is but a few cycles per day. When first observed in vacuum tubes, this noise was called "flicker effect," probably because of the flickering observed in the plate current. Many different names are used some of them uncomplimentary. In the literature, names like excess noise, pink noise, semiconductor noise, low-frequency noise, current noise, and contact noise will be seen. These all refer to the same thing. The term "red noise" is applied to a noise power spectrum that varies as $1/f^2$.

The noise *power* typically follows a $1/f^\alpha$ characteristic with α usually unity, but has been observed to take on values from 0.8 to 1.3 in various devices. The major cause of $1/f$ noise in semiconductor devices is traceable to properties of the surface of the material. The generation and recombination of carriers in surface energy states and the density of surface states are important factors. Improved surface treatment in manufacturing has decreased $1/f$ noise, but even the interface between silicon surfaces and grown oxide passivation are centers of noise generation.

As pointed out by Halford [5] and Keshner [6], $1/f$ noise is quite common. Not only is it observed in vacuum tubes, transistors, diodes, and resistors, but it is also present in thermistors, carbon microphones, thin films, and light sources. The fluctuations of a membrane potential in a biological system have been reported to have flicker noise. No electronic amplifier has been found to be free of flicker noise at the lowest frequencies. Halford points out that $\alpha = 1$ is the most common value, but there are other mechanisms with different values of α. For example, fluctuations of the frequency of rotation of the earth have an α of 2 and the power spectral density of galactic radiation noise has $\alpha = 2.7$.

Since $1/f$ noise power is inversely proportional to frequency, it is possible to determine the noise content in a frequency band by integration over the range of frequencies in which our interest lies. The result is

$$N_f = K_1 \int_{f_l}^{f_h} \frac{df}{f} = K_1 \ln \frac{f_h}{f_l} \tag{1-45}$$

where N_f is the noise power in watts, K_1 is a dimensional constant also in watts, and f_h and f_l are the upper and lower frequency limits of the band being considered. Now consider the noise power present in any decade of frequency such that $f_h = 10f_l$. Equation 1-45 then simplifies to

$$N_f = 2.3K_1 \tag{1-46}$$

This shows that $1/f$ noise results in equal noise power in each decade of frequency. In other words, the noise power in the band from 10 to 100 Hz is equal to that of the band from 0.01 to 0.1 Hz. Since the noise in each of these intervals is uncorrelated, the mean square values must be added. Total noise power increases as the square root of the number of frequency decades.

Since noise power is proportional to the mean square value of the corresponding noise voltage, then the spectral density of the noise voltage for $1/f$ noise is

$$S_f(f) = E_f^2/f \qquad \text{in V}^2/\text{Hz} \tag{1-47}$$

Suppose we know that there is 1 μV of $1/f$ noise in a decade of frequency. This would cause us to write

$$(1\ \mu\text{V})^2 = \int_{f_l}^{10f_l} S_f(f)\, df = \int_{f_l}^{10f_l} \frac{E_f^2\, df}{f} = 2.3E_f^2 \tag{1-48}$$

Consequently, for this example the spectral density of the $1/f$ noise reduces to

$$S_f(f) = \frac{(1\ \mu\text{V})^2}{2.3f} \tag{1-49}$$

Because $1/f$ noise power continues to increase as the frequency is decreased, we might ask the question, "Why is the noise not infinite at dc?" Although the noise voltage in a 1-Hz band may theoretically be infinite at dc or zero frequency, there are practical considerations that keep the total noise manageable for most applications. The noise power per decade of bandwidth is constant, but a decade such as that from 0.1 to 1 Hz is narrower than the decade from 1 to 10 Hz. But, when considering the $1/f$ noise in a dc amplifier, there is a lower limit to the frequency response set by the length of time the amplifier has been turned on. This low-frequency cutoff attenuates frequency components with periods longer than the "on" time of the equipment.

Example 1-2 A numerical example may be of assistance. Consider a dc amplifier with upper cutoff frequency of 1000 Hz. It has been on for 1 day. Since 1 cycle/day corresponds to about 10^{-5} Hz, its bandwidth can be stated as 8 decades. If it is on for 100 days, we add 2 more decades or $\sqrt{10/8} = \sqrt{1.25} = 1.18$ times its 1-day noise. The noise per hertz approaches infinity, but the total noise does not.

A fact to remember concerning a $1/f$ noise-limited dc amplifier is that measurement accuracy cannot be improved by increasing the length of the measuring time. In contrast, when measuring white noise, the accuracy increases as the square root of the measuring time.

1-11 SHOT NOISE

In transistors, diodes, and vacuum tubes, there is a noise current mechanism called shot noise. Current flowing in these devices is not smooth and continuous, but rather it is the sum of pulses of current caused by the flow of carriers, each carrying one electronic charge. Consider the case of a simple forward-biased silicon diode with electrons and holes crossing the potential barrier. Each electron and hole carries a charge q, and when they arrive at the anode and cathode, respectively, an impulse of current results. This pulsing flow is a granule effect, and the variations are referred to as shot noise. The rms value of the shot noise current is given by

$$\boxed{I_{\text{sh}} = \sqrt{2qI_{\text{DC}}\,\Delta f}} \tag{1-50}$$

where q is the electronic charge (1.602×10^{-19} Coulombs), I_{DC} is the direct current in amperes, and Δf is the noise bandwidth in hertz. We note that the shot noise current is proportional to the square root of the noise bandwidth. This means that it is white noise containing constant noise power per hertz of bandwidth.

One example of shot noise is a heavy rain on a tin roof. The drops arrive with about equal energy, the inches per hour rate corresponds to the current I_{DC} and the area of the roof relates to the noise bandwidth Δf.

Shot noise is associated with current flow across a potential barrier. Such a barrier exists in every *pn* junction in semiconductor devices and in the charge-free space in a vacuum tube. No barrier is present in a simple conductor; therefore, no shot noise is present. The most important barrier is the emitter–base junction in a bipolar transistor and the gate–source junction in a junction field effect transistor (JFET). The V–I behavior of the base–emitter junction is described by the familiar diode equation

$$I_E = I_S(e^{qV_{BE}/kT} - 1) \tag{1-51}$$

where I_E is the emitter current in amperes, I_S is the reverse saturation current in amperes, and V_{BE} is the voltage between the base and emitter. Suppose that we consider separately the two currents that make up I_E in Eq. 1-51:

$$I_E = I_1 + I_2 \tag{1-52}$$

where $I_1 = -I_S$ and $I_2 = I_S \exp(V_{BE}/kT)$.

Current I_1 is caused by thermally generated minority carriers, and current I_2 represents the diffusion of majority carriers across the junction. Each of these currents has full shot noise, and even though the direct currents they represent flow oppositely, their mean square noise values are added.

Under reverse biasing, $I_2 \cong 0$ and the shot noise current of I_1 dominates. On the other hand, when the diode is strongly forward-biased, the shot noise current of I_2 dominates. For zero bias, there is no external direct current, and I_1 and I_2 are equal and opposite. The mean square value of shot noise is twice the reverse-bias noise current:

$$I_{\text{sh}}^2 = 4qI_S\,\Delta f \tag{1-53}$$

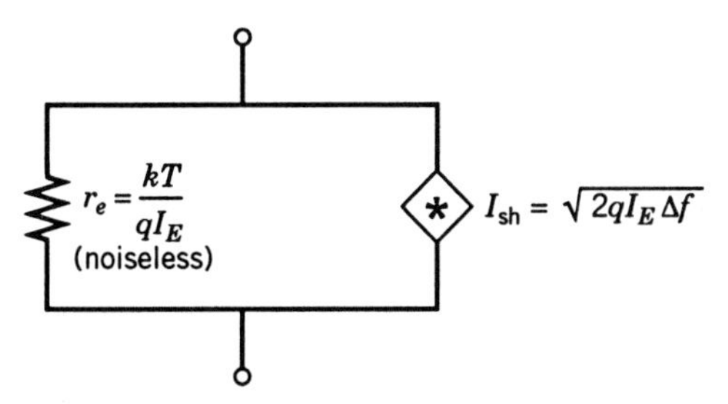

Figure 1-12 Shot noise equivalent circuit for forward-biased *pn* junction.

The equivalent circuit representation for a shot noise source is a current generator as previously noted in Eq. 1-50. For the case of the forward-biased *pn* junction, a noiseless resistance parallels this current generator. By differentiating Eq. 1-51 with respect to V_{BE}, we obtain a conductance. The reciprocal of that conductance is referred to as the Shockley emitter resistance r_e and is given by

$$r_e = kT/qI_E \tag{1-54}$$

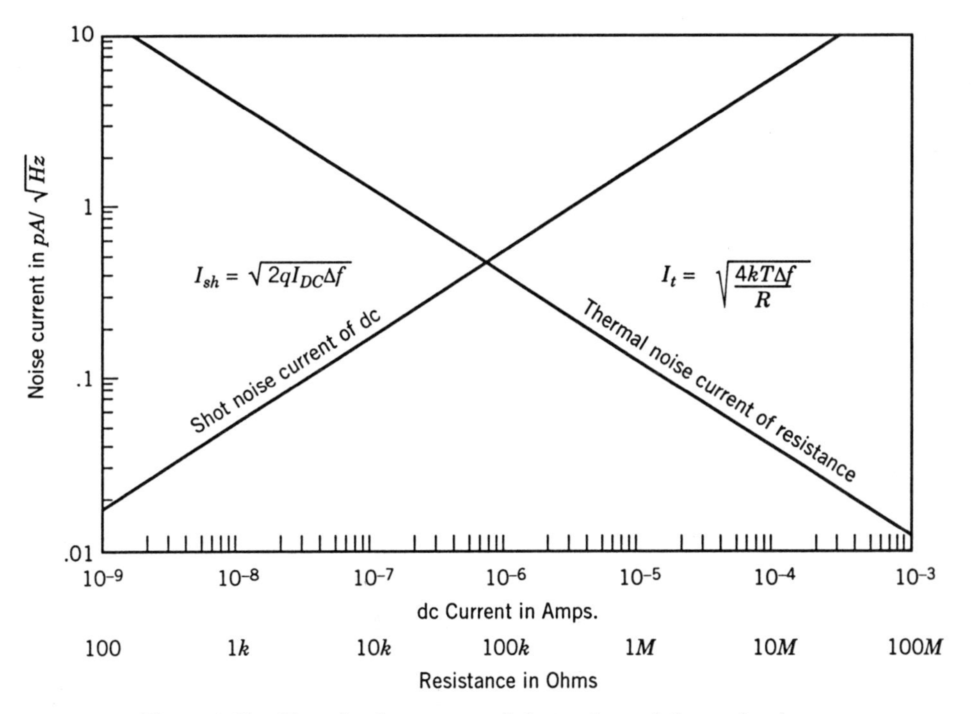

Figure 1-13 Plot of noise current of shot noise and thermal noise.

At room temperature $r_e = 0.025/I_E$. The element r_e is not a thermally noisy component, for it is the dynamic effect of the junction, and not a bulk or material characteristic.

An equivalent circuit representing shot noise at a forward-biased *pn* junction is shown in Fig. 1-12. Mathematically, the mean square value of shot noise is equal to the thermal noise for an unbiased junction, and equal to one-half of the resistive thermal noise voltage at a forward-biased junction. The noise voltage E_{sh} of a forward-biased junction is the product of the shot noise current I_{sh} and the diode resistance r_e:

$$E_{\text{sh}} = \frac{0.025}{I_E}\sqrt{2qI_E\,\Delta f} = 1.42 \times 10^{-11}\sqrt{\frac{\Delta f}{I_E}} \tag{1-55}$$

which shows that the shot noise voltage E_{sh} will decrease by the square root of the diode current. Further discussion is available in the literature [7, 8].

The shot noise current I_{sh} of a diode and the thermal noise current I_t of a resistor are compared in the plot of Fig. 1-13.

1-12 CAPACITANCE SHUNTING OF THERMAL NOISE: kT/C NOISE

The thermal noise expression $E_t = (4kTR\,\Delta f)^{1/2}$ predicts that an open circuit (infinite resistance) generates an infinite noise voltage. This is not observed in a practical situation since there is always some shunt capacitance that limits the voltage. Consider the actual noisy resistance–shunt capacitance combination as shown in Fig. 1-14.

The thermal noise voltage E_t increases as the square root of the resistance. Low-frequency noise from E_t directly affects the output noise voltage E_{no}. Higher-frequency components from E_t are more effectively shunted by the capacitor C. Increasing R increases the noise voltage, but decreases the cutoff frequency and consequently the noise bandwidth. A plot of the resulting noise voltage versus frequency is shown in Fig. 1-15 for one value of capacitance and resistance values of R, $4R$, and $9R$. The noise voltage increases as the square root of the resistance and the bandwidth decreases proportional to the resistance but the integrated mean square noise under each curve is equal. The total output noise voltage which would be measured

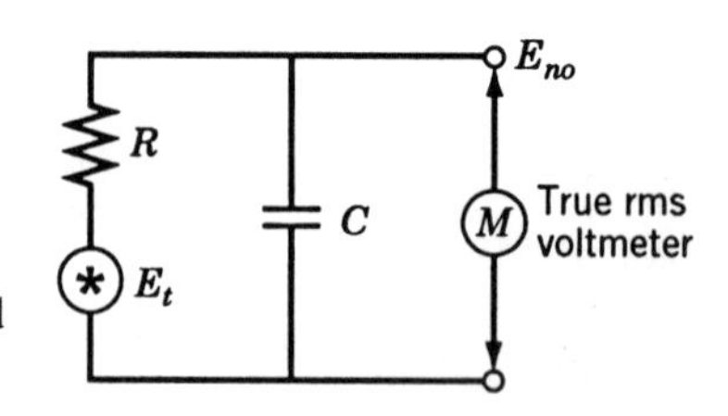

Figure 1-14 Thermal noise of a resistor shunted by a capacitance.

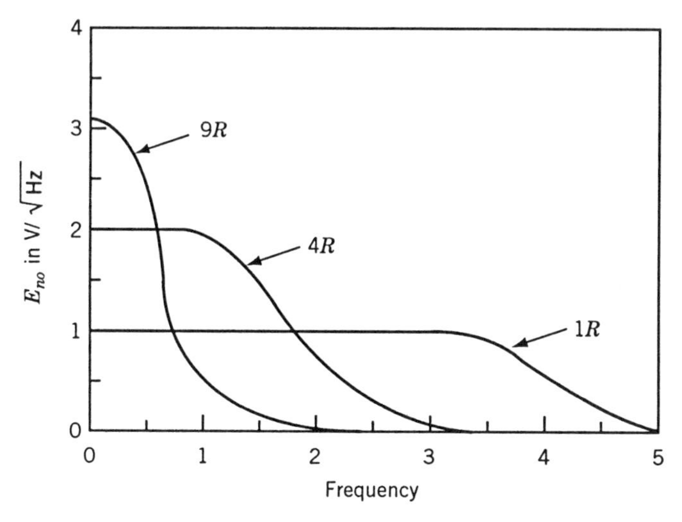

Figure 1-15 Noise spectral density for a resistance shunted by a capacitance.

by a true rms voltmeter having an infinite bandwidth is found from

$$E_{no}^2 = \int_0^{\infty} E_t^2 \left| \frac{1/j\omega C}{R + 1/j\omega C} \right|^2 df = \int_0^{\infty} \frac{E_t^2\, df}{1 + (\omega RC)^2} \tag{1-56}$$

Now, making a change of variable so we can perform the integration, we let $f = f_2 \tan\theta$, $f_2 = 1/2\pi RC$, $df = f_2 \sec^2\theta\, d\theta$, and change the upper limit of integration to $\pi/2$. The output noise voltage now becomes

$$E_{no}^2 = \int_0^{\pi/2} \frac{E_t^2 f_2 \sec^2\theta\, d\theta}{1 + \tan^2\theta} = \int_0^{\pi/2} E_t^2 f_2\, d\theta \tag{1-57}$$

Next, substituting for the thermal noise voltage E_t, we have

$$E_{no}^2 = \int_0^{\pi/2} 4kTRf_2\, d\theta = 2\pi kTRf_2$$

$$E_{no}^2 = kT/C \qquad \text{in mean squared volts} \tag{1-58}$$

Note that the output rms noise voltage is independent of the source resistance and only depends on the temperature and capacitance. The majority of the energy is contained in the very low frequency region because the shunt capacitance attenuates high frequencies. This noise limit is often referred to simply as kT/C noise. It becomes important in applications where sample and hold circuits are utilized such as with analog-to-digital converters and switched capacitor circuits.

Example 1-3 As a numerical example if $C = 1$ pF of stray shunting capacitance, $E_{no}^2 = 4 \times 10^{-9}$ V^2, or $E_{no} \approx 64$ μV. This is a significant noise voltage level. To minimize noise, reduce the system bandwidth to that which is absolutely necessary for properly processing the desired signals.

SUMMARY

a. Noise is any unwanted disturbance that obscures or interferes with a signal.

b. Thermal noise is present in every electrical conductor, with rms value:

$$E_t = \sqrt{4kTR\,\Delta f}$$

When evaluated this yields 4 nV for 1000 Ω and $\Delta f = 1$ Hz.

c. The noise bandwidth Δf is the area under the $|A_v(f)|^2$ curve divided by A_{vo}^2, the reference or maximum value of gain squared.

d. For circuit analysis a noisy resistance can be replaced by a noise voltage generator in series with a noiseless resistance, or a noise current generator in parallel with a noiseless resistance.

e. Noise quantities can be added according to

$$E^2 = E_1^2 + E_2^2 + 2CE_1E_2$$

where C is the correlation coefficient, $-1 \le C \le +1$. Usually $C = 0$.

f. Excess noise is generated in most components when direct current is present.

g. $1/f$ noise is especially troublesome at low audio frequencies.

h. Shot noise is present when direct current flows across a potential barrier:

$$I_{\text{sh}} = \sqrt{2qI_{\text{DC}}\,\Delta f}$$

i. The total thermal noise energy in a resistance is finite and is limited by the effective capacitance across its terminals and the absolute temperature. In the limiting case $E_{no}^2 = kT/C$.

PROBLEMS

1-1. Determine the rms thermal noise voltage of resistances of 1 kΩ, 50 kΩ, and 1 MΩ for each of the following noise bandwidths: 50 kHz, 1 MHz, and 20 MHz. Consider $T = 290$ K.

1-2. Calculate the rms thermal noise voltage of a 100-mH inductance in a 1-Hz noise bandwidth. Consider that the inductor has an impedance of 8 kΩ and that the frequency band is centered around 10 kHz.

1-3. Determine the noise bandwidth of a circuit with $|A_v|^2$ frequency response described as follows:

$$0 \leq f \leq 1 \text{ kHz} \qquad |A_v|^2 = f$$

$$1 \text{ kHz} \leq f \leq 20 \text{ kHz} \qquad |A_v|^2 = 1000$$

$$20 \text{ kHz} \leq f \leq 1000 \text{ kHz} \qquad |A_v|^2 = 1000 - 0.0125(f - 20{,}000)$$

$$100 \text{ kHz} < f \qquad |A_v|^2 = 0$$

1-4. The frequency response of the magnitude of the voltage gain for a certain amplifier is

$$A_v(f) = \begin{cases} 3\sin(2\pi f/400) & \text{for } 0 \leq f \leq 200 \text{ Hz} \\ 0 & \text{elsewhere} \end{cases}$$

Determine the noise bandwidth for this amplifier.

1-5. Calculate E_t for a noisy resistance of 500 kΩ. Transform this into the noise current generator form and determine I_t. Let $T = 290$ K and $\Delta f = 10^5$ Hz.

1-6. The resistor in the previous problem is connected in parallel with another noisy resistor of 250 kΩ. Determine the mean square and rms values of the noise voltage present at the terminals of the pair.

1-7. Find the resistance of a *pn* junction that exhibits 200 nV rms of shot noise. Assume $\Delta f = 1$ MHz and $I_{\text{DC}} = 10$ mA. Compare your answer with the value of r_e predicted by Eq. 1-54. What conclusions can you reach?

1-8. Find the noise bandwidth for a cascade of three identical low-pass filter stages which are buffered by ideal amplifiers. Each single stage has a −3-dB cutoff frequency of f_2.

1-9. Determine the noise bandwidth Δf for the filter whose frequency characteristics are shown in Fig. P1-9.

(a) First find Δf for the asymptotic response.

(b) Then find Δf for the exact response. Compare the two values of Δf.

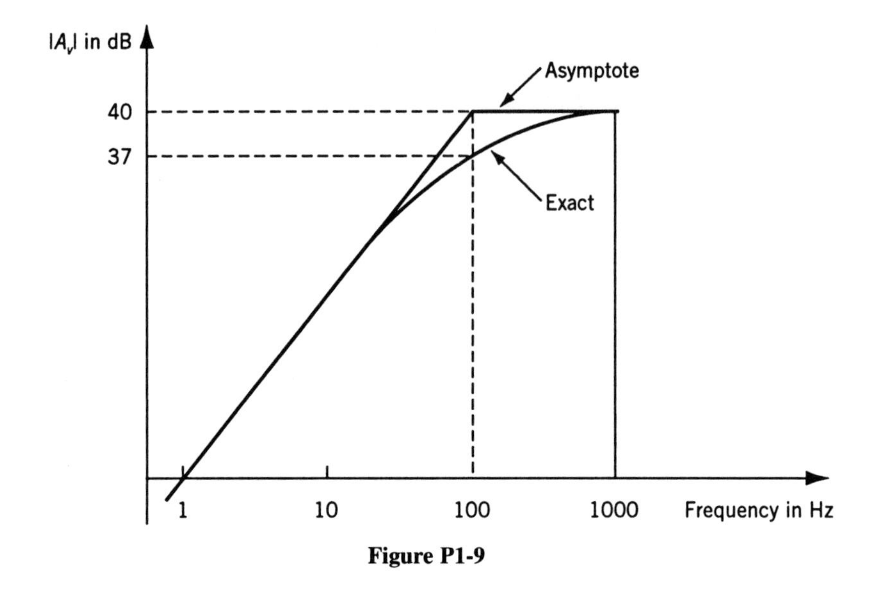

Figure P1-9

1-10. A certain noise source is known to have $1/f$ spectral density. The noise voltage in one decade of bandwidth is measured to be 1 μV rms. How many decades would be involved to produce a total noise voltage of 3 μV?

1-11. Determine the noise bandwidth Δf for the circuit shown in Fig. P1-11. The 20 kΩ represents the input resistance of the amplifier which has a voltage gain of 100.

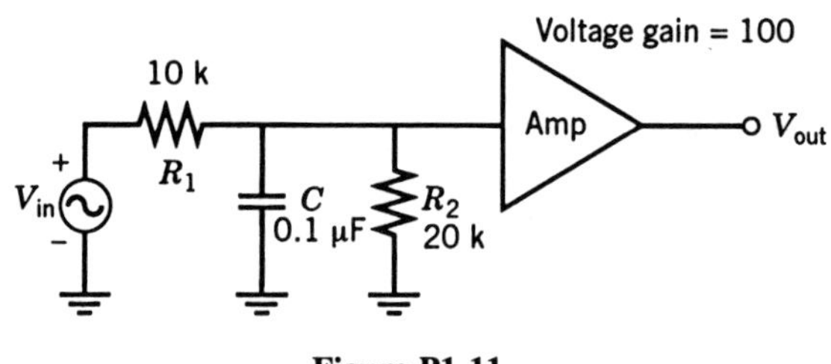

Figure P1-11

1-12. Determine the noise bandwidth Δf for the circuit shown in Fig. P1-12.

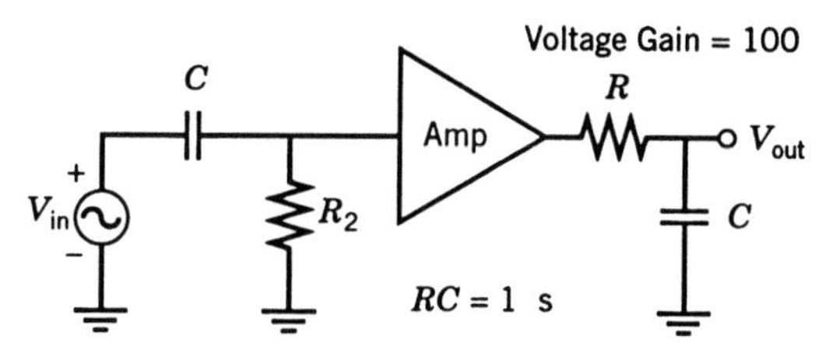

Figure P1-12

1-13. The transfer function for a second-order bandpass active filter is given by

$$T(s) = \frac{GBs}{s^2 + Bs + \omega_o^2}$$

where G is the gain at the radian center frequency, ω_o, and B is the -3-dB radian bandwidth. Prove that the noise bandwidth is given by $\Delta f = B/4 = (\text{BW})\pi/2$, where BW is the -3-dB bandwidth in hertz.

1-14. The transfer function of an active bandpass amplifier is known to be

$$T(s) = \frac{-10sR_2C_1}{(1 + sR_1C_1)(1 + sR_2C_2)}$$

where $R_1 = 15.9\ \text{k}\Omega$, $R_2 = 31.8\ \text{k}\Omega$, $C_1 = 0.1\ \mu\text{F}$, and $C_2 = 0.01\ \mu\text{F}$. Find the equivalent noise bandwidth, Δf, in units of hertz for this amplifier.

1-15. It is desired to replace the diode in Fig. P1-15*a* with the resistor R_x as in Fig. P1-15*b*.

(a) Determine the value of R_x which will produce the same amount of noise voltage as the diode produces. Assume $\Delta f = 1$ Hz.

(b) Now suppose V_{in} is a 1-mV peak amplitude sinusoidal signal at a frequency of 1000 Hz and it is applied to both circuits. Determine the output signal voltages, V_{o1} and V_{o2}, for both circuits which a true rms voltmeter would measure.

(c) Which circuit gives the better signal-to-noise ratio at the output? Explain your answer. Consider only the effects of substituting the resistor R_x for the diode.

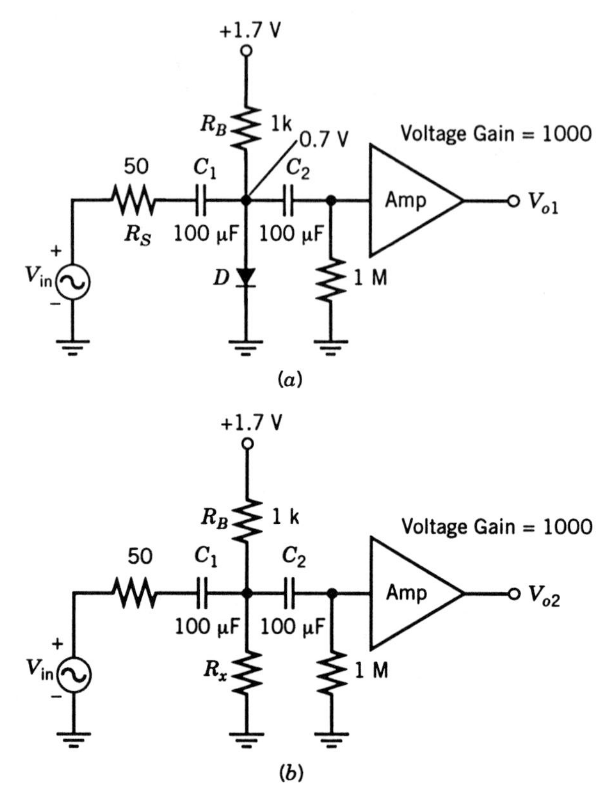

Figure P1-15

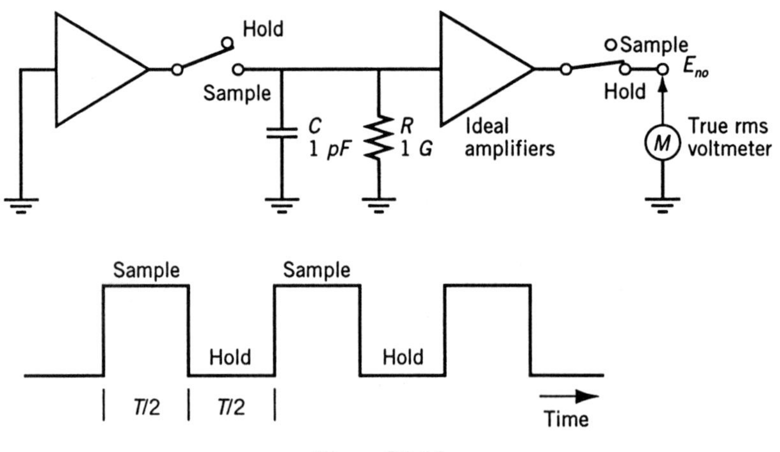

Figure P1-16

1-16. The sample-hold circuit shown in Fig. P1-16 is clocked by a square wave. Determine the output noise voltage, E_{no}, which would be recorded by a true rms voltmeter having infinite bandwidth if (a) $T =$ 1 μs, (b) $T = 1$ ms, and (c) $T = 1$ s. For all practical purposes, the 1-GΩ resistor can be considered to be an open circuit.

1-17. **(a)** The statistically expected number of maxima per second in white noise with upper and lower frequency limits f_2 and f_1 is [9]:

$$\left[\frac{3(f_2^5 - f_1^5)}{5(f_2^3 - f_1^3)}\right]^{1/2}$$

Find the maxima for $f_1 = 0$ and $f_2 = 10^6$ MHz.

(b) The expected total number of zero crossings per second is given by

$$\left[\frac{4(f_2^2 + f_1 f_2 + f_1^2)}{3}\right]^{1/2}$$

If $f_1 = 0$ and $f_2 = 10$ MHz, evaluate the number of zero crossings. Show that for narrowband noise the assumption that $f_1 \cong f_2$ yields

$$(f_1 + f_2)$$

REFERENCES

1. Bennett, A. R., *Electrical Noise*, McGraw-Hill, New York, 1960, p. 42.
2. Van der Ziel, A., *Noise*, Prentice-Hall, Englewood Cliffs, NJ, 1954, pp. 8–9.
3. DeFelice, L. J., "$1/f$ Resistor Noise," *J. Applied Physics*, **47**, (January 1976), 350–352.
4. Firle, J. E., and H. Winston, *Bull. Amer. Phys. Soc.*, **30**, 2 (1955).
5. Halford, D., "A General Model for Spectral Density Random Noise with Special Reference to Flicker Noise $1/f$," *Proc. IEEE*, **56**, 3 (March 1968), 251.
6. Keshner, M. S., "$1/f$ Noise," *Proc. IEEE*, **70**, 3 (March 1982), 212–218.
7. Thornton, R. D., D. DeWitt, E. R. Chenette, and P. E. Gray, *Characteristics and Limitations of Transistors*, Wiley, New York, 1966, pp. 138–145.
8. Baxandall, P. J., "Noise in Transistor Circuits," *Wireless World*, **74**, 1397–1398 (November 1968), 388–392 (December 1968), 454–459.
9. Rice, S. O., "Mathematical Analysis of Random Noise," *Bell System Technical Journal*, **23**, 4 (July 1944).

CHAPTER 2

AMPLIFIER NOISE MODEL

Since every electrical component is a potential source of noise, a network such as an amplifier that contains many components could be difficult to analyze from a noise standpoint. Therefore, a noise model is helpful to simplify noise analysis. The E_n–I_n amplifier model discussed in this chapter contains only two noise parameters. The parameters are not difficult to measure.

The concept of noise figure is introduced, and it is shown that optimization of the noise figure is possible. From a study of the noise contributions of the stages of a cascaded network, it can be concluded that the major noise source is the first signal processing stage. If a high level of power amplification is available from that stage, noise contributions from other portions of the electronics will be negligible.

2-1 THE NOISE VOLTAGE AND CURRENT MODEL

There are universal noise models for any two-port network [1]. The network is considered as a noise-free black box, and the internal sources of noise can be represented by two pair of noise generators (four generators) located at the input or the output or both, usually the input. It turns out that an amplifier can be represented as a voltage generator, a current generator, and a complex correlation coefficient to provide the four generators. Usually this can be simplified to two generators because the correlation coefficient is one. This noise model, shown in Fig. 2-1, is used to represent any type of

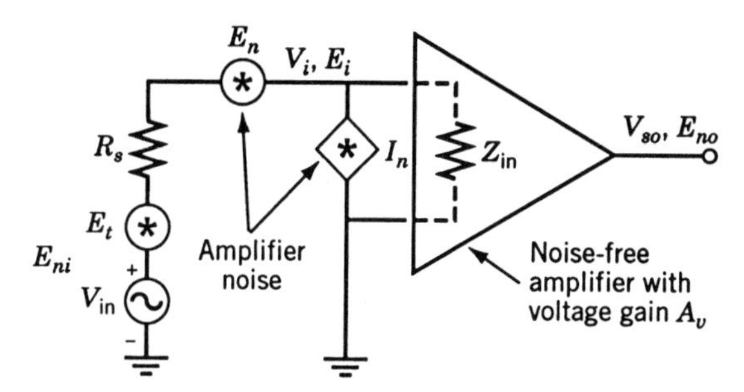

Figure 2-1 Amplifier noise and signal source.

amplifier. It can also apply to passive circuits, single transistors, tunnel diodes, integrated circuit (IC) amplifiers, and so on. The figure also includes the signal source V_{in} and a noisy source resistance R_s.

Amplifier noise is represented completely by a zero impedance voltage generator E_n in series with the input port, an infinite impedance current generator I_n in parallel with the input, and by a complex correlation coefficient C (not shown). Each of these terms typically are frequency dependent. The thermal noise of the signal source is represented by the noise generator E_t.

In a practical design, we are usually concerned with the signal-to-noise ratio at the output of the system. That is where we are using the signal for processing, display, level detecting, or driving a load. Because we are considering signal and noise in an electronic system that has stages of amplification, frequency response shaping, and so forth, it is usually quite difficult to evaluate the results of even minor circuit modifications on the signal-to-noise ratio. By referring all noise to the input port and considering the amplifier to be noise free, it is easier to appreciate the effects of such changes on both the signal and the noise.

Both E_n and I_n parameters are required to represent adequately an amplifier.

2-2 EQUIVALENT INPUT NOISE

Although we have reduced the number of noise sources to three in the system shown in Fig. 2-1 by using the E_n–I_n model for the electronic circuitry, additional simplifications are welcomed. *Equivalent input noise, E_{ni}, will be used to represent all three noise sources.* This parameter refers all noise sources to the signal source location. Since both the signal and the noise equivalent are then present at that point in the system, the S/N can be easily evaluated. We proceed to determine the signal voltage gain and the equivalent input noise voltage, E_{ni}.

The levels of signal voltage and noise voltage that reach Z_{in} in the circuit are multiplied by the noiseless voltage gain A_v. We will determine those levels. For the signal path, the transfer function from input signal source to output port is called *system gain* K_t. By definition,

$$K_t = V_{so}/V_{\text{in}} \tag{2-1}$$

Note that K_t *is different from the voltage gain* A_v. It is dependent on both the amplifier's input impedance and the signal generator's source resistance and it varies with frequency. The rms output voltage signal can be expressed by

$$V_{so} = \left| \frac{A_v V_{\text{in}} Z_{\text{in}}}{R_s + Z_{\text{in}}} \right| \tag{2-2}$$

Substituting Eq. 2-1 into Eq. 2-2 gives an expression for the system gain K_t in terms of network parameters:

$$K_t = \left| \frac{A_v Z_{\text{in}}}{R_s + Z_{\text{in}}} \right| \tag{2-3}$$

For the signal voltage, linear voltage and current division principles can be applied. However, for evaluating noise, we must sum each contribution in mean square values. The total noise at the output port is

$$E_{no}^2 = A_v^2 E_i^2 \tag{2-4}$$

The noise at the input to the amplifier is

$$E_i^2 = \left(E_t^2 + E_n^2\right) \left| \frac{Z_{\text{in}}}{Z_{\text{in}} + R_s} \right|^2 + I_n^2 |Z_{\text{in}} \parallel R_s|^2 \tag{2-5}$$

Note that the $\parallel$ symbol refers to the parallel combination of Z_{in} and R_s. Therefore,

$$E_{no}^2 = \left(E_n^2 + E_t^2\right) |A_v|^2 \left| \frac{Z_{\text{in}}}{Z_{\text{in}} + R_s} \right|^2 + I_n^2 |A_v|^2 |Z_{\text{in}} \parallel R_s|^2 \tag{2-6}$$

The total output noise given in Eq. 2-6 divided by the square of the magnitude of the system gain given in Eq. 2-3 yields an expression for equivalent input noise, E_{ni}^2

$$\boxed{E_{ni}^2 = E_t^2 + E_n^2 + I_n^2 R_s^2} \tag{2-7}$$

This equation is very important for the analysis of many noise problems; it can be applied to systems using any type of active device. Note the simplicity of its terms. The mean square equivalent input noise is the sum of the mean square values of the three noise voltage generators. This single noise source located at V_{in} can be substituted for all sources of system noise. In other words, the single noise source E_{ni}^2 inserted in series with the signal source V_{in} will produce the same total output noise, E_{no}^2, as calculated by Eq. 2-6.

Note that E_{ni}^2 is independent of the amplifier's gain and its input impedance. It is this independence of gain and impedance that makes E_{ni}^2 the most useful index upon which to compare the noise characteristics of various amplifiers and devices. Since amplifier input resistance and capacitance are not factors in the equivalent input noise expression, we can omit them whenever we need to determine E_{ni}^2. However, they must be considered for determining K_t and E_{no}.

As previously mentioned, the amplifier's noise voltage and current generators may not be completely independent. To be absolutely correct, we must introduce the correlation coefficient C as discussed in Sec. 1-7. A modified form of Eq. 2-7 results:

$$E_{ni}^2 = E_t^2 + E_n^2 + I_n^2 R_s^2 + 2CE_n I_n R_s \tag{2-8}$$

The correlation term can be approached, if desired, from the standpoint that it is another noise generator in the system of Fig. 2-1. It could be represented as a voltage generator with rms value $(2CE_n I_n R_s)^{1/2}$ in series with E_n or a current generator $(2CE_n I_n / R_s)^{1/2}$ in parallel with I_n.

2-3 MEASUREMENT OF E_n AND I_n

Another reason for the wide acceptance of the E_n–I_n model is the ease of measurement of its parameters. The thermal noise of the source resistance, E_t, can be easily calculated from Eq. 1-6:

$$E_t = \sqrt{4kTR_s\,\Delta f} \tag{1-6}$$

It can be observed that if R_s is purposely made to equal zero, two terms in Eq. 2-7 drop out, and the resulting equivalent input noise is simply the noise generator E_n. A measurement of total output noise, the $R_s = 0$ condition therefore equals $A_v E_n$. Division of total output noise by A_v gives a value for E_n.

We now have determined two of the three components of the equivalent input noise expression. The third component, $I_n R_s$, is most easily determined by making the source resistance very large. The source thermal noise contribution is proportional to the square root of source resistance, whereas the $I_n R_s$ term is proportional to the first power of resistance and dominates at a sufficiently large value of source resistance. To determine I_n then, we measure the total output noise with a large source resistance and divide by the system gain as measured with this source resistance in series with the input. This gives E_{ni}, which is mostly the $I_n R_s$ term. Dividing by R_s yields the I_n component. If there is still a contribution from the thermal noise of the source, it can be subtracted from E_{ni}.

The values of E_n and I_n vary with frequency, operating point, and the type of amplifier input device. Values have been measured and are compiled in the appendixes.

Equivalent input noise, E_{ni}, applies to a specific frequency band. If it is necessary to operate over a wide bandwidth, the noise at discrete frequencies can be calculated or measured and a mean square summing made. In effect, this process is integrating the equivalent input noise versus frequency expression over the bandwidth of interest. Direct numerical integration using PSpice or other simulation and computer programs is another way to determine the total equivalent input noise voltage over a wide bandwidth. This will be discussed further in Chap. 4.

2-4 INPUT NOISE EXAMPLES

Plots of equivalent input noise voltage versus source resistance are given in Fig. 2-2. Also shown in these plots are values of the components E_n and $I_n R_s$ for two amplifiers and the behavior of source resistance thermal noise E_t.

In each graph, the equivalent input noise E_{ni} is bounded by three separate lines. Each line corresponds to a term in Eq. 2-7. At low values of source resistance, E_n alone is important. As the source resistance increases, the thermal noise of the source becomes significant. At sufficiently high values of R_s, the total equivalent input noise is equal to the $I_n R_s$ term.

In Fig. 2-2*a* the E_n–I_n noise dominates the thermal noise throughout the total range of values of R_s. In Fig. 2-2*b* the noise current, I_n, is an order of magnitude smaller. The total noise is limited therefore by thermal noise through a part of the range. Lowering the values of E_n and I_n widens the thermal noise limited region. System noise limited by sensor thermal noise is an ideal case.

Curves of the type shown in Fig. 2-2 apply to all kinds of active devices. The levels, however, differ. With a bipolar or CMOS operational amplifier,

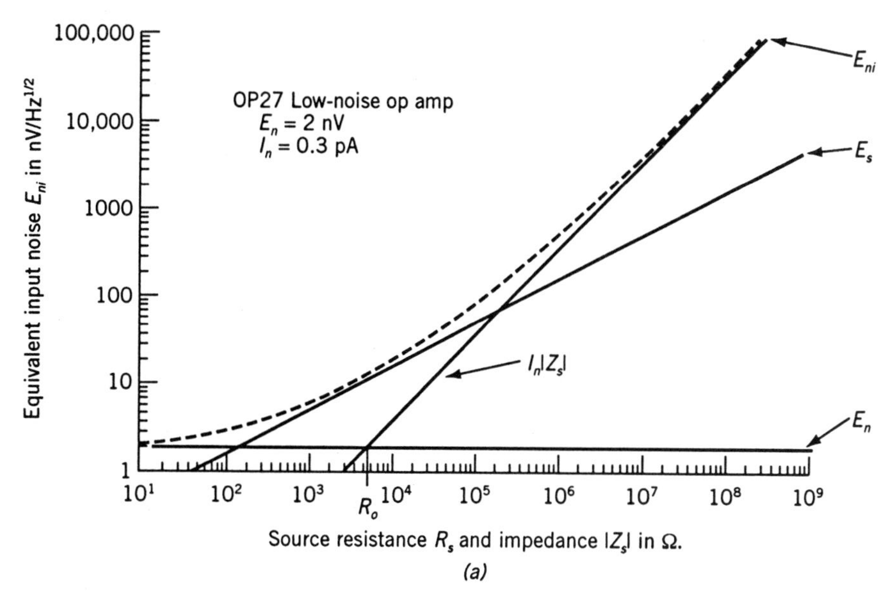

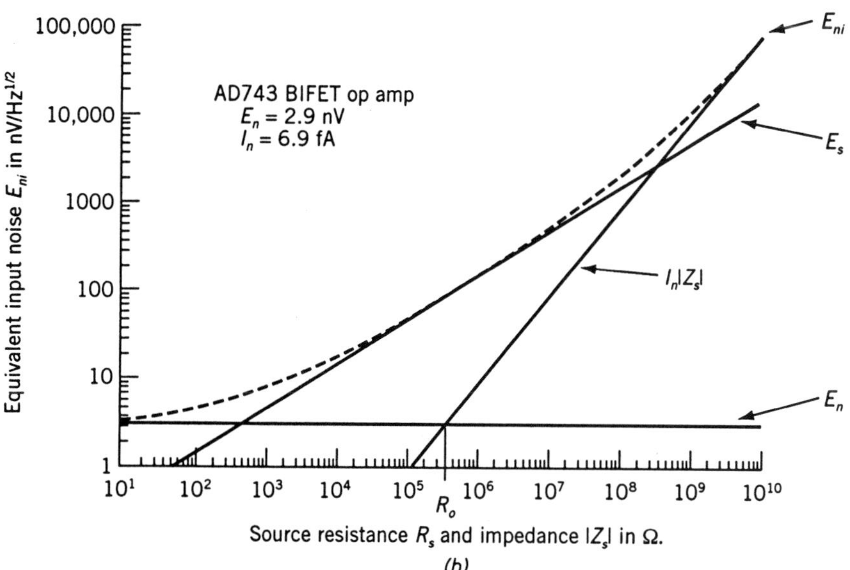

Figure 2-2 Plots of E_{ni} versus R_s.

we expect E_n to be typically as shown at frequencies substantially above the $1/f$ noise corner. For FET devices or amplifiers with FET first stages, the noise current I_n can be typically only 1/100 of the values given in the figure.

2-5 NOISE FIGURE (NF) AND SIGNAL-TO-NOISE RATIO (SNR)

The *noise factor* F is a figure-of-merit for a device or a circuit with respect to noise. According to IEEE standards, *the noise factor of a two-port device is the ratio of the available output noise power per unit bandwidth to the portion of that noise caused by the actual source connected to the input terminals of the device, measured at the standard temperature of* 290 K [2]. This definition of the noise factor in equation form is

$$F = \frac{\text{Total available output noise power}}{\text{Portion of output noise power caused by } E_t \text{ of source resistance}} \tag{2-9}$$

An equivalent definition of the noise factor is

$$F = \frac{\text{Input signal-to-noise ratio}}{\text{Output signal-to-noise ratio}} = \frac{S_i/N_i}{S_o/N_o} \tag{2-10}$$

The noise factor, since it is a power ratio, can be expressed in decibels. When expressed in decibels, this ratio is referred to as the *noise figure*. The logarithmic expression for the noise figure NF is

$$\text{NF} = 10 \log F \tag{2-11}$$

The noise figure is a measure of the signal-to-noise degradation attributed to the amplifier. For a perfect amplifier, one that adds no noise to the thermal noise of the source, the ideal noise factor is $F = 1$ and the noise figure NF = 0 dB.

The noise figure NF can be defined in terms of E_n and I_n. Thus

$$\text{NF} = 10 \log \frac{E_{ni}^2}{E_t^2} = 10 \log \frac{E_t^2 + E_n^2 + I_n^2 R_s^2}{E_t^2} \tag{2-12}$$

This equation shows that the noise figure can also be expressed as the ratio of the total mean square equivalent input noise to the mean square thermal noise of the source.

An illustration of the noise factor can be obtained from Fig. 2-2*b*. The noise factor is proportional to the square of the ratio of total equivalent input noise (dotted curve) to thermal noise (solid curve). At low resistances the ratio of total noise to thermal noise is very large and the noise figure is large, representing poor performance. As the source resistance increases, thermal noise increases, but total input does not. The noise factor therefore decreases. The plot of total input noise E_{ni} is closest to thermal noise when $E_n = I_n R_s$. This is the point of minimum noise factor. As we go to higher source resistances, total noise follows the $I_n R_s$ curve, and the noise factor again becomes large.

The definition of NF used here is based on a reference temperature of 290 K. When this definition is applied to sensors that are cooled, apparent negative values of NF can result.

The term spot noise factor F_o, when used to describe system noise, is simply a narrowband F. Often F_o is defined for $\Delta f = 1$ Hz, and a test frequency such as 1000 Hz can be used. The spot noise factor is clearly a function of frequency and is sometimes simply written as $F(f)$.

The principal value of the concept of noise factor F is to compare amplifiers or amplifying devices; it is not necessarily the appropriate indicator for optimizing noise performance. Because of the definition, F can be reduced by an increase in the thermal noise of the source resistance. Since a change of this type has little bearing on amplifier design, F is not as useful for system optimization as quantities such as E_{ni} or S_o/N_o.

2-6 OPTIMUM SOURCE RESISTANCE

The point at which the total equivalent input noise approaches closest to the thermal noise curve in Fig. 2-2*b* is significant. At this point, the amplifier adds minimum noise to the thermal noise of the source. The noise figure reaches a minimum value. This optimum source resistance is called R_{opt} or R_o and may be obtained from

$$R_o = E_n/I_n \qquad \text{where} \quad E_n = I_n R_s \tag{2-13}$$

The value of the noise factor at this point can be called F_{opt}. A rearrangement of Eq. 2-13 yields

$$F_{\text{opt}} = 1 + (E_n I_n / 2kT\Delta f) \tag{2-14}$$

Noise figure variations are illustrated in Fig. 2-3.

The minimum value of the noise figure occurs at $R_s = R_o$. As the product of E_n and I_n increases, not only does F_{opt} increase, but NF is more highly sensitive to source resistance variations. The lowest curve gives a good noise figure over a wide range of source resistance, whereas the upper curve represents poor operation when the source is not equal to R_o. From an engineering standpoint, if the noise figure is less than 3 dB, there may be little advantage to be gained from continued noise reduction effort because half the noise is from the source.

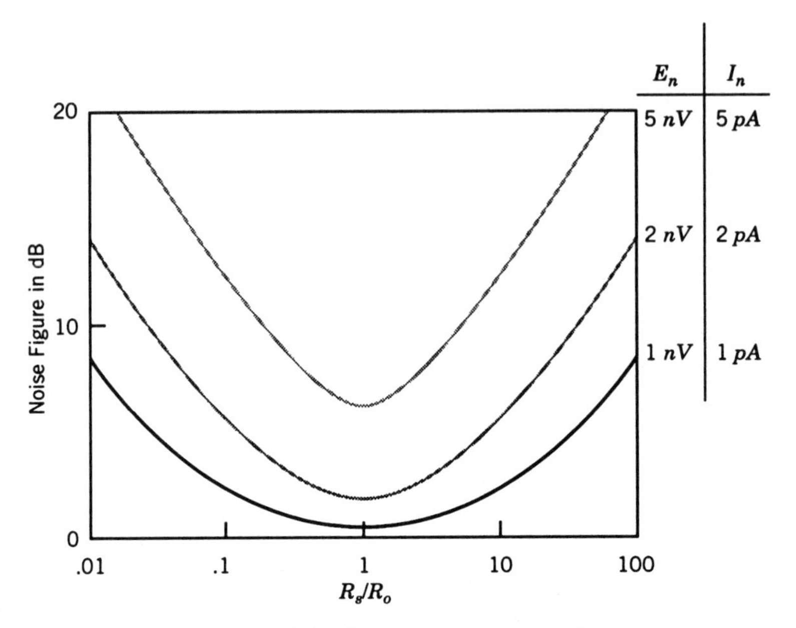

Figure 2-3 Noise figure versus source resistance.

Optimum source resistance, R_o, is not the resistance for maximum power transfer. There is no direct relation between R_o and the amplifier input impedance Z_i. R_o is determined by the amplifier noise mechanisms and has a bearing on the maximum signal-to-noise ratio. Optimum power transfer is based on maximizing the signal only. The input impedance of an amplifier is strongly affected by circuit conditions such as feedback, but the noise of the amplifier is unaffected by feedback, except insofar as the feedback resistors generate noise.

2-7 NOISE RESISTANCE AND NOISE TEMPERATURE

Noise resistance R_n is the value of a resistance that would generate thermal noise of value equal to the amplifier noise. An expression for R_n is desired. Equating thermal noise to amplifier noise gives

$$4kTR_n \Delta f = E_n^2 + I_n^2 R_s^2 \tag{2-15}$$

Therefore,

$$R_n = \left(E_n^2 + I_n^2 R_s^2\right)/4kT \Delta f \tag{2-16}$$

This equivalent noise resistance is not related to the amplifier input resistance, nor does it bear any relation to source resistance.

Noise temperature T_s is the value of the temperature of the source resistance that generates thermal noise equal to the amplifier noise. Proceeding as before, we equate terms:

$$4kT_sR_s\,\Delta f = E_n^2 + I_n^2R_s^2 \tag{2-17}$$

Therefore,

$$T_s = \left(E_n^2 + I_n^2R_s^2\right)/4kR_s\,\Delta f \tag{2-18}$$

Equation 2-18 yields T_s in kelvins. R_n and T_s are most useful for device and circuit applications where $E_n \gg I_nR_s$.

2-8 NOISE IN CASCADED NETWORKS

We now consider the problem of locating the important noise sources within a system. If we derive a useable expression for the noise factor of cascaded networks in terms of the characteristics of each network, we will then be able to predict for design purposes ways of minimizing system noise.

The system to be analyzed is shown in Fig. 2-4. It consists of a signal source with internal thermal noise and two cascaded networks. Equation 2-10 gives the noise factor as the quotient of input S/N to output S/N. The available power gain of the system is represented by G_a, and the available thermal noise power is $N_i = E_t^2/4R_s$. Therefore, an alternate expression for F is

$$F = N_o/G_akT\,\Delta f \tag{2-19}$$

where N_o is the available noise power at the load terminals. Equation 2-19 is not a useful design equation in its present form because we do not know N_o and G_a at this point.

The available noise power at the input to network 2, N_{i2}, is

$$N_{i2} = N_{o1} = F_1G_1kT\,\Delta f \tag{2-20}$$

Equation 2-20 is simply a rearrangement of Eq. 2-19. The second stage,

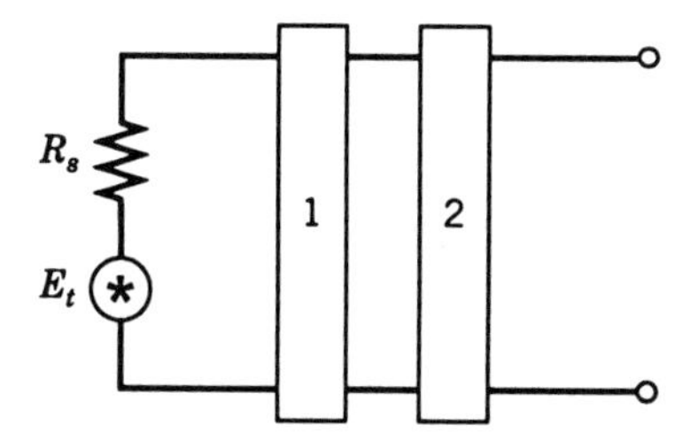

Figure 2-4 Cascaded networks.

considered separately, behaves according to

$$F_2 = N_{o2}/G_2 kT\,\Delta f \tag{2-21}$$

The noise originating in the second stage is $N_{o2} - G_2 kT\,\Delta f$, or from Eq. 2-21 it is

$$F_2 G_2 kT\,\Delta f - G_2 kT\,\Delta f = (F_2 - 1)G_2 kT\,\Delta f \tag{2-22}$$

The $G_2 kT\,\Delta f$ term represents the thermal noise power in the hypothetical source resistance for network 2.

The total output noise N_{oT} is given by the sum of terms from Eqs. 2-20 and 2-22:

$$\begin{aligned} N_{oT} &= G_2(F_1 G_1 kT\,\Delta f) + (F_2 - 1)G_2 kT\,\Delta f \\ &= (F_1 G_1 G_2 + F_2 G_2 - G_2)kT\,\Delta f \end{aligned} \tag{2-23}$$

The noise factor of the cascaded pair is

$$F_{12} = N_{oT}/G_1 G_2 kT\,\Delta f \tag{2-24}$$

By substitution of Eq. 2-23 into Eq. 2-24, we obtain

$$F_{12} = F_1 + (F_2 - 1)/G_1 \tag{2-25}$$

If the analysis is extended to three stages, we obtain the classical relation developed by Friis [3]:

$$F_{123} = F_1 + (F_2 - 1)/G_1 + (F_3 - 1)/G_1 G_2 \tag{2-26}$$

One concludes then that the noise factor of a cascaded network is primarily influenced by first-stage noise, provided that the gain of that stage is large.

When network 1 is a combination of passive circuit elements, for example, a coupling or equalizing network, its available power gain is less than unity, and the overall system noise is severely influenced by noise contributions represented by F_2.

SUMMARY

a. A universal noise model for amplifiers uses generators E_n and I_n at the input port.
b. All noise sources in a sensor–amplifier system can be represented by equivalent input noise E_{ni} (or I_{ni}), a voltage (or current) generator located in series (in parallel) with the signal source.

c. Equivalent input noise for a simple sensor–amplifier system is

$$E_{ni}^2 = E_t^2 + E_n^2 + I_n^2 R_s^2$$

d. From data on E_{ni} versus R_s, both E_n and I_n can be determined. When R_s is zero, E_n is the only noise source. When R_s is very large, then I_n is dominant.

e. The noise factor is useful in comparing amplifiers and devices:

$$F = \frac{S_i/N_i}{S_o/N_o}$$

It is often expressed in decibels as the noise figure.

f. The optimum source resistance ($R_o = E_n/I_n$) is defined for $E_n = I_n R_s$. This results in the optimum noise factor.

g. For source resistance thermal noise equal to amplifier noise, the noise resistance and noise temperature are defined.

h. For cascaded networks,

$$F = F_1 + (F_2 - 1)/G_1 + (F_3 - 1)/G_1 G_2$$

System noise is predominantly first-stage noise when the gain of that stage is high.

PROBLEMS

2-1. Derive Eq. 2-14 from Eq. 2-12.

2-2. Calculate the noise figure and the noise factor for the following values of $(E_n^2 + I_n^2 R_s^2)/E_t^2$: (a) 0, (b) 1, and (c) 2.

2-3. Find the noise figure in decibels for a system with noise temperature equal to ambient (290 K).

2-4. Derive Eq. 2-21.

2-5. A system is composed of two noisy resistances in series. Resistance R_1 is the signal source and R_2 is the load or output. Determine the noise figure for this network. Can you conclude that the broadband noise figure and the narrowband or spectral density noise figure are identical for this system?

2-6. Find the equivalent input and output noises E_{ni} and E_{no} for the circuit shown in Fig. P2-6.

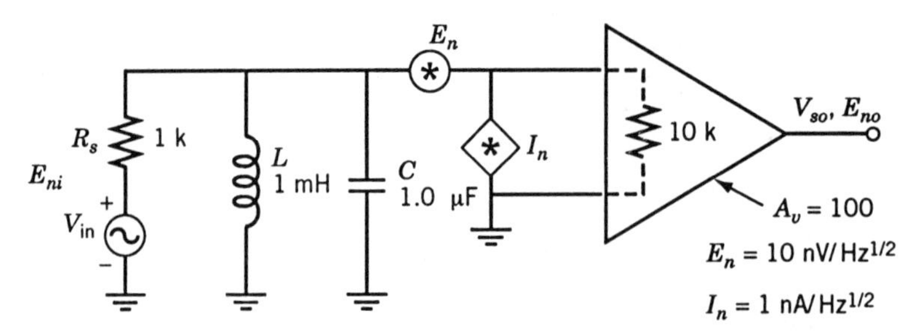

Figure P2-6

(a) Assume that the operating frequency is 10 kHz and we are interested in a 1-Hz bandwidth. Evaluate each contribution to E_{ni}^2 separately. Then determine E_{ni} (with appropriate units) for the entire system.

(b) Repeat part (a) for an operating frequency of 5033 Hz.

2-7. The noise figure of an op amp is 5 dB with a source resistance of 10 kΩ.

(a) Determine E_{ni}^2 as a spectral density for this amplifier.

(b) Determine the noise temperature in degrees celsius for this amplifier.

(c) Determine the noise resistance for this amplifier.

2-8. Consider the amplifier system shown in Fig. P2-8. The operating frequency of interest is 1.0 kHz and the bandwidth is 1.0 Hz. The amplifier's input impedance at 1.0 kHz is 2.8 kΩ at an angle of $-45°$ and the magnitude of its voltage gain is 20. The amplifier contributes noise through $E_n = 4\ \mathrm{nV/Hz}^{1/2}$ and $I_n = 8\ \mathrm{pA/Hz}^{1/2}$. E_n and I_n are not correlated. Find E_{ni}^2 for this amplifier system by *separately identifying and evaluating* the terms which contribute to E_{ni}^2. For example,

$$E_{ni}^2 = A + B + C + \cdots$$

Identify and evaluate A, B, C, and so on separately. Then sum to find E_{ni}^2.

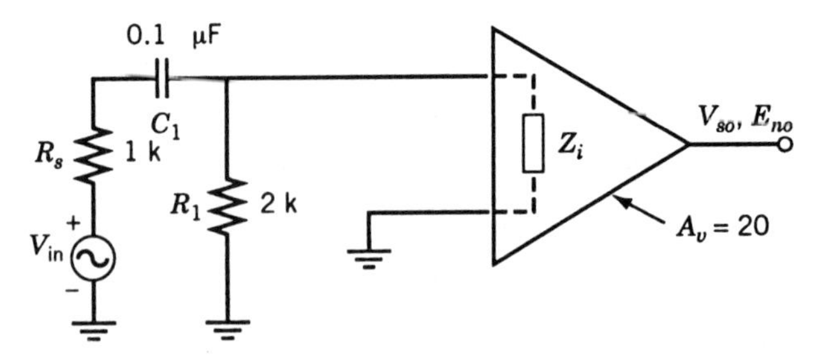

Figure P2-8

2-9. Find the mean square equivalent input noise, E_{ni}^2, for the system shown in Fig. P2-9. The coupling network is a high-pass filter composed of R_1–C_1. Consider that the operating frequency of interest is 1.0 kHz and the bandwidth is 1.0 Hz. Assume the amplifier's input impedance is infinite and its voltage gain is +1.0. However, the amplifier does contribute noise through $E_n = 4\ \text{nV/Hz}^{1/2}$ and $I_n = 8\ \text{pA/Hz}^{1/2}$ which are not correlated.

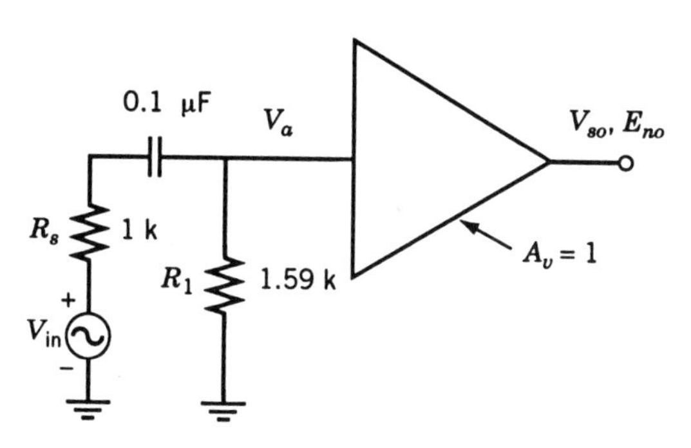

Figure P2-9

2-10. Find the K_t for the circuit in Fig. P2-10 at an operating frequency of 10 kHz.

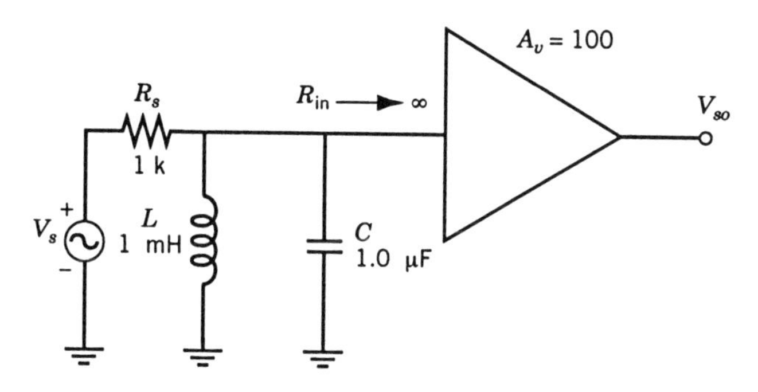

Figure P2-10

2-11. Determine the individual contributions to E_{ni}^2 and E_{no}^2 from the source resistor and the amplifier's E_n and I_n noises in Fig. P2-10. Let $E_n = 2\ \text{nV/Hz}^{1/2}$ and $I_n = 1\ \text{pA/Hz}^{1/2}$. Consider a noise bandwidth of 10 Hz centered at an operating frequency of 10 kHz.

2-12. Consider the amplifier system shown in Fig. P2-12. The bandwidth of interest is 1 Hz centered at an operating frequency of 100 kHz. At this frequency, the amplifier's input impedance is 5 kΩ at an angle of −60°. The amplifier consists of the portion of Fig. P2-12 outlined and labeled in the box. This amplifier contributes noise through $E_n = 10\ \text{nV/Hz}^{1/2}$ and $I_n = 12\ \text{pA/Hz}^{1/2}$. E_n and I_n are not correlated.

(a) Find the numerical value for K_t.

(b) Determine the output noise voltage, E_{no}^2 for this circuit. Separately identify and evaluate all the individual terms which contribute to E_{no}^2. Then find the total E_{no}^2.

(c) Calculate the equivalent input noise voltage E_{ni}^2 for this same circuit. Again separately identify and evaluate all the individual terms which contribute to E_{ni}^2.

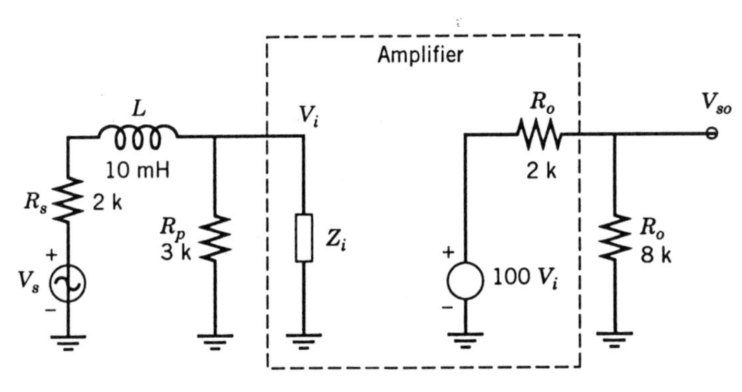

Figure P2-12

REFERENCES

1. "Representation of Noise in Linear Twoports," *Proc. IRE*, **48**, 1 (January 1960), 69–74.
2. "IRE Standards on Methods of Measuring Noise in Linear Twoports, 1959," *Proc. IRE*, **32**, 7 (July 1944), 419–422.
3. Friis, H. F., "Noise Figures of Radio Receivers," *Proc. IRE*, **32**, 7 (July 1944), 419–422.

CHAPTER 3

NOISE IN FEEDBACK AMPLIFIERS

The previous chapter demonstrated how noise sources can be reflected to the amplifier's input and output terminals under open-loop conditions, that is, in the absence of feedback. This chapter extends these reflection processes to amplifiers utilizing feedback. The additional noises introduced by the feedback elements are derived and included in the determinations of E_{ni} and E_{no}. The analysis techniques illustrate how to determine E_{ni} and E_{no} for virtually any network topology with or without feedback.

3-1 NOISE AND SOME BASIC FEEDBACK PRINCIPLES

Feedback is a handy way to change gains, impedance levels, and frequency response, reduce distortion, and alter many other properties of an electronic circuit. Circuit and system textbooks are filled with derivations which typically demonstrate that critical performance indexes are improved by the factor $1 + A\beta$ every time negative feedback is properly used. It is often tempting to think that correctly using negative feedback can reduce the noise level in a circuit or system by the same $1 + A\beta$ factor. *Would that it were so!* As will be shown, feedback does not increase or decrease the equivalent input noise, but the added feedback resistive elements themselves will add noise.

To see how noise is affected by feedback, consider the block diagram shown in Fig. 3-1 [1]. The desired input voltage is V_{in} and all the E's represent contaminating noise voltages being injected at various critical points in the system. Blocks A_1 and A_2 represent amplifiers with voltage gains and β represents the feedback network. The output voltage V_o is a

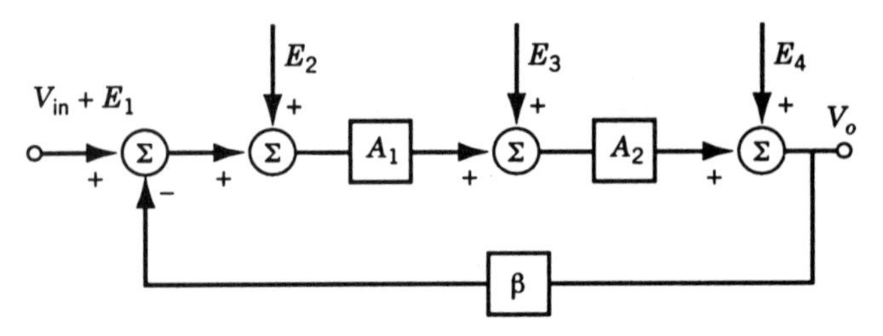

Figure 3-1 Two-stage amplifier with feedback for determining the effects of noise.

function of all five inputs according to

$$V_o = E_4 + A_2[E_3 + A_1(E_2 + V_{in} + E_1 - \beta V_o)] \tag{3-1}$$

After rearranging, the output voltage becomes

$$V_o = \frac{A_1 A_2}{1 + A_1 A_2 \beta}(V_{in} + E_1 + E_2) + \frac{A_2 E_3}{1 + A_1 A_2 \beta} + \frac{E_4}{1 + A_1 A_2 \beta} \tag{3-2}$$

Now for comparison purposes, consider an open-loop system as shown in Fig. 3-2. There are still two amplifier gain blocks of A_1 and A'_2, and the output voltage is simply

$$V_o = A_1 A'_2(V_{in} + E_1 + E_2) + A'_2 E_3 + E_4 \tag{3-3}$$

Now we impose the condition that for a meaningful comparison, the voltage gain from V_{in} must be the same for both the open- and closed-loop cases. We accomplish this by setting the A'_2 gain to be

$$A'_2 = A_2/(1 + A_1 A_2 \beta) \tag{3-4}$$

With A'_2 set to this value, Eq. 3-3 for the open-loop case becomes

$$V_o = \frac{A_1 A_2}{1 + A_1 A_2 \beta}(V_{in} + E_1 + E_2) + \frac{A_2 E_3}{1 + A_1 A_2 \beta} + E_4 \tag{3-5}$$

Comparing Eqs. 3-2 and 3-5, we see that feedback does not give any

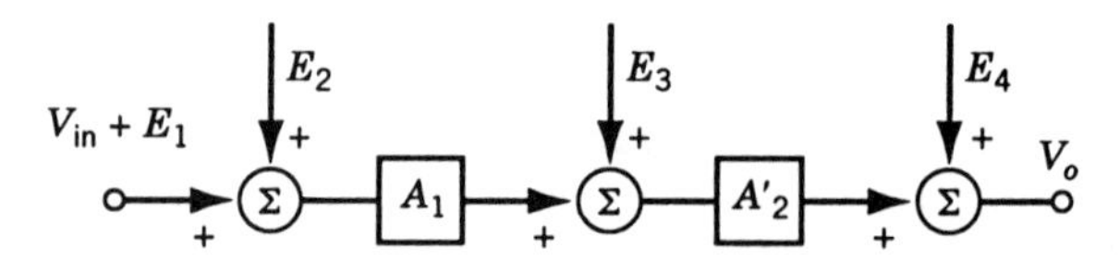

Figure 3-2 Open-loop amplifier used for comparison.

improvement for any noise source introduced at the input to either amplifier regardless of whether this noise source exists before or after the summer. Noise injected at the amplifier's output is attenuated in the feedback amplifier. Unintentionally, noise might be added directly at the output for example by adding an additional loading element with its own noise. This additional noise will have different effects on E_{ni} depending upon whether feedback is present or not.

For cases where noise is introduced at the amplifier's input or within the feedback loop, feedback gives no improvement at all. In fact, as we will demonstrate in the next section, providing feedback will actually increase the output noise level due to added thermal noise from the feedback resistors. *Consequently, the old standby principle of using feedback to improve some amplifier performance index by the* $1 + A\beta$ *factor does not work for noise.*

3-2 AMPLIFIER NOISE MODEL FOR DIFFERENTIAL INPUTS

The overwhelming majority of operational amplifiers are configured with differential inputs. The user can configure the feedback network and input signal so as to produce a noninverting amplifier, an inverting amplifier, or a true differential amplifier. Therefore, any generalized op amp model having equivalent noise sources must be able to handle all of these different configurations. The basic amplifier noise model previously introduced in Fig. 2-1 is now expanded in Fig. 3-3 to function with fully differential inputs. Noise sources E_{n1} and I_{n1} are noise contributions from the amplifier reflected to the inverting input terminal referenced to ground potential. Likewise, noise sources E_{n2} and I_{n2} are the noise contributions reflected to the noninverting input terminal. It will be shown later how these four E_n–I_n sources are reduced to simpler models with the E_n–I_n sources. The two input voltages and the one output voltage are referenced to the common ground [2].

Consider now the typical differential amplifier circuit shown in Fig. 3-4. If the op amp is ideal, the output voltage is a function of the two signal input

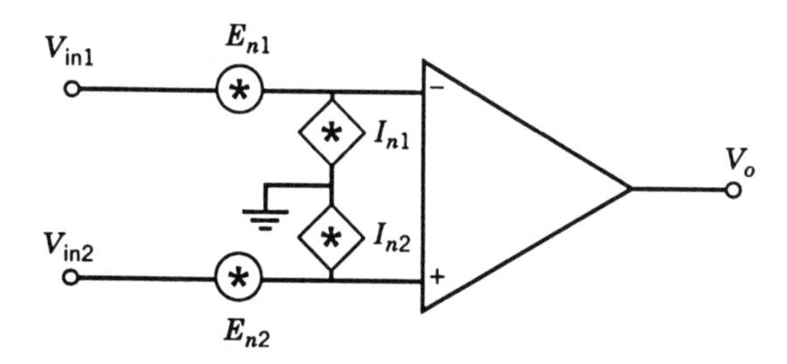

Figure 3-3 Amplifier noise and signal source.

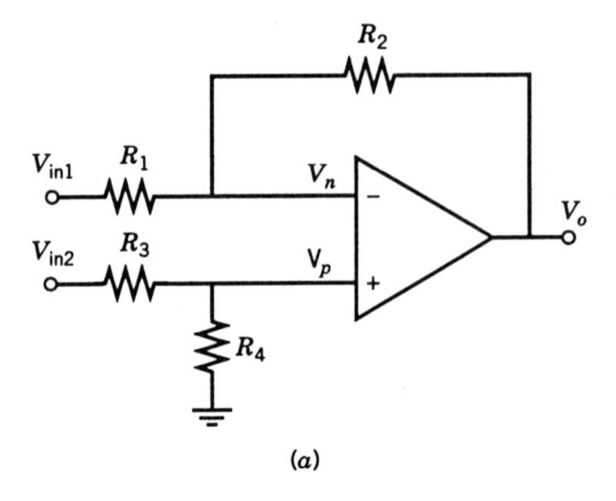

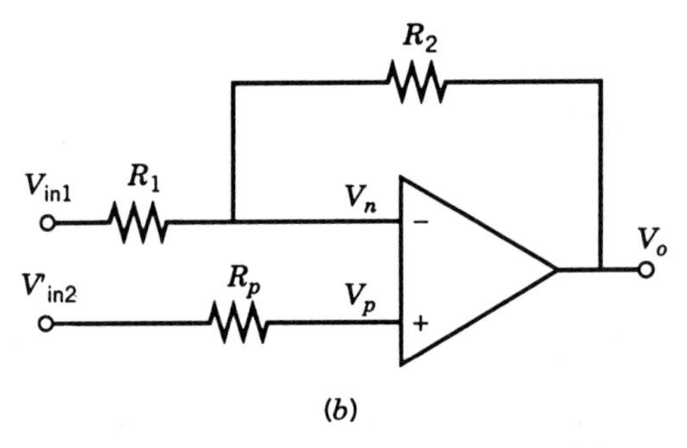

Figure 3-4 Differential amplifier using one op amp: (*a*) complete circuit and (*b*) reduced circuit.

voltages. Voltages V_p and V_n are the voltages at the respective positive and negative inputs to the amplifier referenced to ground.

The output voltage is a function of the two signal voltages according to

$$V_o = \left(\frac{R_4}{R_3 + R_4}\right)\left(\frac{R_1 + R_2}{R_1}\right)V_{\text{in2}} - \left(\frac{R_2}{R_1}\right)V_{\text{in1}} \tag{3-6}$$

An ideal differential amplifier occurs when we make the coefficients of V_{in1} and V_{in2} have identical magnitudes and opposite signs. This condition is satisfied by choosing the resistors such that $R_1R_4 = R_2R_3$ or, alternatively,

$$R_2/R_1 = R_4/R_3 \tag{3-7}$$

When Eq. 3-7 is satisfied, the output voltage becomes

$$V_o = (R_2/R_1)(V_{\text{in2}} - V_{\text{in1}}) \tag{3-8}$$

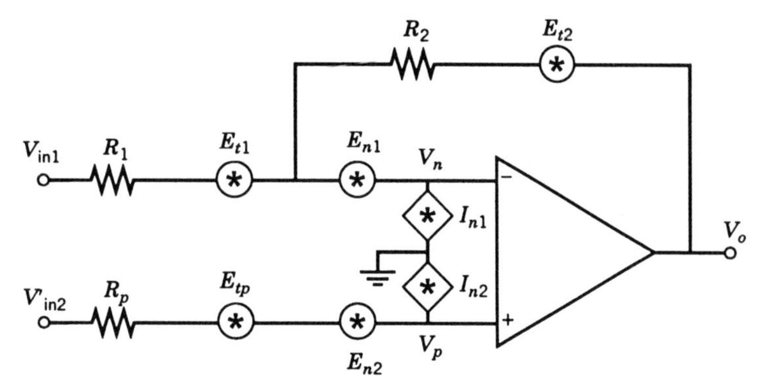

Figure 3-5 Differential amplifier with all noise sources in place.

Thus the ideal difference mode voltage gain is R_2/R_1. To examine the noise behavior of the differential amplifier, first form a Thevenin equivalent circuit at the noninverting input as shown in Fig. 3-4*b* where

$$R_p = R_3 \parallel R_4 \qquad \text{and} \qquad V'_{\text{in}2} = (R_4 V_{\text{in}2})/(R_3 + R_4) \tag{3-9}$$

Next insert noise voltage and current sources for the op amp and Thevenin equivalent noise sources for the resistors as shown in Fig. 3-5 [3, 4].

We could write the necessary equations to determine the effects of both input signal sources plus the seven noise sources on the output voltage. However, this would be quite complicated because of the rms effects of all sources and the need to square all terms. Instead we replace all noise sources with uncorrelated and independent signal sources as shown in Fig. 3-6. Here

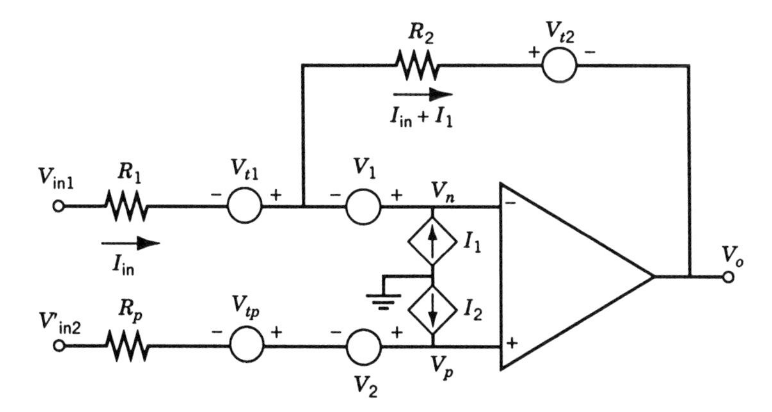

Figure 3-6 Differential amplifier with signal sources in place.

the seven signal sources have arbitrary polarities as indicated. To better relate to a practical limitation, consider that the operational amplifier has a finite open-loop voltage gain of A but is ideal otherwise.

The four defining equations for this circuit are

$$V_o = A(V_p - V_n) \tag{3-10}$$

$$V_p = V'_{\text{in}2} + R_p I_2 + V_{tp} + V_2 \tag{3-11}$$

$$V_n = V_{\text{in}1} - R_1 I_{\text{in}} + V_{t1} + V_1 \tag{3-12}$$

$$V_{\text{in}1} - R_1 I_{\text{in}} + V_{t1} = V_o + V_{t2} + R_2(I_{\text{in}} + I_1) \tag{3-13}$$

Combining and simplifying these four equations gives

$$V_o\left(\frac{1}{A} + \frac{R_1}{R_1 + R_2}\right) = V'_{\text{in}2} - V_{\text{in}1} + V_2 - V_1 + V_{tp} - V_{t1} + R_p I_2$$
$$+ \left(\frac{R_1}{R_1 + R_2}\right)(V_{\text{in}1} + V_{t1} - V_{t2} - I_1 R_2) \tag{3-14}$$

Now we can easily produce the case for the ideal operational amplifier by taking the limit as $A \to \infty$ and rearranging Eq. 3-14 to give

$$V_o = \left(1 + \frac{R_2}{R_1}\right)(V'_{\text{in}2} + V_2 + V_{tp} + I_2 R_p - V_1)$$
$$- \frac{R_2}{R_1}(V_{\text{in}1} + V_{t1}) - V_{t2} - I_1 R_2 \tag{3-15}$$

The coefficient of each term represents the transfer gain (respective K_t's—either voltage gains or transresistance gains) to the output where V_o is produced.

Previously, for clarity, we substituted voltage and current signal sources for corresponding noise sources. The gain to the output will be the same for both signal sources and noise sources from the same circuit position. Therefore, we pair up the corresponding signal and noise sources in Figs. 3-5*b* and 3-6. Then we modify Eq. 3-15 to express the square of the equivalent output noise voltage, E_{no}^2, contributed by each noise source. This result is

$$E_{no}^2 = \left(1 + \frac{R_2}{R_1}\right)^2 \left(E_{n2}^2 + E_{tp}^2 + I_{n2}^2 R_p^2 + E_{n1}^2\right)$$

$$+ \left(\frac{R_2}{R_1}\right)^2 \left(E_{t1}^2\right) + E_{t2}^2 + I_{n1}^2 R_2^2 \qquad (3\text{-}16)$$

This equation shows how each noise source contributes to the total squared output noise. Specifically, both equivalent input noise voltages and the noise from R_p are reflected to the output by the square of the noninverting voltage gain, $(1 + R_2/R_1)^2$. The positive input noise current "flows through" R_p establishing a noise voltage which, in turn, is reflected to the output by the same gain factor, $(1 + R_2/R_1)^2$. The noise from R_1 is reflected to the output by the square of the inverting gain factor, $(R_2/R_1)^2$. The negative input noise current "flows through" the feedback resistor R_2 establishing a noise voltage directly at the output. Finally, the noise contribution due to R_2 appears directly at the output.

When we attempt to determine E_{ni}, we must first decide which terminal will be the reference. This is critical since the K_t's are different (by one) for the inverting and noninverting inputs unless the ideal differential amplifier condition of Eq. 3-7 is satisfied. Under this special condition the differential voltage gain from either terminal is $\pm R_2/R_1$.

First reflect E_{no}^2 to the *inverting* input by dividing Eq. 3-16 by $(R_2/R_1)^2$ to obtain

$$E_{ni1}^2 = \left(1 + \frac{R_1}{R_2}\right)^2 \left(E_{n2}^2 + E_{tp}^2 + E_{n1}^2\right)$$

$$+ R_1^2 I_{t2}^2 + E_{t1}^2 + I_{n1}^2 R_1^2 + I_{n2}^2 R_p^2 \left(1 + \frac{R_1}{R_2}\right)^2 \qquad (3\text{-}17)$$

where

$$R_1^2 I_{t2}^2 = R_1^2 \frac{E_{t2}^2}{R_2^2}$$

Note that the two amplifier noise voltages plus E_{tp}^2 are all increased at the input by the $(1 + R_1/R_2)^2$ factor. Usually $R_1 \ll R_2$ for a typical high-gain amplifier application, so the first three noise voltage sources essentially contribute directly to E_{ni1}^2 as does E_{t1}^2. The noise current of the feedback resistor R_2 is multiplied by R_1^2. In effect, the noise current generator of resistor R_2 is in parallel with the input. The I_{n1} noise current "flows through" R_1 creating a direct contribution to E_{ni1}^2. The I_{n2} noise current "flows through" R_p to produce a noise voltage and then is reflected to the inverting input by the same $(1 + R_1/R_2)^2$ factor as the first three noise voltages.

When reflected to the *noninverting* input, we divide Eq. 3-10 by $(1 + R_2/R_1)^2$ and obtain

$$E_{ni2}^2 = (E_{n2}^2 + E_{tp}^2 + E_{n1}^2) + \left(\frac{R_1}{R_1 + R_2}\right)^2 (E_{t2}^2) + \left(\frac{R_2}{R_1 + R_2}\right)^2 (E_{t1}^2) + I_{n1}^2 (R_1 \parallel R_2)^2 + I_{n2}^2 R_p^2 \qquad (3\text{-}18)$$

Here the two amplifier noise voltages as well as the noise voltage from R_p contribute directly to E_{ni2}^2. The noise voltage in the feedback resistor is divided (or reduced) by the square of feedback factor. The noise in R_1 is slightly diminished, but essentially unchanged when $R_1 \ll R_2$. The inverting noise current "flows through" the parallel combination of R_1 and R_2 and then contributes directly to E_{ni2}^2. The noninverting noise current "flows through" R_p and contributes directly.

Next consider the special and desired case where the ideal differential amplifier is produced by choosing resistors according to Eq. 3-7. Under this condition, $K_{t1} = -R_2/R_1$, $K_{t2} = R_2/R_1$, and $K_{t1}^2 = K_{t2}^2 = K_t^2$. Furthermore,

$$E_{ni1}^2 = E_{ni2}^2 = E_{ni}^2 = E_{no}^2/K_t^2 \qquad (3\text{-}19)$$

After some mathematical manipulations, Eqs. 3-17 and 3-18 simplify to

$$E_{ni}^2 = \left(1 + \frac{R_1}{R_2}\right)^2 (E_{n1}^2 + E_{tp}^2 + E_{n2}^2) + \left(\frac{R_1}{R_2}\right)^2 (E_{t2}^2) + E_{t1}^2 + I_{n1}^2 R_1^2 + I_{n2}^2 R_p^2 \left(1 + \frac{R_1}{R_2}\right)^2 \qquad (3\text{-}20)$$

which is identical to the inverting amplifier in Eq. 3-17.

Example 3-1 In the differential amplifier of Fig. 3-4*a*, find the total output noise voltage, the total noise referenced to the *inverting* input, and the minimum amplified rms signal which occurs when this signal is equal to the rms value of the total input reference noise. This latter condition defines a signal-to-noise ratio of 1. Assume typical 741 op amp noise parameters of $E_n = 20\ \text{nV/Hz}^{1/2}$ and $I_n = 0.5\ \text{pA/Hz}^{1/2}$. Neglect the $1/f$ noise contributions of the amplifier and resistors. Assume that the 741 op amp is ideal except for noise. The resistor values are $R_1 = R_3 = 1\ \text{k}\Omega$ and $R_2 = R_4 = 50\ \text{k}\Omega$

Solution Note that Eq. 3-7 is satisfied for producing an ideal differential amplifier. First we find $R_p = R_3 \parallel R_4 = 980\ \Omega$, and the difference mode gain to be 50. Using Eqs. 3-16 and 3-17, we tabulate the noise contributions as follows.

Noise Source	Noise Value	Gain Multiplier	Output Noise Contribution	Input Noise Contribution
R_1	$4\ \text{nV/Hz}^{1/2}$	50	$200\ \text{nV/Hz}^{1/2}$	$4\ \text{nV/Hz}^{1/2}$
R_2	$28.3\ \text{nV/Hz}^{1/2}$	1	$28.3\ \text{nV/Hz}^{1/2}$	$0.566\ \text{nV/Hz}^{1/2}$
R_p	$3.96\ \text{nV/Hz}^{1/2}$	51	$202\ \text{nV/Hz}^{1/2}$	$4.04\ \text{nV/Hz}^{1/2}$
E_{n1}	$14.14\ \text{nV/Hz}^{1/2}$	51	$721\ \text{nV/Hz}^{1/2}$	$14.4\ \text{nV/Hz}^{1/2}$
E_{n2}	$14.14\ \text{nV/Hz}^{1/2}$	51	$721\ \text{nV/Hz}^{1/2}$	$14.4\ \text{nV/Hz}^{1/2}$
I_{n1}	$0.5\ \text{pA/Hz}^{1/2}$	50k	$25\ \text{nV/Hz}^{1/2}$	$0.5\ \text{nV/Hz}^{1/2}$
I_{n2}	$0.5\ \text{pA/Hz}^{1/2}$	49.98k	$25\ \text{nV/Hz}^{1/2}$	$0.5\ \text{nV/Hz}^{1/2}$
Total Noise Contributions			$1059.5\ \text{nV/Hz}^{1/2}$	$21.16\ \text{nV/Hz}^{1/2}$

Clearly, in this example the E_n noise is dominant in both E_{no} and E_{ni} calculations. If we assume a 1-MHz gain bandwidth product for the 741 op amp, the -3-dB corner frequency will be approximately 1 MHz/50 = 20 kHz. The noise bandwidth, assuming a single pole response, is 20 kHz($\pi/2$) = 31.42 kHz. Therefore, $E_{no} = 1.06\ \mu\text{V/Hz}^{1/2} \times (31.42\ \text{kHz})^{1/2} = 188\ \mu\text{V}$ and $E_{ni} = 21.16\ \text{nV/Hz}^{1/2} \times (31.42\ \text{kHz})^{1/2} = 3.75\ \mu\text{V}$. Hence, for a signal-to-noise ratio of 1, the input signal would be equal to 3.75 μV.

As a general rule, a balanced differential stage with matched transistors and equal bias currents is normally used as the first stage in most op amps. Under this condition, $E_{n1} = E_{n2}$. If a data sheet value for E_n is given and you wish to produce an exact noise model, then divide the given E_n value by $\sqrt{2}$ and insert each part in your circuit model for the individual E_{n1} and E_{n2} noise voltage sources. On the other hand, in the inverting configuration, it is often simpler to use the specified data sheet value for E_n as a single noise voltage source and incorporate it into the circuit as shown in Fig. 3-7 [5]. Here the *output* noise voltage contributed by the op amp, neglecting the

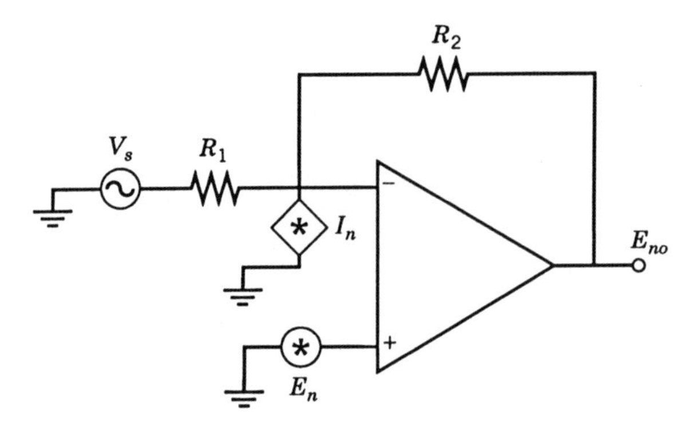

Figure 3-7 Simplified inverting amplifier with noise sources in place.

thermal noise in the resistors, is simply

$$E_{no}^2 = (1 + R_2/R_1)^2 E_n^2 + R_2^2 I_n^2 \tag{3-21}$$

This is the noise present at the output from the amplifier's noise sources alone as affected by the gain setting resistors. This output noise does not depend upon whether the amplifier is used in the inverting or noninverting configuration. The respective E_{ni} voltages will be different because the two voltage gains are different.

It is interesting to note that in the simplified circuit of Fig. 3-7, the noise resistance, R_o, is obtained when equal noise contributions occur at the *output* from the E_n and I_n sources. From Eq. 3-21, R_o is found as

$$R_o = E_n/I_n = R_1R_2/(R_1 + R_2) = R_1 \parallel R_2 \tag{3-22}$$

At impedance levels of source resistance R_1 that are below R_o, E_n noise is dominant; for levels above R_o, I_n noise dominates. In high-gain applications, $R_1 \ll R_2$ so that $R_o \approx R_1$. Note that by choosing $R_1 = R_o$, we do not guarantee minimum circuit noise. This only occurs for zero source resistance which maximizes the signal-to-noise ratio. The minimum noise factor (but not minimum noise) is achieved with $R_1 = R_o$, where equal noise contributions come from E_n and I_n.

One novel method for measuring I_n is to use the circuit shown in Fig. 3-8. Although this method is simple, it may be difficult to use with very low noise amplifiers because of pickup. In that case, use the method of Ch 15. Initially, both switches are closed (shorted) and E_{no} is measured. Since R_1 is omitted (infinite), the measured output noise will be the total amplifier noise voltage, E_n, plus additional noise from a necessary second stage because the voltage gain of the stage under test is only unity. Then the switch paralleling R_2 is opened and E_{no} measured again. With reference to Eq. 3-16, we see that E_{no}

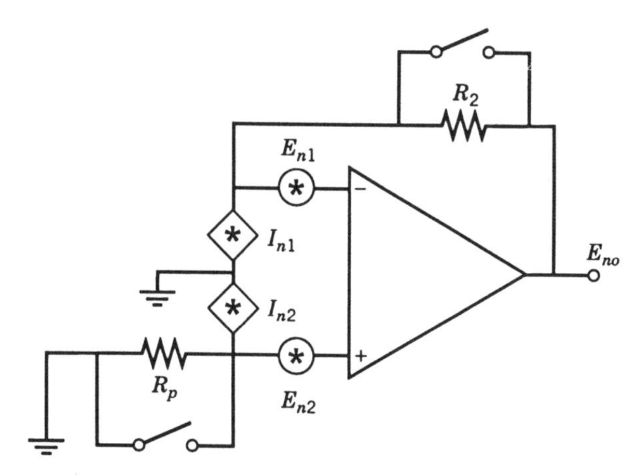

Figure 3-8 Circuit for measuring current noise sources.

now includes contributions from E_n, $I_{n1}R_2$, and E_{t2}. If R_2 is large, one should first check to make sure that the operating point of the op amp does not change significantly. The thermal noise from R_2 can be calculated, leaving only the I_{n1} term as the unknown. Finally, both switches are opened and E_{no} remeasured. Again recheck the operating point. The new terms added to this E_{no} measurement are the thermal noise from R_p (which can be calculated) and $I_{n2}R_p$. Thus I_{n2} can be determined and its value should be included in the inverting op amp noise model whenever R_p is not zero.

When $R_p = 0$, the I_{n2} noise source is essentially shorted out and only I_{n1} contributes noise to E_{ni1}. When measuring I_n, the "source" resistor R_1 should be made very large such that $I_{n1}^2 R_1^2$ becomes the dominant term as derived in Sec. 15-2-1. When measured in this way, the I_n specified on a data sheet is the noise current associated with the inverting input terminal. There is no need to include an I_{n2} term or a noise source unless R_p is not very small or equal to zero. Under this condition, one should add a second noise current source for I_{n2} equal in value to the data sheet amplifier's I_n value.

3-3 INVERTING NEGATIVE FEEDBACK

The inverting amplifier configuration with resistive negative feedback is the most widely used stage configuration. It is easily obtained by grounding $V'_{\text{in}2}$ in Fig. 3-5, replacing R_1 with the signal source resistance R_s. The input offset voltage due to bias current will be canceled by making R_p a single resistor equal to the parallel combination of R_s and R_2. These changes are shown in Fig. 3-9.

All noise sources are now reflected to the $V_{\text{in}1}$ input and we look for ways to simplify this circuit. Grouping and rearranging the noise voltage and

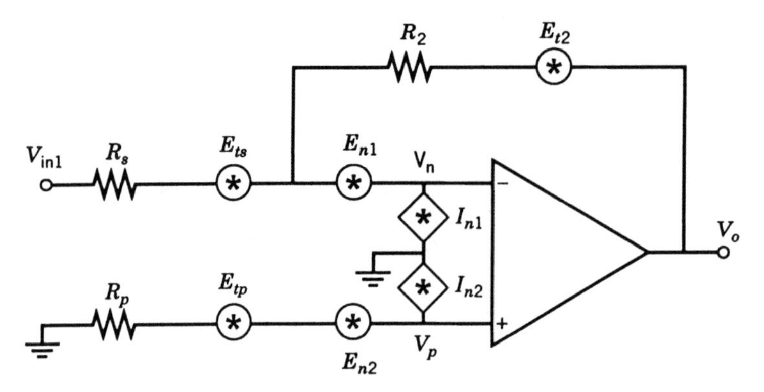

Figure 3-9 Simplified closed-loop inverting amplifier.

current terms from Eq. 3-17 gives

$$E_{ni1}^2 = \left(1 + \frac{R_s}{R_2}\right)^2 \left(E_{n1}^2 + E_{n2}^2 + E_{tp}^2 + I_{n2}^2 R_p^2\right) + \left(\frac{R_s}{R_2}\right)^2 \left(E_{t2}^2\right) + E_{ts}^2 + I_{n1}^2 R_s^2$$

$$E_{ni1}^2 = E_{ts}^2 + R_s^2\left(I_{n1}^2 + I_{t2}^2\right) + \left(1 + \frac{R_s}{R_2}\right)^2 \left(E_{n1}^2 + E_{n2}^2 + E_{tp}^2 + I_{n2}^2 R_p^2\right) \tag{3-23}$$

where $I_{t2} = E_{t2}/R_2$. An op amp specification sheet will normally provide E_n and I_n values which are related to the preceding amplifier according to

$$E_n = \sqrt{E_{n1}^2 + E_{n2}^2} \tag{3-24}$$

$$I_n = I_{n1} = I_{n2} \tag{3-25}$$

Thus we see that the single E_n noise voltage typically specified on a manufacturer's data sheet is really the rms sum of the two equal input noise voltages from each separate input terminal. Equation 3-23 simplifies further to

$$E_{ni}^2 = E_{ni1}^2 = E_{ts}^2 + I_n^2 R_s^2 + \left(1 + \frac{R_s}{R_2}\right)^2 \left(E_n^2 + E_{tp}^2 + I_{n2}^2 R_p^2\right) + I_{t2}^2 R_s^2 \tag{3-26}$$

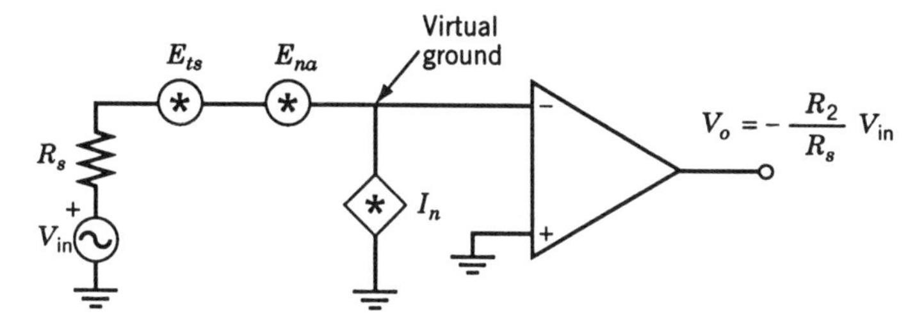

Figure 3-10 Simplified open-loop inverting amplifier with noise sources in place.

Now we define a new equivalent amplifier noise voltage E_{na}^2 as

$$E_{na}^2 = \left(1 + \frac{R_s}{R_2}\right)^2 \left(E_n^2 + E_{tp}^2 + I_{n2}^2 R_p^2\right) + I_{t2}^2 R_s^2 \qquad (3\text{-}27)$$

So the equivalent input noise simplifies to

$$E_{ni}^2 = E_{ts}^2 + E_{na}^2 + I_n^2 R_s^2 \qquad (3\text{-}28)$$

The closed-loop inverting amplifier can now be equivalently represented by the simplified open-loop circuit shown in Fig. 3-10. Note that all the noise due to the R_2 feedback and the R_p offset adjust resistors as well as the amplifier E_n noise are contained in the equivalent E_{na} noise source.

Often the R_p resistor on the noninverting input terminal is omitted, especially in op amps having a MOSFET first stage, since this resistor's main function is to eliminate the offset voltage due to input bias current. In addition, low noise amplifiers often require closed-loop gains of 30 or more which makes $R_2 \gg R_s$. For this typical condition where $R_2 \gg R_s \gg R_p$, then Eq. 3-28 for the equivalent input noise of an inverting feedback amplifier simplifies to

$$E_{ni}^2 = E_{ts}^2 + E_n^2 + \left(I_n^2 + I_{t2}^2\right) R_2^2 \qquad (3\text{-}29)$$

This demonstrates that the thermal noise current of the feedback resistor R_2 is effectively in parallel with the input. To reduce it's noise contribution, R_2 must be large.

3-4 NONINVERTING NEGATIVE FEEDBACK

The noninverting op amp feedback configuration is obtained from Fig. 3-5 by shorting the V_{in1} source to ground and applying the input signal directly as the V'_{in2} source. The resistor R_p becomes the source resistance, R_s, for the V'_{in2} source now simply designated as V_{in} in Fig. 3-11.

Now we show how to reduce the noninverting feedback circuit to a simpler, open-loop equivalent network. Using Eq. 3-16 as a starting point, the

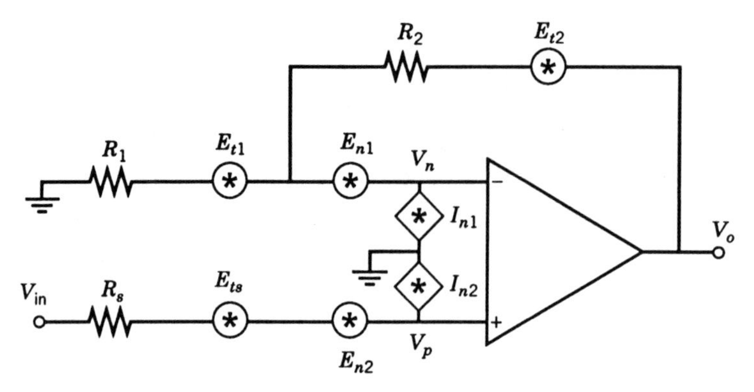

Figure 3-11 Simplified closed-loop noninverting amplifier.

output squared noise for Fig. 3-11 is

$$E_{no}^2 = \left(1 + \frac{R_2}{R_1}\right)^2 (E_{n1}^2 + E_{n2}^2 + E_{ts}^2 + I_{n2}^2 R_s^2)$$

$$+ \left(\frac{R_2}{R_1}\right)^2 (E_{t1}^2) + E_{t2}^2 + I_{n1}^2 R_2^2 \tag{3-30}$$

Using the normal E_n relationship of Eq. 3-24, E_{no}^2 simplifies to

$$E_{no}^2 = \left(1 + \frac{R_2}{R_1}\right)^2 (E_n^2 + E_{ts}^2 + I_{n2}^2 R_s^2)$$

$$+ \left(\frac{R_2}{R_1}\right)^2 (E_{t1}^2) + E_{t2}^2 + I_{n1}^2 R_2^2 \tag{3-31}$$

Next we reflect the output noise to the noninverting input by dividing E_{no}^2 by the squared noninverting gain factor of $(1 + R_2/R_1)^2$ giving

$$E_{ni}^2 = E_{ts}^2 + E_n^2 + \left(\frac{R_1}{R_1 + R_2}\right)^2 E_{t2}^2 + \left(\frac{R_2}{R_1 + R_2}\right)^2 E_{t1}^2$$

$$+ I_{n1}^2 (R_1 \parallel R_2)^2 + I_{n2}^2 R_s^2 \tag{3-32}$$

Using the normal situation where $I_n = I_{n1} = I_{n2}$, we can define a new

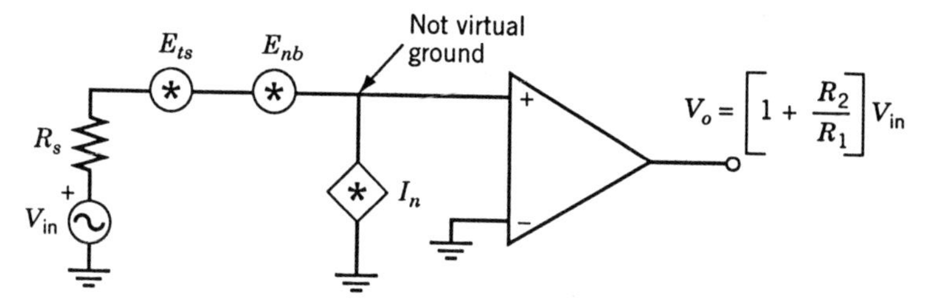

Figure 3-12 Simplified open-loop noninverting amplifier with noise sources in place.

equivalent noninverting amplifier noise source according to

$$E_{nb}^2 = E_n^2 + \left(\frac{R_1}{R_1 + R_2}\right)^2 (E_{t2}^2) + \left(\frac{R_2}{R_1 + R_2}\right)^2 (E_{t1}^2) + I_{n1}^2 (R_1 \parallel R_2)^2 \tag{3-33}$$

The previous two equations produce the simplified open-loop equivalent amplifier circuit shown in Fig. 3-12 where the total equivalent input noise is given by

$$E_{ni}^2 = E_{ts}^2 + E_{nb}^2 + I_n^2 R_s^2 \tag{3-34}$$

Note that all the noise due to both feedback resistors and the amplifier's voltage noise are contained in the equivalent E_{nb} noise source. If the closed-loop voltage gain is greater than 30, the E_{nb} noise of the R_2 feedback resistor becomes negligible. Then E_{nb} contains the amplifier E_n, the thermal noise of R_1, and the current noise from I_n "flowing through" R_1. Thus remember that the R_1 resistor is essentially in series with the source resistor as both a source of thermal noise at the input and as a path through which I_n can flow.

3-5 POSITIVE FEEDBACK

Many circuits and systems have positive feedback, either deliberately introduced as in oscillators or gain-boosting amplifiers or accidental as with unstable oscillations in high-gain amplifiers or as latched-up phenomena where some critical internal node is saturated to a power supply rail [6]. Positive feedback does not alter the basic principles of noise regardless of how this feedback is incorporated into the circuit. The concepts and techniques for analyzing and designing with noise in positive feedback systems are the same as those presented earlier in this chapter for negative feedback systems. For example, when SPICE is used to analyze a network, its algorithms establish the circuit equations without a priori knowing or caring whether positive feedback is present. Solutions to the network equations are found by numerical iterating techniques. If a solution exists, SPICE will find it, barring some special case control statements or tricky networks. The solution obtained will do the same noise analysis and determine noise results regardless of the type of feedback actually present.

Sometimes one is required to design a low-noise sinusoidal oscillator. Usually there is a design specification that the phase noise level must be down a certain number of decibels from the fundamental frequency at a frequency spacing of so many hertz. The usual question is, "How does one use the noise and feedback principles to design such an oscillator?" The secret lies in the realization that a low-noise oscillator is produced by first designing a low-noise amplifier and a low-noise feedback network. The noise produced in the amplifier and feedback elements will cause random fluctuations in the output frequency which will appear as jitter or as a "skirt" on a spectrum analyzer display. In Sec. 3-1 we showed that feedback affects the signal and noise at the input to the amplifier in the same way. Therefore, the way to reduce the noise jitter at the oscillator's output is to ensure a low equivalent noise level at the input summing junction where the feedback is connected and use a low-noise power supply.

Another very important consideration in feedback systems is stability. Is the amplifier stable? There are many well-established analytical techniques to test for stability such as using Nyquist plots, Routh–Herwitz criteria, and gain and phase margin determinations using Bode plots. Each approach requires either a knowledge or determination of the poles and zeros in the amplifier and feedback network. Often the determination of stability may be extremely tedious and time consuming. However, there are two easy ways to determine stability when using SPICE [7]. First, when the circuit is simulated, the transient response of an unstable network will show an unbounded output. Of course, to see this effect may require several simulations over long time intervals. Second, positive feedback is present in any system whenever the closed-loop gain is larger than the forward gain. Positive feedback does not always mean the system is unstable. However, it should raise a warning flag to do further investigations.

Example 3-2 Consider an amplifier which has an open-loop voltage gain of 80 dB and poles at 1, 6, and 22 MHz. Determine if this amplifier will be stable when negative feedback is used to set the closed-loop voltage gain to 40 dB.

Solution Using the modeling techniques in [8], we can produce a simple macromodel for the amplifier. We conduct a simulation as shown in Fig. 3-13 to verify the correct open-loop response. Note, in general, that one could already have the frequency response simulation completed for an amplifier, and may just wish to determine whether it is stable by performing and correctly interpreting the following simulation results.

Next we connect the feedback network around the amplifier and close the loop. Simulation results of this network are also shown in the figure. Note that at about 12 MHz the magnitude of the closed-loop response exceeds the

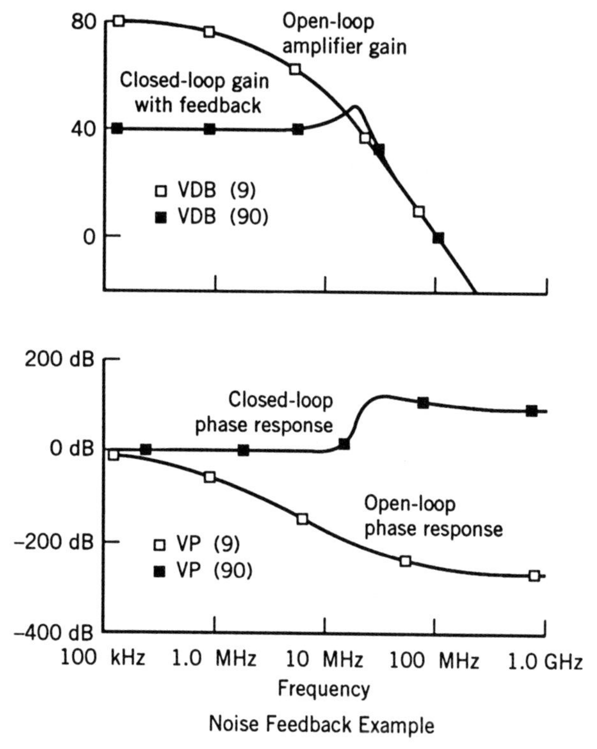

Figure 3-13 Effect of positive feedback as shown by PSpice.

open-loop gain. This means that positive feedback is present since the only way the closed-loop gain can be larger than the forward amplifier gain is for the denominator of the closed-loop gain expression, namely $|1 + A\beta|$, to be less than 1. Examination of the phase responses shows that the closed-loop phase shift becomes positive or leads for frequencies of about 12 MHz and greater. A leading phase means that the 40-dB closed-loop amplifier configuration is unstable. Performing a transient analysis at the output of the feedback amplifier will confirm this instability. If so instructed, PSpice will perform a normal noise analysis on this feedback amplifier, calculating the noise contribution from each resistor at the specified frequencies even though this amplifier is unstable due to positive feedback.

SUMMARY

a. Feedback cannot be used to reduce the equivalent input noise of an op amp circuit.

b. The four op amp noise sources in a differential amplifier as well as all the resistor noise sources can be treated as signal sources to determine their effects on the total output noise.

c. In general, the reflected equivalent input noises to the two inputs of an op amp with feedback will be different only because the voltage gains of the two configurations are different.

d. Noise in circuits and systems having positive feedback is determined exactly the same way as for circuits with negative feedback.

PROBLEMS

3-1. Calculate E_{na} for Example 3-1 using Eq. 3-25.

3-2. Consider the feedback amplifier shown in Fig. P3-2a. The operational amplifier can be considered ideal except for its noise parameters of $E_n = 30\ \text{nV/Hz}^{1/2}$ and $I_n = 0.3\ \text{pA/Hz}^{1/2}$. Resistor R_1 has its associated noise source E_{t1}. In Fig. P3-2b the amplifier and R_1 noise sources have been replaced by signal sources.

(a) The output voltage can be expressed as a function of the signal sources according to $V_o = AV_s + BV_1 + CV_{t1} + DI_1$, where A, B, C, and D are constants. Determine numerical values for these constants.

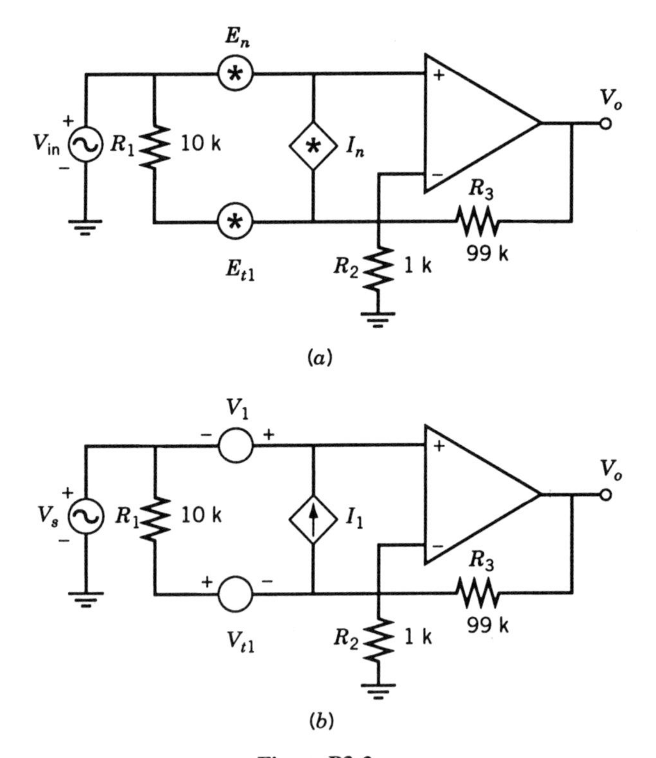

Figure P3-2

(b) Find the individual spectral density noise contributions to E_{no}^2 due to the three resistors plus E_n and I_n. Evaluate each contribution to E_{no}^2 in a 1-Hz bandwidth. Then determine the total E_{no}^2.

(c) Find E_{ni}^2 relative to the signal source V_s.

3-3. Consider the op amp with feedback shown in Fig. P3-3. The op amp can be considered ideal in every respect except for noise.

(a) Determine the noise contributions at the output (V_o) in a 1-Hz band centered at 1 kHz due to the thermal noise of the four resistors R_1, R_2, R_F, and R_L.

(b) Now consider that the op amp has noise parameters of $E_n = 9\ \text{nV/Hz}^{1/2}$ and $I_n = 0.15\ \text{pA/Hz}^{1/2}$, both of which are constant with frequency. Connect the E_n source between ground and the noninverting input to the op amp. Connect two I_n sources between ground and each op amp input. Determine the contributions to E_{no}^2 from these three noise sources. Express your answers in volts^2 per hertz.

(c) Estimate the total output noise voltage that would be indicated by a true rms voltmeter. The noise bandwidth of the voltmeter is three times as large as the noise bandwidth of the amplifier.

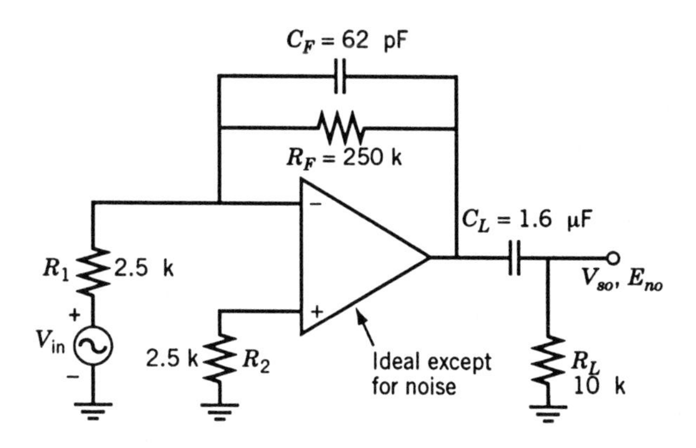

Figure P3-3

3-4. Repeat part (b) of the previous problem if one I_n noise source is connected between the two inputs to the op amp.

3-5. Find the characteristic source resistance for the circuit in Fig. P3-3.

3-6. Consider the op amp with feedback shown in Fig. P3-6. The op amp can be considered ideal in every respect except for noise.

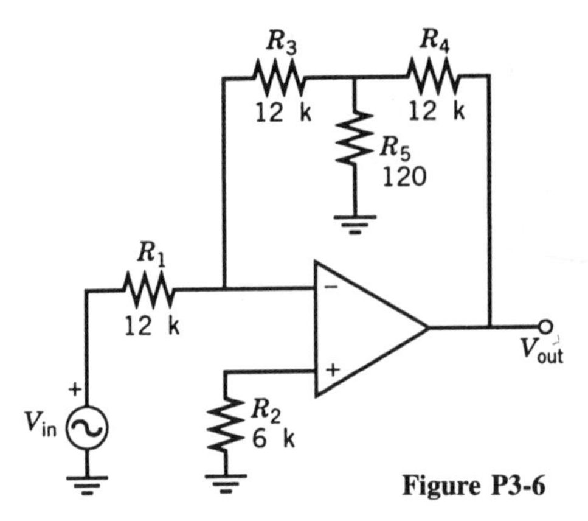

Figure P3-6

(a) Determine the noise contributions at the output due to the thermal noise of the five resistors.

(b) Find the noise contributions at the output due to op amp noise parameters of $E_n = 10\ \mathrm{nV/Hz^{1/2}}$ and $I_n = 1\ \mathrm{pA/Hz^{1/2}}$. Neglect any $1/f$ noise effects.

3-7. The op amp in Fig. P3-7 is biased using a single +15-V power supply. Assume that the op amp has flat noise spectral densities of $E_n = 32\ \mathrm{nV/Hz^{1/2}}$ and $I_n = 0.3\ \mathrm{pA/Hz^{1/2}}$. To set the dc output voltage at $+15\ \mathrm{V}/2 = 7.5\ \mathrm{V}$, a Zener diode is used which has a reference voltage of 7.5 V and a dynamic resistance of $r_Z = 10\ \Omega$. This Zener diode generates avalanche noise according to

$$E_Z^2 = (5 \times 10^{-20} V_Z^4)/I_Z \qquad \mathrm{V^2/Hz}$$

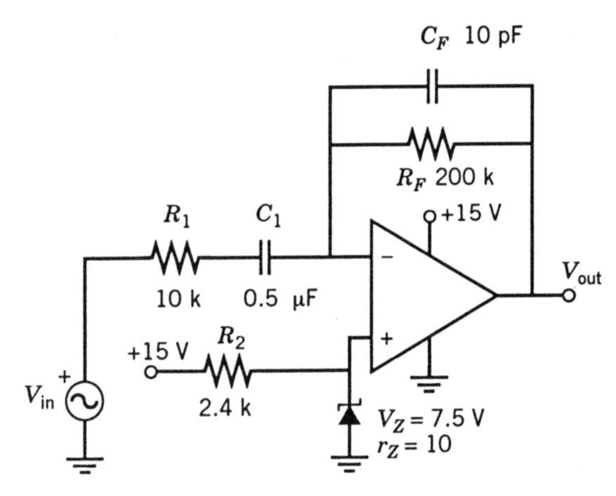

Figure P3-7

(a) Determine the total midband equivalent input noise, E_{ni}, due to three dominant noise sources. Find the contributions for each source separately before doing a summation to determine E_{ni}^2.

(b) Using your answer to part (a), find the total rms value of E_{no} which would be measured by a true rms voltmeter.

3-8. Determine the input and output noise contributions from all resistors and from the E_n and I_n noise sources for the operational amplifier configuration shown in Fig. P3-8. Consider a single E_n noise source in series with the noninverting input and having a value of 10 nV/Hz$^{1/2}$ and two I_n noise sources = 0.3 pA/Hz$^{1/2}$ connected between ground and the amplifier's inputs. Assume the op amp is ideal except for noise.

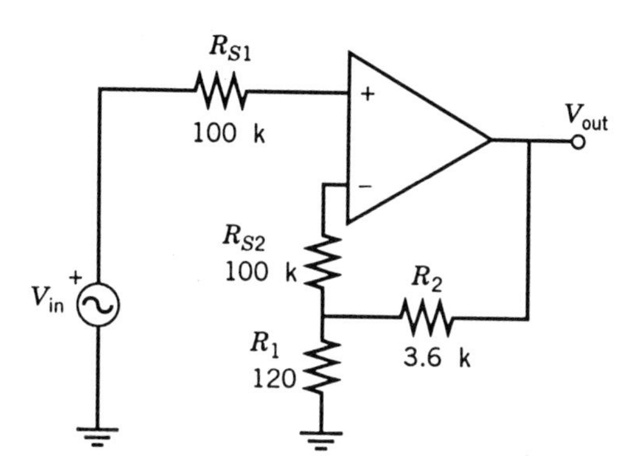

Figure P3-8

3-9. Bridge amplifiers serve as sensing circuits for many low-noise applications. Consider the circuit shown in Fig. P3-9 which is very useful

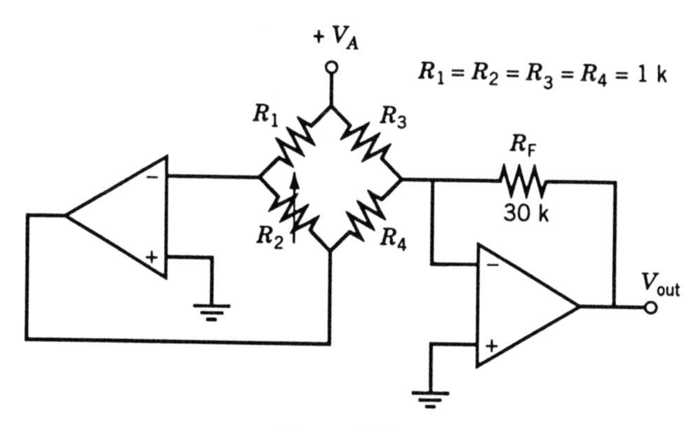

Figure P3-9

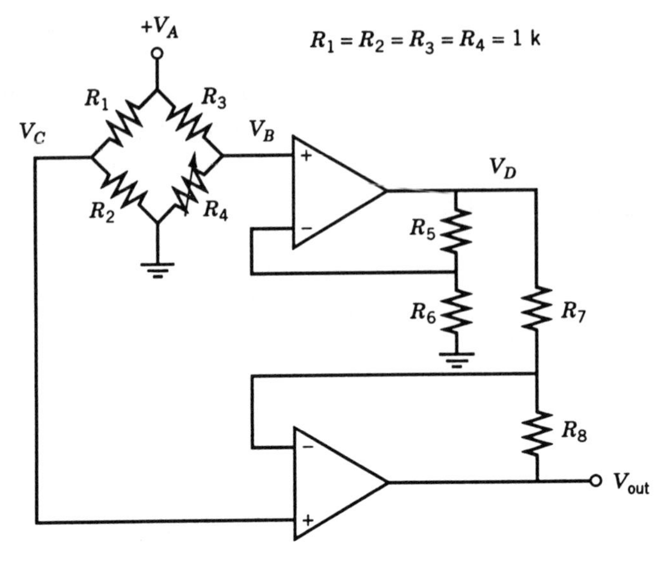

Figure P3-10

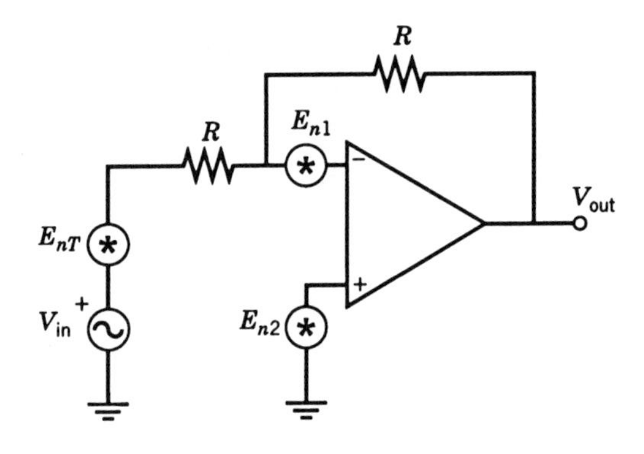

(*a*) Inverting amplifier

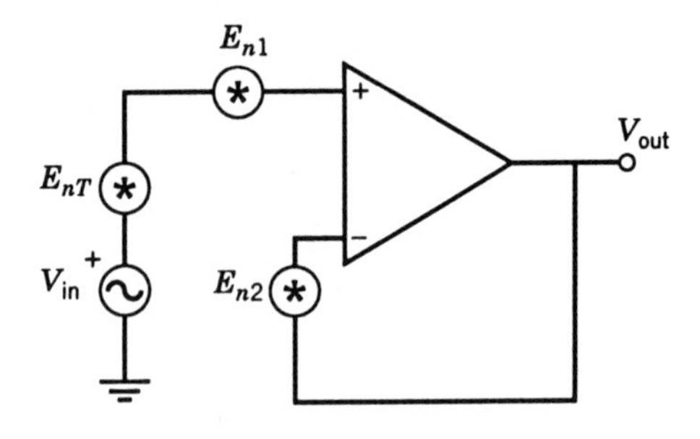

(*b*) Noninverting amplifier

Figure P3-11

because its output voltage is a linear function of the R_2 resistor. Assume both op amps are ideal except for $E_n = 5\ \text{nV/Hz}^{1/2}$ and $I_n = 20\ \text{pA/Hz}^{1/2}$. Find the total output noise due to all noise sources.

3-10. Another bridge amplifier is shown in Fig. P3-10. Here the resistors have values of $R_5 = R_7 = 20\ \text{k}\Omega$ and $R_6 = R_8 = 250\ \Omega$. Assume both op amps are ideal except for $E_n = 5\ \text{nV/Hz}^{1/2}$ and $I_n = 20\ \text{pA/Hz}^{1/2}$. Find the total output noise due to all noise sources.

3-11. Both amplifiers in Fig. P3-11 produce unity voltage gain. Assume that the op amps are ideal. In both circuits it is desired to replace the amplifiers' two noise voltage sources with a single equivalent E_{nT} source in series with the signal source as shown. Neglect all other noise sources except for the E_n sources. Derive and compare E_{nT} for both circuits. What important conclusions can you draw from this comparison?

REFERENCES

1. Rosenstark, S., *Feedback Amplifier Principles*, Macmillan, New York, 1986, pp. 5–9.
2. Robe, T., "Taming Noise in IC Op Amps," *Electronic Design*, **15** (July 19, 1974), 64–70.
3. Trofimenkoff, F. N., and O. A. Onwuachi, "Noise Performance of Operational Amplifier Circuits," *IEEE Trans. Education*, **32**, 1 (February 1989), 12–17.
4. Agouridis, D. C., "Comments on 'Noise Performance of Operational Amplifier Circuits,'" *IEEE Trans. Education*, **35**, 1 (February 1992), 98.
5. Smith, L., and D. H. Sheingold, "Noise and Operational Amplifier Circuits," *The Best of Analog Dialog*, Analog Devices, Inc., Cambridge, MA, 1991, pp. 19–31.
6. Cheah, J. Y. C., "Analysis of Phase Noise in Oscillators," *RF Design*, (November 1991), 99–104.
7. Hageman, S. C., "Spice Techniques Facilitate Analysis of Feedback Circuits," *EDN*, **33**, 20 (September 29, 1988), 173–182.
8. Connelly, J. A., and P. Choi, *Macromodeling with SPICE*, Prentice-Hall, Englewood Cliffs, NJ, 1992, pp. 6–10.

PART II

NOISE MODELING

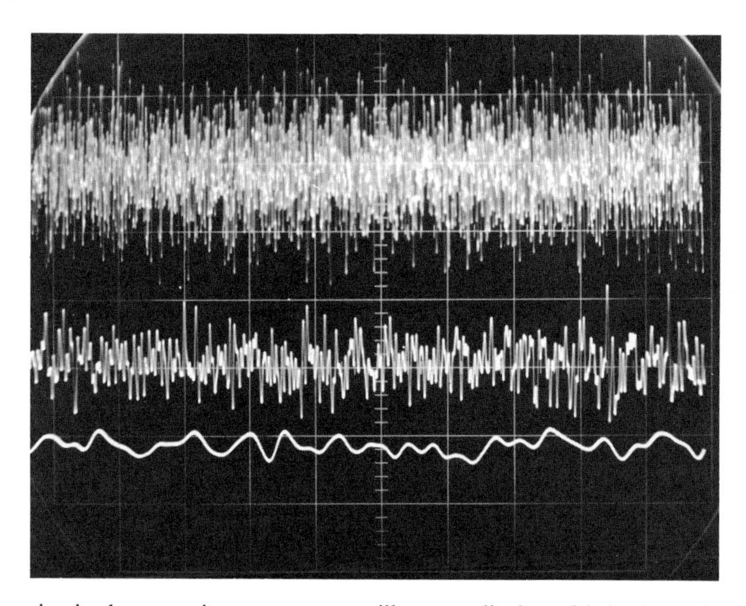

White noise is shown as it appears on oscilloscope display with horizontal sweep of 1 ms/cm. Top waveform, bandwidth is dc to 200 kHz; center waveform, bandwidth is dc to 20 kHz; bottom waveform, bandwidth is dc to 2 kHz. Note that the bandwidth reduction affects both the peak amplitude and the rms value.

CHAPTER 4

CAD FOR NOISE ANALYSIS

Thus far we have modeled noise in amplifiers with simple noise sources positioned at the input terminals of the amplifier. Our analysis has been restricted to "hand calculations" in order to understand the causes and effects of many noise sources generated throughout the circuit. As you have noted, these calculations are complicated and very time-consuming. In addition, network and sensor impedances are complex, and the transistor noise sources are all functions of frequency. To predict the total noise in a usable bandwidth it is necessary to calculate the noise at many frequencies and integrate the mean-square noise signal over the bandwidth of interest. A digital computer can be used to simplify the calculations. In this section we focus on modeling techniques for amplifier blocks using the SPICE computer program and its extensions. Furthermore, we include discussions on the modeling of discrete devices such as resistors, diodes, and transistors.

One word of caution: Sometimes, after a little practice and some successes using the computer to perform the complicated noise calculations, one may become complacent and tempted to rely on the computer to do more and more analysis without carefully cross-checking its output data. This invariably leads to disastrous results because the computer, while extremely fast and accurate, only performs exactly what it is programmed to do. Simple errors in modeling, in the input control file, or in choosing the options control will produce nonsensical results. Here is where the old reliable "hand calculations" must be resurrected and used for cross-checking. *Not to do so is only inviting disaster!*

4-1 HISTORY OF SPICE [1]

SPICE is an extremely powerful, general-purpose circuit analysis program which simulates analog circuits. SPICE is an acronym for **S**imulation **P**rogram with **I**ntegrated **C**ircuit **E**mphasis. Today this program is by far the most popular analog circuit simulation program being used by both practicing engineers and students. SPICE was developed by the Integrated Circuit Group of the Electronics Research Laboratory and the Department of Electrical Engineering and Computer Sciences at the University of California, Berkeley, in the late 1960s and was released to the public in 1972 [2]. The person credited with originally developing SPICE is Dr. Lawrence Nagel, whose Ph.D. thesis describes the algorithms and numerical methods used in SPICE.

Over the years, SPICE has gone through many upgrades, which are still continuing today. The most significant early improvement came with SPICE2, where the kernal algorithms were upgraded to support advanced integrated system methods, many of which relate to IC performance. SPICE2 has replaced SPICE1 as the SPICE choice. SPICE2 has been ported to numerous types of mainframe computers, personal computers, and various operating systems. The development of SPICE2 was supported with public funds, so this software is in the public domain and may be used freely by the citizens of the United States. SPICE2 has become an industry standard and is simply referred to as SPICE. It is very large (over 17,000 lines of FORTRAN source code), powerful, and extremely versatile as the industry-standard program for circuit analysis and IC design.

Recently, SPICE2 was upgraded to SPICE3. In this newer version, the program was converted from FORTRAN to C language for easier portability. Also several devices were added to the program library, such as a varactor, a semiconductor resistor, and lossy *RC* transmission line models. However, the kernel algorithms were not changed, and the added components are not that significant because SPICE2 can simulate all the devices built into SPICE3 using external device modeling techniques.

Today there are more than 35 SPICE derivative programs, many known by associated acronyms such as HSPICE and RAD-SPICE (from Meta-Software), IG-SPICE (from A. B. Associates), I-SPICE (from NCSS timesharing), PSpice (from MicroSim), IS_Spice (from intusoft), SLICE (from Harris), ADVICE (from AT&T Bell Laboratories), Precise (from Electronic Engineering Software), and ASPEC (from Control Data Corporation) [3–5].

In 1984 MicroSim introduced two versions of SPICE, called PSpice, which run on an IBM personal computer. The evaluation version, also called the educational or student version, is distributed free by MicroSim. This has made PSpice available to vast numbers of students and has caused rethinking the way classes and laboratories are being taught. The student version of PSpice that is PC-based is limited to circuits with approximately 10 transistors or less. However, the commercial (or professional or production) version can simulate a circuit with up to 200 bipolar transistors or 150 MOSFETs.

SPICE and other simulation programs will be around and will be widely used for many years to come. Students will find SPICE an important tool for learning circuit analysis and design, and for testing electronic circuits in ways not easily done in most college laboratories.

4-2 SPICE CAPABILITIES

SPICE contains built-in models for passive elements (resistors, capacitors, inductors, and transmission lines), semiconductor devices (diodes, bipolar transistors, JFETs, and MOSFETs), independent voltage and current sources, and dependent voltage and current sources including linear and nonlinear types (VCVS, VCCS, CCVS, and CCCS). By including control lines in an input file, SPICE can be made to perform many kinds of analyses of a circuit:

1. Nonlinear dc analysis, which determines the dc operating point of the circuit (.DC).
2. Linear small-signal ac analysis, which calculates the frequency response of the circuit (.AC).
3. Transient analysis, which determines the response as a function of time over a specified time interval (.TRAN).
4. Small-signal dc transfer function analysis of a circuit from a specified input to a specified output (input resistance, output resistance, and transfer function) (.TF).
5. dc small-signal sensitivity analysis of one or more specified output variables with respect to every parameter in the circuit (.SENS).
6. Distortion analysis with ac analysis (.DISTO).
7. Noise analysis with ac analysis, which determines the equivalent output and input noise at specified output and input nodes (.NOISE).
8. Fourier analysis of an output variable when done with a transient analysis (.FOUR).
9. Temperature analysis (.TEMP).

4-3 SPICE DESCRIPTION

Noise analysis is performed in SPICE by using the .NOISE statement in the input control file. To perform noise analysis, the .AC option must also be included in the control file. SPICE will calculate a thermal noise spectral density for every resistor in the circuit. Furthermore, SPICE calculates shot and flicker noise spectral densities for every diode and transistor in the circuit.

The specific commands, examples, and explanations of the program statements which must be included in the SPICE input control file for noise analysis are as follows:

```
.NOISE    OUTV    INSRC    NUMS
```

Example

```
.NOISE    V(5)    VIN    10
```

Explanation

.NOISE	initiates the noise analysis routine
OUTV	output voltage which defines the summing point or node
INSRC	independent voltage or current source which is the input noise reference
.NUMS	the summing interval

In the example, SPICE will calculate the noise produced by every resistor, diode, and transistor in the circuit at the output node 5. Then SPICE will reflect this output noise to the independent voltage source identified elsewhere in the file as VIN. The analysis will be calculated for every tenth frequency point as set by the NUMS index.

Examples

```
.AC    DEC    ND     FSTART    FSTOP
.AC    DEC    10     1         10K

.AC    OCT    NO     FSTART    FSTOP
.AC    OCT    10     1         1.048575MEG

.AC    LIN    NP     FSTART    FSTOP
.AC    LIN    100    1         100HZ
```

Explanation

.AC	initiates the AC analysis routine
DEC	sets the sweep for decade frequency variations
OCT	sets the sweep for octave frequency variations
LIN	sets the sweep for linear frequency variations
ND	number of points per decade of frequency

NO	number of points per octave of frequency
NP	total number of points for linear frequency sweep
FSTART	the starting frequency
FSTOP	the final frequency

In the first example, the frequency will be swept in decades from 1 Hz to 10 kHz, and 10 AC calculations will be performed in each decade. However, noise calculations will be done for every tenth frequency, that is, 1 Hz, 10 Hz, 100 Hz, 1 kHz, and 10 kHz in conjunction with the NUMS control statement. In the second example, the frequency will be swept in octaves of the starting frequency, 1 Hz in this case to a final frequency of about 1.05 MHz, equivalent to 20 octaves. Here 10 points per octave are calculated for AC and only 1 point per octave for NOISE. In the third example, the frequency is swept linearly from 1 to 100 Hz in 1-Hz steps. Here noise calculations will be performed at frequencies of 1, 11, 21, 31, 41, 51, 61, 71, 81, and 91 Hz.

Sometimes we may desire a tabulated listing of the input and output noises in the circuit under investigation. This is accomplished by including the following optional command in the input SPICE control file:

```
.PRINT    NOISE    INOISE    ONOISE
```

INOISE is the noise reflected to the input, the independent voltage, or current source defined as the INSRC in the .NOISE statement, and ONOISE is the output noise as defined by OUTV.

4-4 AMPLIFIER NOISE SOURCES

Often for modeling purposes it is advantageous to lump all the noise produced by an amplifier or other functional block into two noise sources, E_n and I_n, as shown in Fig. 4-1. There are several ways to implement this type of an amplifier model in SPICE. The easiest way is to use two resistors as noise sources and reflect their noises to the appropriate nodes using dependent voltage and current sources. Another way is to use diodes as sources of shot

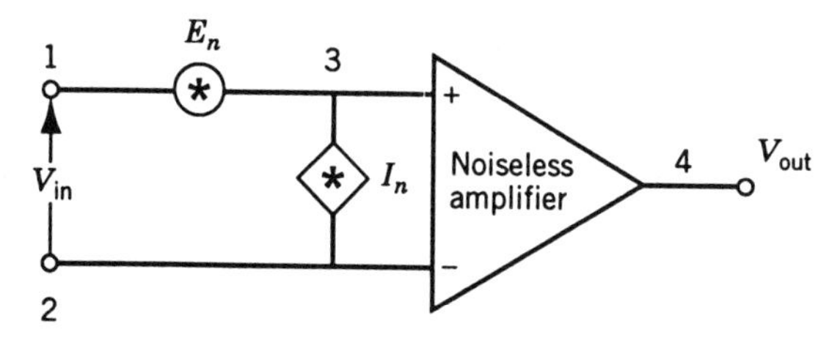

Figure 4-1 Voltage and current noise sources in an amplifier.

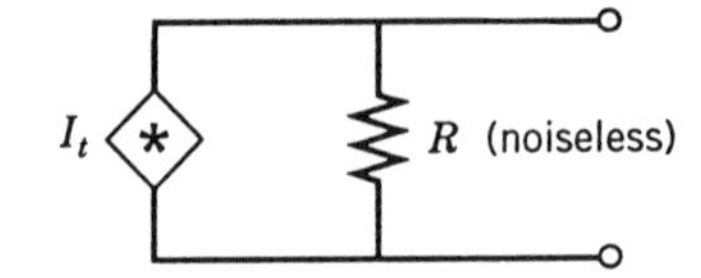

Figure 4-2 SPICE representation of thermal noise in a resistor.

and flicker noise and again reflect these noises to the appropriate nodes in the circuit. First we consider the resistor modeling approach.

SPICE calculates the spectral density of the thermal noise current in a resistor according to

$$I_t^2 = 4kT/R \qquad A^2/\text{Hz} \tag{4-1}$$

where k is Boltzmann's constant (1.38×10^{-23} W-s/K) and T is the temperature in kelvins. The default temperature is 27°C or 300 K. This resistor noise is modeled in SPICE as a current source in parallel with a noiseless resistor as shown in Fig. 4-2.

Example 4-1 Create E_n and I_n noise voltage and current sources of 3 nV/Hz$^{1/2}$ and 10 pA/Hz$^{1/2}$, respectively.

Solution First we create two thermal noise current standards of 1 pA/Hz$^{1/2}$ by using two resistors of value

$$R_n = \frac{4(1.38 \times 10^{-23})(300)}{10^{-24}} = 16.56 \text{ k}\Omega \tag{4-2}$$

Two separate sources are necessary since it is normally assumed that the E_n and I_n amplifier noise sources are uncorrelated. If E_n and I_n are 100% correlated, only one standard source would be needed. Next we sense the noise currents produced in these resistors by using independent (dummy) voltage sources set to 0 V. Finally, we reflect each standard noise current to the appropriate nodes by using current-controlled dependent sources as shown in Fig. 4-3.

Polarity does not matter for any of the sources because noise is a squared quantity. The multipliers needed for the H and F dependent sources are (3 nV/Hz$^{1/2}$)/(1 pA/Hz$^{1/2}$) = 3k and (10 pA/Hz$^{1/2}$)/(1 pA/Hz$^{1/2}$) = 10, respectively.

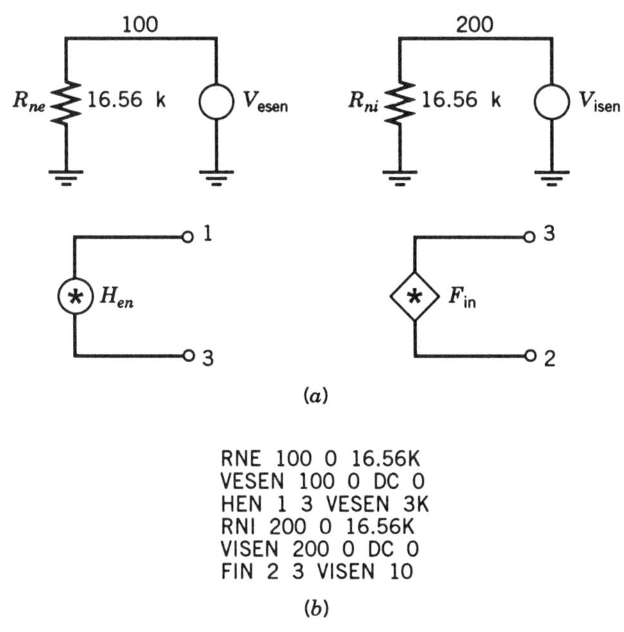

Figure 4-3 Modeling E_n and I_n noises: (*a*) circuit configuration and (*b*) SPICE code to reflect noise sources to the amplifier nodes of Fig. 4-1.

If we desire to include the amplifier's input impedance into our noise model, an extra noise source will be unintentionally introduced through the resistive component of the input impedance. For example, suppose we wanted to model the amplifier's differential input impedance as the parallel combination of $R_{\text{in}} = 100\ \text{k}\Omega$ and $C_{\text{in}} = 100$ pF. Also let the amplifier have a voltage gain of 750. Figure 4-4 shows the more complete amplifier model.

Extra noise will be added through R_{in} unless we replace this resistor with a SPICE element that has the same circuit effect as the resistor but does not produce thermal noise. We can use a voltage-controlled current source (VCCS) to model R_{in} as a noiseless resistor. This substitution is shown in

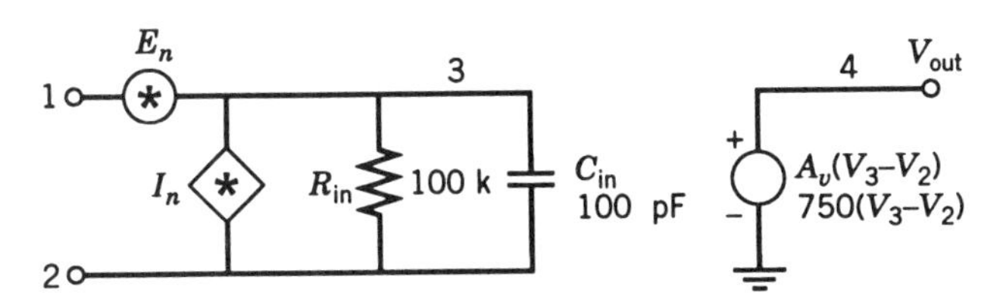

Figure 4-4 Amplifier model with noise sources, input impedance, and voltage gain.

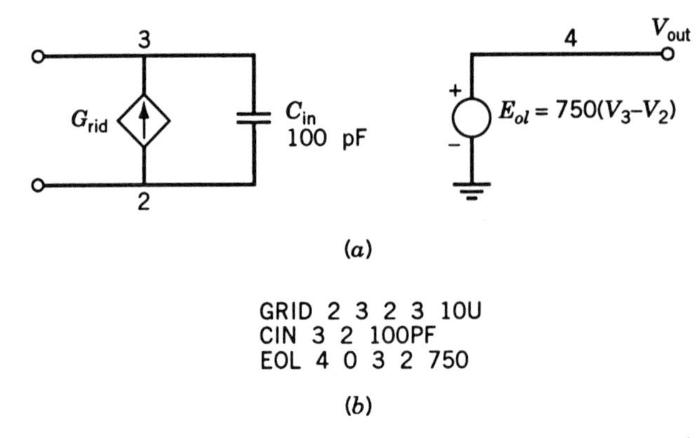

Figure 4-5 Adding a noiseless input resistance to the amplifier model: (*a*) equivalent circuit and (*b*) SPICE code.

Fig. 4-5 together with its SPICE code. The 10U term in GRID is the conductance of R_{in}, that is, the reciprocal of the 100 kΩ input resistance.

4-5 MODELING 1 / *f* NOISE

When it becomes necessary to include the frequency effects due to $1/f$ noise in E_n and I_n, the resistors must be replaced with diodes [6–8]. SPICE calculates a noise current in a diode according to

$$I_d^2 = \frac{KF(I_{\text{dc}})^{AF}}{f} + 2qI_{\text{dc}} \tag{4-3}$$

where *KF* is the flicker noise coefficient in amps, *AF* is the flicker noise exponent, q is the electronic charge (1.602×10^{-19} C), f is the frequency in hertz, and I_{dc} is the dc bias current through the diode.

Example 4-2 Model E_n noise of 20 nV/Hz$^{1/2}$ with a noise corner frequency $f_{nce} = 200$ Hz and I_n noise of 0.5 pA/Hz$^{1/2}$ with a noise corner frequency $f_{nci} = 500$ Hz. Use independent current sources to set a dc diode current which will produce a standard noise reference level of 1 pA/Hz$^{1/2}$ in each diode.

Solution The dc diode current is calculated according to

$$I_{\text{dc}} = \frac{10^{-24}}{2 \times 1.602 \times 10^{-19}} = 3.122\ \mu\text{A} \tag{4-4}$$

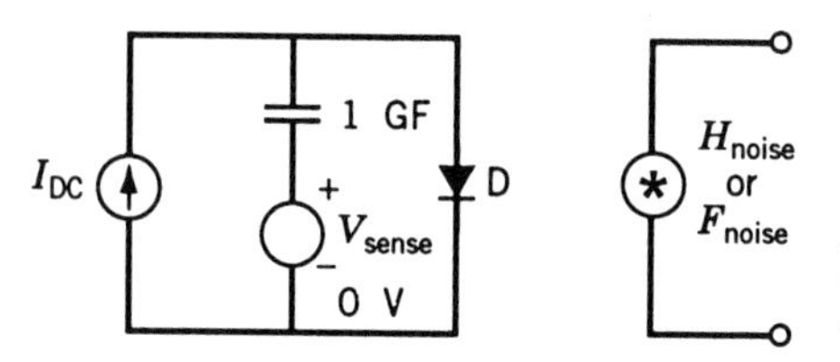

Figure 4-6 Circuit to produce noise with a $1/f$ frequency component.

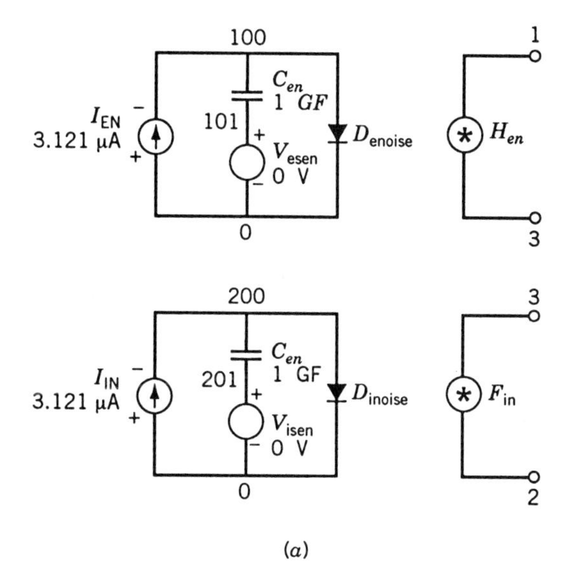


```
IEN 0 100 3.121UA
CEN 100 101 1GF
VESEN 101 0 DC 0
DENOISE 100 0 DENOISE
.MODEL DENOISE D (KF = 6.408E-17, AF = 1)
HEN 1 3 VESEN 20K
IIN 0 200 3.121UA
CIN 200 201 1GF
VISEN 201 0 DC 0
DINOISE 200 0 DINOISE
.MODEL DINOISE D (KF = 1.602E-16, AF = 1)
FIN 2 3 VISEN 0.5
```

(*b*)

Figure 4-7 E_n and I_n noise sources having $1/f$ components for inclusion into the amplifier of Fig. 4-1: (*a*) equivalent circuit and (*b*) SPICE code.

Next the flicker noise term is set equal to the 1-pA/Hz$^{1/2}$ reference noise current at the f_{nc} frequency. For the E_n source with $AF = 1$, KF is calculated by equating the two right-hand terms of Eq. 4-3, or

$$KF = 2qf_{nce} = 6.408 \times 10^{-17} \text{ A} \tag{4-5}$$

In a like manner, the KF for I_n is determined to be 1.602×10^{-16} A.

Figure 4-6 shows how the bias current is established in the diode. The 1-GF capacitor blocks the dc bias current and permits the ac short-circuit noise current to be sensed by the V_{sense} independent voltage source. Figure 4-7 shows the complete macromodel and the SPICE code for the E_n and I_n noise sources with $1/f$ variations positioned to match the nodes in the amplifier previously cited in Fig. 4-1.

4-6 MODELING EXCESS NOISE

Resistors are known to produce more noise than that attributable solely to temperature (see Chap. 12 of this book and Chap. 13 of Motchenbacher and Fitchen [9]). This additional noise called excess noise is a function of the dc biasing voltage, frequency, wattage rating, and the composition of the material making up the resistor. Excess noise has a $1/f$ frequency spectrum and is specified according to a noise index (NI) in decibels or in microvolt/V_{dc}/decade of frequency. The square of the spectral density of excess noise in volts^2 per hertz is given by

$$E_{\text{ex}}^2 = \left(\frac{10^{\text{NI}/10} \times 10^{-12}}{\ln 10} \right) \left(\frac{V_{\text{dc}}^2}{f} \right) \quad \text{V}^2/\text{Hz} \tag{4-6}$$

V_{dc} is the dc voltage across the resistor, f is the frequency in hertz, and NI is the noise index in decibels. As a numerical example, a resistor with a NI of -10 dB biased with 5 V dc produces an excess spectral density noise voltage of 33 nV/Hz$^{1/2}$ at a frequency of 1000 Hz.

Excess and thermal noises can be modeled in SPICE using the circuit shown in Fig. 4-8 where R is the resistor of interest. The dependent current source, F_{noise}, produces the excess noise current. The independent voltage source, V_{isense}, senses the dc current through R. This current is reflected to the diode subcircuit to control the dependent current source which, in turn, biases one of the diodes, D_1 or D_2, depending upon polarity. The gain of this dependent current source is equal to R so that the diode current has a magnitude equal to V_{dc}, the dc voltage developed across the resistor R. In the diode model, KF is set equal to

$$KF = \left(\frac{10^{\text{NI}/10} \times 10^{-12}}{\ln 10} \right) \quad \text{A} \tag{4-7}$$

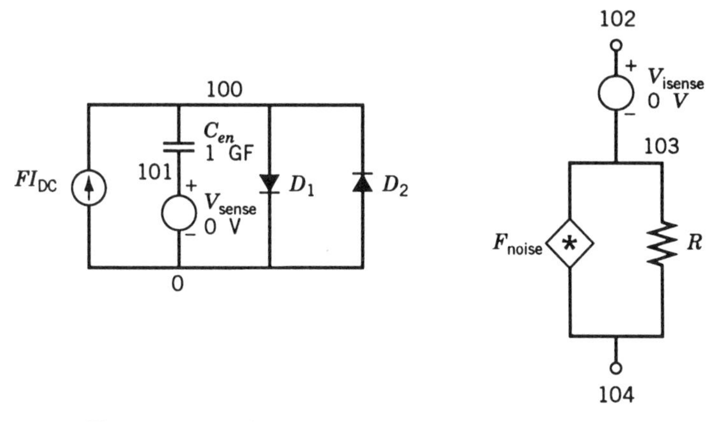

Figure 4-8 Resistor current noise model for SPICE.

and AF is set equal to 2. Then the diode spectral density expression of Eq. 4-3 yields an expression whose first term is that of Eq. 4-6. For typical NI values of -40 dB < NI < $+10$ dB, the shot noise term of the diode spectral density is negligible. The magnitude of the diode flicker noise term is the noise voltage in R caused by resistor current noise. Since the noise element F_{noise} is a current source, the gain of this dependent source must be $1/R$ when the noise is reflected to the main circuit.

Example 4-3 Consider the circuit shown in Fig. 4-9*a* where the resistor R_{excess} has a noise index of 10 dB. Develop a noise model for the excess noise in this resistor.

Solution For simplicity, we assume that the excess noises of all the other resistors are negligibly small. Next we calculate KF from Eq. 4-7 to be 4.343×10^{-12}. Finally, we use the excess noise model from Fig. 4-8 to produce the complete circuit of Fig. 4-9*b* with its associated SPICE code. Edited SPICE simulation results shown in Fig. 4-10 match the results obtained by hand calculations using Eq. 4-6.

The excess noise model of Fig. 4-8 must be modified when it is used with active devices where both dc biasing and ac frequency response simulations are required. As an example consider the single transistor circuit shown in Fig. 4-11.

The bias resistor, R_B, is to be modeled with a noise index of $+10$ dB through the F_{noise} current-controlled current source (CCCS). The current through R_B is reflected to the model circuit of Fig. 4-11*b* via the CCCS, FI_{dc}, which has a multiplier of R_B. Any ac component of FI_{dc} is shorted to ground by C_B. This current flows through one of the D_{excess} diodes which produces a

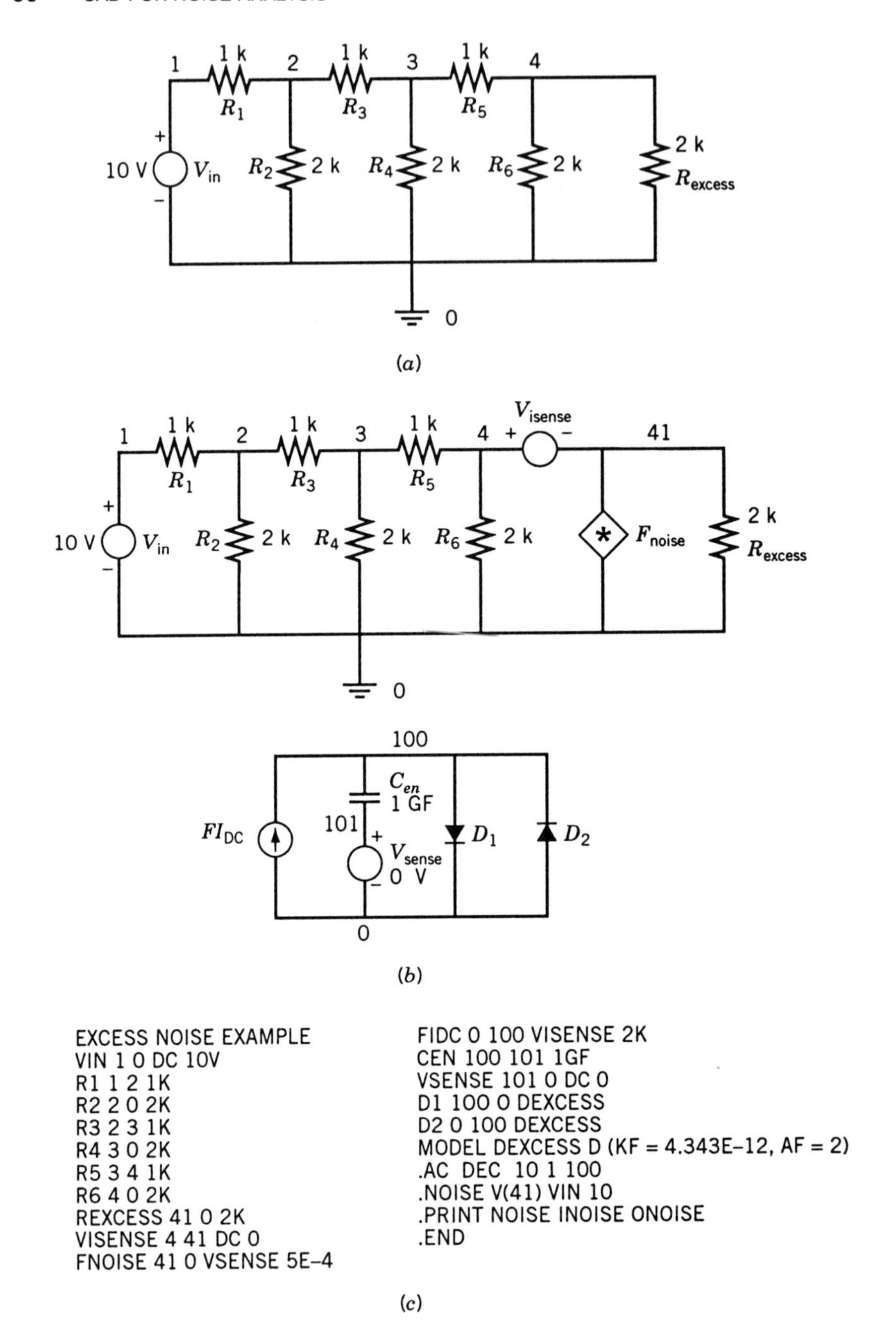


```
EXCESS NOISE EXAMPLE
VIN 1 0 DC 10V
R1 1 2 1K
R2 2 0 2K
R3 2 3 1K
R4 3 0 2K
R5 3 4 1K
R6 4 0 2K
REXCESS 41 0 2K
VISENSE 4 41 DC 0
FNOISE 41 0 VSENSE 5E–4
FIDC 0 100 VISENSE 2K
CEN 100 101 1GF
VSENSE 101 0 DC 0
D1 100 0 DEXCESS
D2 0 100 DEXCESS
MODEL DEXCESS D (KF = 4.343E–12, AF = 2)
.AC DEC 10 1 100
.NOISE V(41) VIN 10
.PRINT NOISE INOISE ONOISE
.END
```

(c)

Figure 4-9 Example of modeling excess noise: (*a*) example circuit, (*b*) circuit with added excess noise source, and (*c*) SPICE control file.

```
EXCESS NOISE EXAMPLE

SMALL SIGNAL BIAS SOLUTION  TEMPERATURE = 27.000 DEG C

NODE VOLTAGE NODE VOLTAGE NODE VOLTAGE NODE VOLTAGE
(1) 10.0000    (2) 5.000      (3) 2.5000  (4) 1.2500
(41) 1.2500 (100)  .8395  (101)  0.0000

NOISE ANALYSIS    TEMPERATURE = 27.000 DEG C

FREQUENCY = 1.000E+00 HZ
DIODE SQUARED NOISE VOLTAGES (SQ V/HZ)

               D1            D2

RS        0.000E+00    0.000E+00
ID        4.005E-19    0.000E+00
FN        6.786E-12    0.000E+00

TOTAL     6.786E-12    0.000E+00

RESISTOR SQUARED NOISE VOLTAGES (SQ V/HZ)

                REXCESS

TOTAL     3.315E-17

TOTAL OUTPUT NOISE VOLTAGE = 6.786E-12 SQ V/HZ
                           = 2.605E-06 V/RT HZ
```

Figure 4-10 Edited SPICE output showing 2.60 $\mu V/Hz^{1/2}$ excess noise voltage at a frequency of 1 Hz.

$1/f$ noise current. The inductor L_B blocks any ac component of the noise current generated by either D_{excess} diode from flowing back through C_B or FI_{dc}. Therefore, any noise current produced by the diode flows through C_{sen} and V_{sense}. This noise current through V_{sense} is then reflected back to the main circuit via F_{noise} and a constant multiplier. The value of this multiplier is critical to producing an accurate value of excess noise at R_B. Calculation of this reflection coefficient, X_C, is dependent upon the Thevenin resistance at node 4. Specifically, this Thevenin resistance is the parallel combination of the ac input resistance looking into the transistor and the source resistance R_s. Assuming a nominal $V_{BE} = 0.7$ V and $\beta = 100$, the emitter current in the transistor is 0.761 mA. The r_π of Q_1 is calculated as

$$r_\pi = \beta r_e = 100(0.026 \text{ V}/0.761 \text{ mA}) = 3.17 \text{ k}\Omega \quad (4\text{-}8)$$

The reflection coefficient for F_{noise} is then found as the reciprocal of the parallel combination of r_π and R_s. Numerically, this is

$$X_C = (r_\pi \parallel R_s)^{-1} = (3.17k \parallel 5k)^{-1} = (2.03k)^{-1} = 492.6 \times 10^{-6} \quad (4\text{-}9)$$

The sample SPICE file for the complete circuit is shown in Fig. 4-11*c*. The circuit was simulated and the voltage gain from node 4 to node 5 was found to be 173.5 as shown in Fig. 4-12. Next the total output noise at 1 Hz was found to be 4.1 $mV/Hz^{1/2}$, corresponding to a noise voltage at node 4 of 4.1 $mV/Hz^{1/2}/173.5 = 23.6$ $\mu V/Hz^{1/2}$. Furthermore, we see from PSpice that the dc voltage across R_B is 11.23 V. Relating these simulation results to the design methodology, we use Eq. 4-6 to calculate the excess noise voltage produced by R_B to be 23.7 $\mu V/Hz^{1/2}$. Figure 4-13 is an edited version of the simulation results obtained from PSpice.

The dependence on the reflection coefficient and the input capacitance was investigated. At 1 Hz the impedance seen looking back toward the signal source from node 4 is $(5k - j159)$ Ω and the contribution from C_{in} is negligible with respect to R_s. Another simulation was done after changing C_{in} to 220 μF which gave an impedance of $(5k - j5k)$ Ω, which is no longer negligible with respect to the 5-kΩ source resistance. The reflection coeffi-

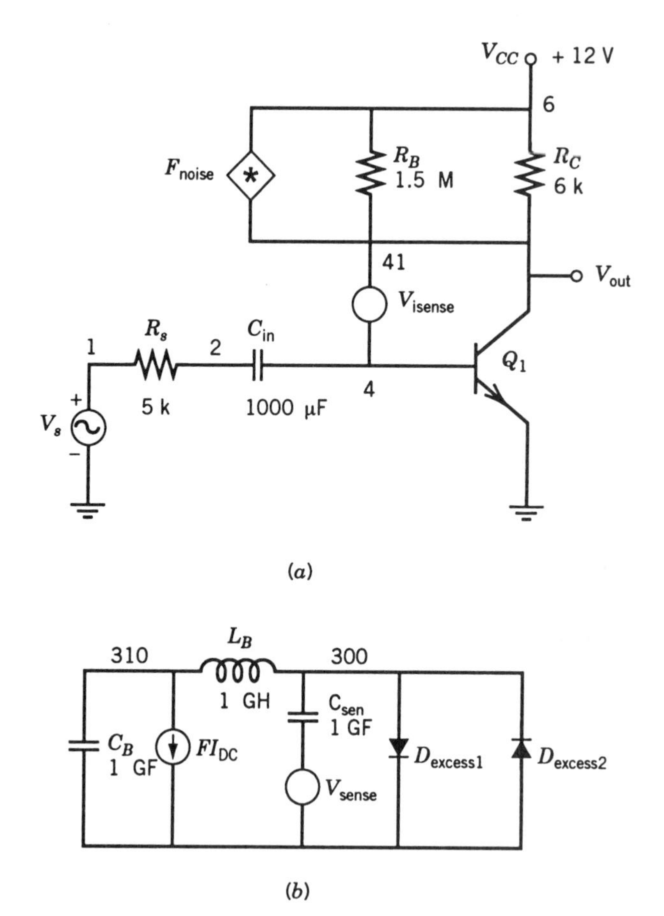

Figure 4-11 Example of modeling excess noise with an active device: (*a*) example circuit, (*b*) circuit with added excess noise source, and (*c*) SPICE control file.

```
EXCESS NOISE EXAMPLE FOR ACTIVE DEVICE
VS 1 0 AC 1V
RS 1 2 5K
CIN 2 4 1000UF
Q1 5 4 0 NPN1
.MODEL NPN1 NPN
VCC 6 0 12VOLTS
RB 6 41 1.5MEG
RC 5 6 6K
*EXCESS NOISE MODEL
VISENSE 41 6 VSENSE 492.6U
FIDC 310 0 VISENSE 1.5MEG
CB 310 0 1GF
LB 310 300 1GH
CSEN 300 301 1GF
VSENSE 301 0 0
DEXCESS1 300 0 D1
DEXCESS2 0 300 D1
*MODEL FOR FLICKER NOISE IN DIODE NI =+10 DB
.MODEL D1 D (KF=4.343E-12, AF=2)
.AC DEC 10 1 100K
.NOISE V(5) VISENSE 10
.PRINTING NOISE INOISE ONOISE
.PROBE
.END
```

(*c*)

Figure 4-11 *(Continued)*

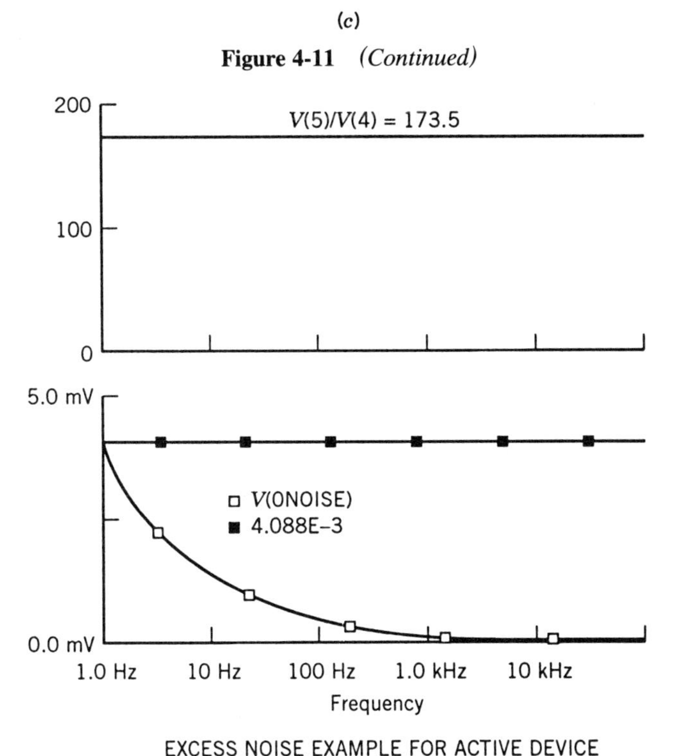

Figure 4-12 Voltage gain and output noise simulation for excess noise model.

```
                     EXCESS NOISE EXAMPLE FOR ACTIVE DEVICE

        * * * * SMALL SIGNAL BIAS SOLUTION  TEMPERATURE = 27.000 DEG C
NODE    VOLTAGE    NODE    VOLTAGE    NODE    VOLTAGE    NODE    VOLTAGE
 (1)     0.0000     (2)     0.0000      (4)    0.7667      (5)    7.5067
 (6)    12.0000    (41)     0.7667    (300)   -0.8963    (301)    0.0000
(310)   -0.8963
* * * * NOISE ANALYSIS      TEMPERATURE = 27.000 DEG C

FREQUENCY = 1.000E + 00 HZ
* * * * DIODE SQUARED NOISE VOLTAGES (SQ V/HZ)

                    DEXCESS1      DEXCESS2

          RS        0.000E + 00   0.000E + 00

          ID        8.857E - 27   1.098E - 13

          FN        0.000E + 00   1.671E - 05

          TOTAL     8.857E - 27   1.671E - 05

* * * * TRANSISTOR SQUARED NOISE VOLTAGES (SQ V/HZ)

                           Q1
             RB            0.000E + 00
             RC            0.000E + 00
             RE            0.000E + 00
             IB            3.016E - 13
             IC            8.639E - 15
             FN            0.000E + 00

             TOTAL         3.102E - 13

* * * * RESISTOR SQUARED NOISE VOLTAGES (SQ V/HZ)

                    RS            RB            RC

         TOTAL   4.162E - 13   1.389E - 15   9.945E - 17


* * * * TOTAL OUTPUT NOISE VOLTAGE = 1.671E - 05 SQ V/HZ
                                   = 4.088E - 03 V/RT HZ


TRANSFER FUNCTION VALUE:
V(5)/VISENSE = 2.363E - 01
EQUIVALENT INPUT NOISE AT VISENSE = 1.730E - 02 V/RT HZ
```

Figure 4-13 Edited PSpice output showing excess noise voltage at a frequency of 1 Hz.

cient remained constant at 497.2×10^{-6}. The output noise was identical to that when $C_{in} = 1000\ \mu F$. Therefore, the reflection coefficient depends only upon the *real* portion of the Thevenin resistance looking back from the modeled resistor.

To conclude, excess noise can be accurately modeled in SPICE. Both the dc and ac analyses are required. The reflection coefficient is the multiplier on the F_{noise}-dependent CCCS and is critical to producing accurate results. This reflection coefficient is dependent on the Thevenin resistance seen by the modeled resistor.

4-7 RANDOM NOISE GENERATOR

Occasionally, we may need to test a circuit or system to see how it would respond to a random input voltage like that produced by Gaussian noise. As we have just seen, the .NOISE function in SPICE performs an ac analysis in the frequency domain, but does not have a built-in way to simulate noise in the time domain. The transient analysis option (.TRAN) can be used with a specially configured time domain noise generator (TDNG).

The first step for constructing a TDNG is to find some means of generating a set of random numbers having a Gaussian distribution. If the mean, μ, and the standard deviation, σ, are not predetermined in the selection process, then calculate them according to

$$\text{Mean} = \mu = \frac{1}{N} \sum_{j=1}^{N} X_j \tag{4-10}$$

$$\text{Variance}(X_1, \ldots, X_N) = \frac{1}{N} \sum_{j=1}^{N} (X_j - \mu)^2 \tag{4-11}$$

$$\text{Standard deviation} = \sigma = \sqrt{\text{Variance}(X_1, \ldots, X_N)} \tag{4-12}$$

If $\sigma \neq 1$ and $\mu \neq 0$, perform the following transformation so that a new mean and standard deviation are found, respectively identified as $\mu' = 0$ and $\sigma' = 1$:

$$\frac{X_i - \mu}{\sigma} \Rightarrow Y_1, Y_2, Y_3, \ldots, Y_n \tag{4-13}$$

Decide upon the time interval of interest and split it up into a number of equal segments equal to the number of random numbers in the set. The program included in App. D can be used to do this [10]. This program automatically produces a set of numbers having an approximate Gaussian distribution. Furthermore, the program automatically generates the SPICE input file for the random noise generator. As an example, consider a set of 50 numbers with a Gaussian distribution spread over a known range. Here we might be interested in the time response from 0 to 5 μs. We would divide up the segments in 100-ns intervals. Finally, we construct a series combination of three independent, piecewise linear, voltage sources, each having voltage values at 20 specific instants of time and zero for all other times. The piecewise linear voltage source model is used as this voltage source. Note that a single PWL source is limited to a maximum of twenty amplitude–time data points in SPICE.

Example 4-4 Construct a 20-nV random noise generator model and show its simulated output noise over the time interval from 0 to 5 μs.

Solution Figure 4-14*a* shows the general approach where PWL source V_1 has its random voltage values for the first 20 time steps, from 0 to 1.9 μs and then is zero at all other times. The second source, V_2, has value for the second set of 20 time steps and is zero elsewhere. This process is repeated for the third PWL source up to the 5 μs limit. If more data points are desired, additional PWL sources as well as additional random numbers need to be added in the TDNG model.

Over the time interval of interest, the output voltage between node 1 and ground in Fig. 4-14 will have approximately a zero mean value and a standard deviation voltage (rms noise voltage) of 1 V. The VCVS E_{ngen} reflects and scales this 1-V noise level to the desired new level. For example, since a noise level of 20 nV is desired as the output level between node 3 and ground, the multiplier is set to 20N as shown. Figure 4-15 shows the output from the TDNG model where the time interval was the specified 0 to 5 μs and the rms noise level was 20 nV.

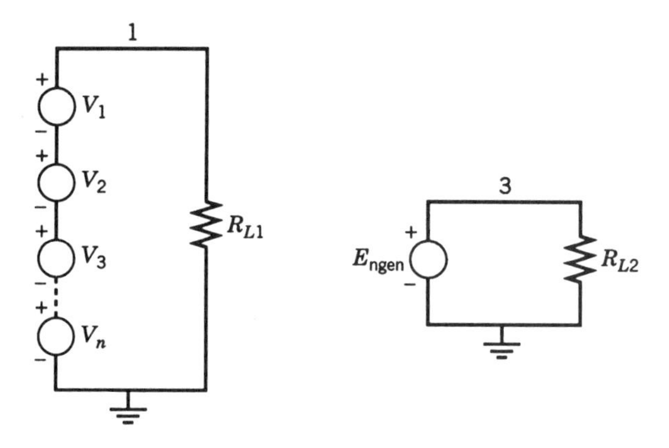

(*a*)

Figure 4-14 Time-dependent random noise generator: (*a*) circuit representation and (*b*) SPICE netlist.

```
Time-Dependent Noise Generator

*

ENGEN 3 0 1 0 (A)    ;(A) is the rms noise level = 20 n v for this example
RL1 1 0 1
RL2 3 0 1T
X1 1 0 TDNG

*

*The mean and standard deviation before transformation are
*mean = 1.69325368409091E-0001 rms = 1.00016888842765E+0000

*

.Subckt TDNG POS NEG
V1 100 NEG pwl(0s 0
+ 1.000E-0007s 0.89034047 2.000E-0007s -1.16319155 3.000E-0007s 2.03930666
+ 4.000E-0007s 0.40463799 5.000E-0007s -0.47319771 6.000E-0007s 0.32083804
+ 7.000E-0007s 0.32537534 8.000E-0007s 0.13314042 9.000E-0007s -0.30955093
+ 1.000E-0006s 1.65501569 1.100E-0006s -1.50768434 1.200E-0006s -0.03783616
+ 1.300E-0006s 0.51317443 1.400E-0006s -1.68355872 1.500E-0006s -0.65928206
+ 1.600E-0006s 0.43142757 1.700E-0006s 0.18312538 1.800E-0006s 1.87252633
+ 1.900E-0006s 0)

*

V2 200 100 pwl(0s 0 1.800E-0006s 0
+ 1.900E-0006s -1.36038170 2.000E-0006s -0.05226307 2.100E-0006s -0.10261622
+ 2.200E-0006s -0.99439241 2.300E-0006s 0.56870622 2.400-0006s 0.49330001
+ 2.500E-0006s -1.19179314 2.600E-0006s 1.38155654 2.700E-0006s 0.66423013
+ 2.800E-0006s -1.07387959 2.900E-0006s -0.10540508 3.000E-0006s -1.28235444
+ 3.100E-0006s 1.01340746 3.200E-0006s -0.60131112 3.300E-0006s 1.07827081
+ 3.400E-0006s 0.32531230 3.500E-0006s 0.13651206 3.600E-0006s -1.10343446
+ 3.700E-0006s 0)

*

V3 POS 200PWL(0s 0 3.600E-0006s 0
+ 3.700E-0006s 1.36542284 3.800E-0006s -0.20886253 3.900E-0006s -0.51574401
+ 4.000E-0006s -0.26872715 4.100E-0006s 1.64804841 4.200E-0006s -0.25428479
+ 4.300E-0006s 0.50392776 4.400E-0006s -0.46641442 4.500E-0006s 0.73564880
+ 4.600E-0006s -0.83540614 4.700E-0006s -2.22394443 4.800E-0006s 1.95317640
+ 4.900E-0006s -0.45252145 5.000E-0006s -1.05763975 5.100E-0006s 0)

*

.ends

*

.PROBE
.TRAN 20NS 5US
.END
```

Figure 4-14 *(Continued)*

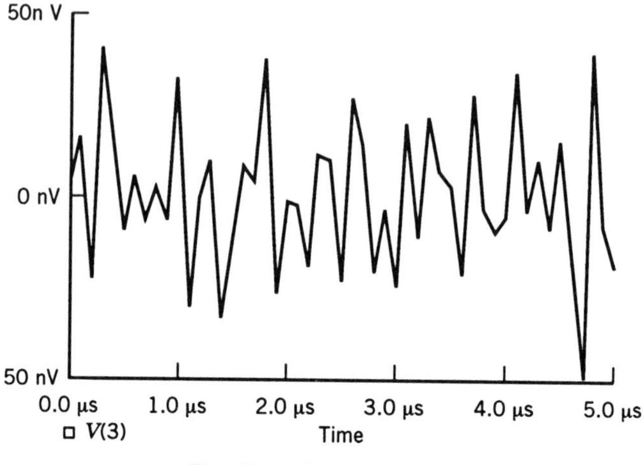

Figure 4-15 Output of time-dependent noise generator.

4-8 FINDING NOISE BANDWIDTH USING PSpice

Some extended SPICE simulation programs such as PSpice have built-in arithmetic postprocessing capabilities. When properly used, these functions enable quick and accurate determination of critical circuit and system performance indexes such as noise bandwidth, noise figure [11], and total input and output noises.

Recall from Eq. 1-9, the noise bandwidth was defined as

$$\Delta f = \frac{1}{A_{vo}^2} \int_0^\infty |A_v(f)|^2 \, df \qquad (4\text{-}14)$$

We can use the PROBE postprocessing capabilities in PSpice to find Δf by integrating the square of the magnitude of voltage gain with respect to frequency from dc (or some very low frequency) to some large frequency and then dividing this result by the square of the peak voltage gain. This is done using the PROBE statement.

```
(1 / (Avo*Avo))*S(VM(output node)*VM(output node))
```

The S function in PROBE performs numerical integration over the limits as set by the x-axis, the frequency sweep in this case. Care should be exercised in how the voltage magnitude is specified. Using VM(node number) appears to always give the correct result while just using V(node number) may give incorrect results especially in 4.03 and later versions of PSpice.

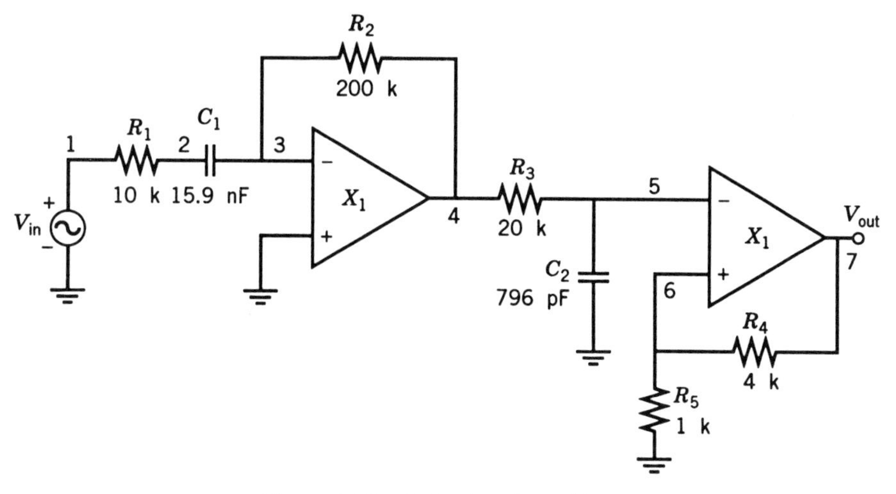

Figure 4-16 Active bandpass filter.

Example 4-5 Find the noise bandwidth using PSpice of the active bandpass filter shown in Fig. 4-16. Assume both op amps are ideal.

Solution Regular circuit analysis shows that the first stage produces a high-pass frequency response with a gain of 26 dB and a single −3-dB frequency of 1 kHz. The second stage is a low-pass filter with 14-dB gain and a high-frequency cutoff of 10 kHz. These conclusions about the filter's operation are easily verified using the following SPICE or PSpice control file.

```
NOISE BANDWIDTH FOR X100 AMPLIFIER
*
VIN 1 0 DC 0 AC 1V
R1 1 2 10K
C1 2 3 15.9NF
R2 3 4 200K
R3 4 5 20K
C2 5 0 796PF
R4 6 7 4K
R5 6 0 1K
*
*OP AMP SUBCIRCUIT
.SUBCKT OPAMP 1 2 3
RIN 1 2 10MEG
EOL 3 0 1 2 100MEG
```

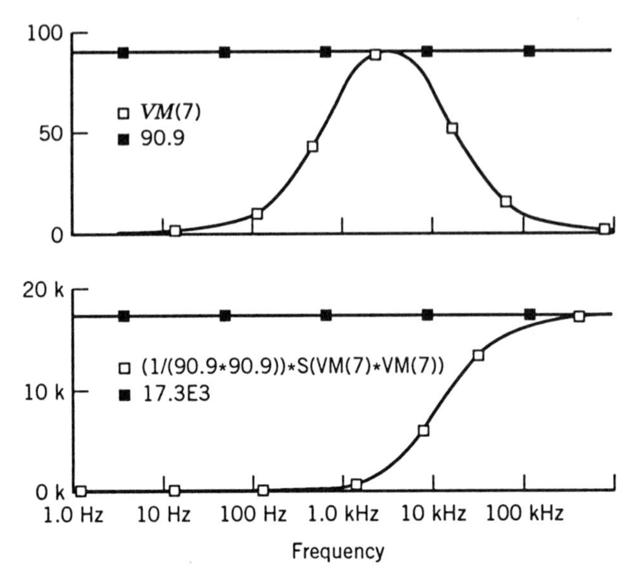

Figure 4-17 PSpice simulation results of bandpass filter: (*a*) frequency response showing peak voltage gain and (*b*) integration to find noise bandwidth.

```
.ENDS OPAMP
*
X1 0 3 4 OPAMP
X2 5 6 7 OPAMP
.AC DEC 10 1HZ 1MEGHZ
.NOISE V(7) VIN 10
.PROBE
.END
```

Simulation results are shown in Fig. 4-17*a* where the maximum voltage gain is 90.9 (39.2 dB) at 3.16 kHz as displayed on the linear voltage axis. (Note that a simple macromodel for both op amps was used which modeled an input resistance of 10 MΩ and an open-loop voltage gain of 10^8.)

Next, the approximate noise bandwidth is found by utilizing the integration and scaling commands in PROBE, namely,

```
(1 / (90.9*90.9))*S(VM(7)*VM(7))
```

Figure 4-17*b* shows the resulting output plot. The curve approaches an asymptotic level which is the noise bandwidth for the bandpass filter, approximately 17.3 kHz. Rigorous mathematical calculations deriving the bandpass filter's transfer function and then its integration produce an exact determination of noise bandwidth to be

$$\Delta f = (\pi f_o/2Q) = (\pi/2)(3.165\ \text{kHz}/0.2875) = 17.3\ \text{kHz} \qquad (4\text{-}15)$$

4-9 INTEGRATING NOISE OVER A FREQUENCY BANDWIDTH

As discussed in Sec. 4-1, the .NOISE command causes SPICE to calculate the spectral density noise contribution for every resistor and semiconductor device in a circuit. However, we often need to know how much wideband noise would be measured by a true rms voltmeter if it were placed at specific nodes in the circuit, usually at the input or output. Again the integration postprocessor in PROBE furnishes an easy way to find the wideband rms noise.

To find the wideband rms noise, include the .NOISE command in the input control file and identify the OUTV (output voltage defining the summing node) and INSRC (independent voltage or current source which is the input noise reference) nodes. Then simulate the circuit to obtain the ac frequency response in the normal way. The spectral densities for the input and output noises can be displayed in PROBE by simply displaying the traces V(ONOISE) and V(INOISE) for the output and input noises, respectively. The rms output and input wideband noises are then found by taking the square root of the integral of the square of V(ONOISE) and V(INOISE).

Example 4-6 Extend the previous example to determine the equivalent input and output noise voltages for the active bandpass amplifier shown in Fig. 4-16. Neglect the noise contributions from the operational amplifiers and consider only the noise from the resistors in the circuit.

Solution After simulating the bandpass filter circuit, the output and input equivalent noise spectral density voltages are displayed in Fig. 4-18. The output noise follows the bandpass response of the filter and peaks at f_o at approximate 1.2 $\mu\text{V/Hz}^{1/2}$. The input noise is flat at approximately 12 $\text{nV/Hz}^{1/2}$ over most of the useable frequency range. This level is due to the 10-kΩ resistance, R_1. The up-turning rises in V(INOISE) at both low and high frequencies is due to the basic definition of E_{ni}, namely,

$$E_{ni} = E_{no}/A_v \qquad (4\text{-}16)$$

Since the voltage gain of the bandpass amplifier approaches zero at dc and infinite frequency, E_{ni} increases without bound at these limits.

The desired wideband noise voltages are calculated and displayed as in Fig. 4-19. The output noise voltage approaches an asymptotic value of approximately 160 μV which is the voltage level a true rms voltmeter with a bandwidth of at least 1 MHz would display. Assuming that R_1 is the dominant noise source, the following quick calculation verifies this output

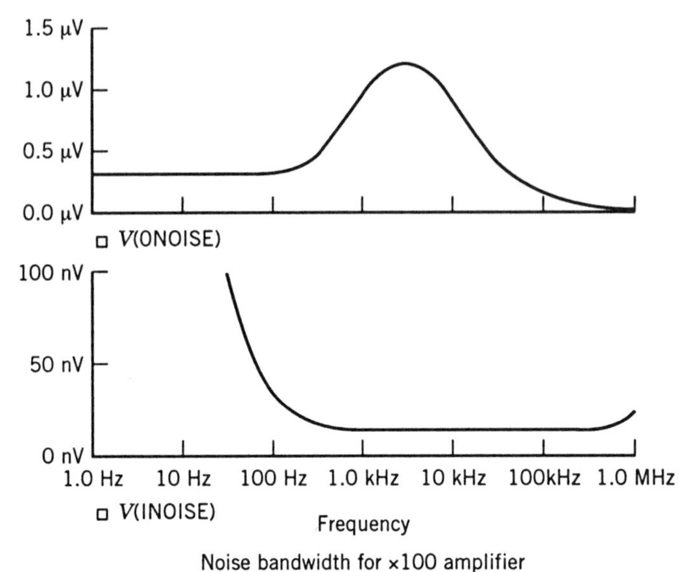

Figure 4-18 Output and input noise simulation results of the bandpass filter.

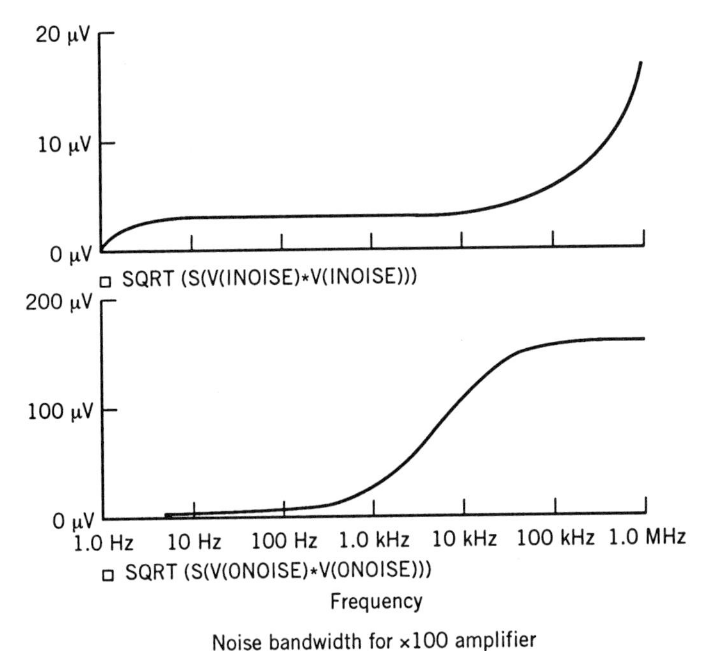

Figure 4-19 PSpice simulation of the wideband noise of the bandpass filter.

wideband noise level:

$$\text{Wideband } E_{no}^2 = \int_0^\infty E_{ni}^2 A_v^2 \, df = \frac{1}{A_{vo}^2} \int_0^\infty A_{vo}^2 E_{ni}^2 A_v^2 \, df = \left(A_{vo}^2\right)(4kTR_1)\,\Delta f$$

$$\text{Wideband } E_{no}^2 = (90.9)^2(1.6 \times 10^{-20} \times 10^4)(17.3 \text{ kHz})$$

$$\text{Wideband } E_{no}^2 = 2.29 \times 10^{-8} \text{ V}^2$$

$$\text{Wideband } E_{no} = 151 \ \mu\text{V} \tag{4-17}$$

Since the wideband input noise voltage is derived from V(ONOISE), E_{ni} increases without limit at high frequencies due to the gain reduction.

4-10 MODEL REDUCTION TECHNIQUES

The signal and noise levels at the output of a sensor–detector–amplifier system depend on the transfer function of the complete amplifier unit. Often a postamplifier and equalizer are necessary parts of the total unit. As we have just seen, it is very easy to use SPICE to simulate a system and determine the input and output noise levels as well as the noise bandwidth. Furthermore, the output file (.OUT) produced after every SPICE simulation reveals the noise contributions of every resistor and semiconductor device in the system. Thus careful examination of the .OUT file will reveal the dominant sources of noise at the frequencies of interest.

Sometimes we may need to reduce the complexity of a sensor–detector–amplifier system by substituting some simple frequency-shaping models or macromodels for the postamplifier, equalizer, or other stages. This may be necessary in order to reduce the simulation time to a reasonable limit or to permit the system to fit within the memory or device size capabilities of the SPICE program. Common filter stages used are low-pass, bandpass, high-pass, lag-lead, and lead-lag networks.

There are many ways to make these circuit simplifications. One easy way is to seek out one of the many textbooks providing examples of passive filter configurations with appropriate equations so that the design specifications can be used to calculate and denormalize element values. If this easy approach is followed, one must be very careful in choosing the location and value of all resistors in the circuit as SPICE will calculate and include their noise contributions. Here is where it is important to examine the .OUT file after simulating the circuit to make sure these resistor noises are not significant.

Another method for system simulation is to utilize behavioral-level continuous-time macromodels as explained in Connelly and Choi [12]. Here one begins with the required mathematical transfer function for the simplified network and synthesizes an equivalent macromodel which produces this frequency characteristic. Feedback around a basic integrator building block is the technique used to produce the required macromodel. The only limit to the order of the filter produced by this technique is the memory capability and resolution of the particular version of SPICE being used. If high-order filters are required ($n > 4$), it is usually best to factor and split the transfer function into two or more separate stages and then cascade them. As with the passive filter approach, one needs to examine the .OUT file after simulating to make sure that the noise added by the resistors in the integrator blocks is insignificant with respect to the other circuit noises.

SUMMARY

a. Simulation alone without independent "hand calculations" for verification invariably leads to disaster.
b. The .NOISE command in SPICE must be used in conjunction with the .AC command.
c. SPICE uses the Norton equivalent representation of the thermal noise in a resistor. If one resistor is used as a noise reference source and then reflected to two or more places in the circuit, all resultant noise sources will be 100% correlated.
d. Diodes must be used to model $1/f$ noise characteristics in SPICE.
e. Excess noise in resistors can be modeled in SPICE. However, the equivalent circuit is complicated and should only be used for the dominant resistive noise source.
f. A random noise generator is useful for modeling noise behavior in the time domain, but not in the frequency domain.
g. The PROBE postprocessor in PSpice is very useful for finding the noise bandwidth and equivalent wideband output noise in a circuit.
h. After using model reduction techniques to simplify a large system, carefully examine the .OUT file to make sure that the models added for simplicity do not contribute significant noise to the total system at all frequencies of interest.

PROBLEMS

4-1. The E_n and I_n noise characteristics of a certain op amp are to be modeled in SPICE. The spectral densities of E_n and I_n versus frequency are given by

$$E_n^2 = (9\ \text{nV/Hz}^{1/2})^2[1 + (200/f)]$$

and

$$I_n^2 = (0.15\ \text{pA/Hz}^{1/2})^2[1 + (500/f)]$$

where f is the frequency in hertz. A partial SPICE input file follows.

```
*NOISE SOURCES
IREF1 0 100 3.125UA
CNE 100 101 1GF
VESEN 101 0 DC 0
DEN 100 0 DMODE
.MODEL DMODE D (KF =____ , AF =____)
HV1F 2 3 VESEN____
*
IREF2 0 200 0.2809UA
CNI 200 201 1GF
VISEN 201 0 DC 0
DIN 200 0 DMODI
.MODEL DMODI D (KF =____ , AF =____)
FI1F 3 4 VISEN____
```

(a) Add the missing information where the blanks appear to the input SPICE file.

(b) Simulate the noise characteristics of this op amp and verify that it produces the correct E_n^2 and I_n^2 noise characteristics with respect to amplitude and frequency.

4-2. The E_n and I_n noise characteristics of an op amp are modeled in SPICE by

```
*NOISE SOURCES
IREF1 0 100 12.48UA
CNE 100 101 1GF
VESEN 101 0 DC 0
DEN 100 0 DMODE
.MODEL DMODE D (KF = 6.408E-17 AF = 1)
HV1F 2 3 VESEN 5K
*
IREF2 0 200 0.2809UA
CNI 200 201 1GF
VISEN 201 0 DC 0
DIN 200 0 DMODI
.MODEL DMODI D (KF = 3.204E-16 AF = 1)
FI1F 3 4 VISEN 4
*
```

Determine the numerical values of E_n and I_n as well as their associated noise corner frequencies which are being modeled.

4-3. Refer to the excess noise resistor example (Example 4-3). What value of noise index must R_{excess} have in order to contribute equal thermal and excess noises over the frequency range from 1 to 10 Hz?

4-4. Change the value of the base biasing resistor, R_B, in Fig. 4-11*a* so that the collector current in the transistor doubles. Determine the effect this will have on the excess noise contributed by R_B. Verify your conclusion by simulation.

4-5. Drive the transistor amplifier circuit of Fig. 4-11 with the 20-nV random noise generator shown in Fig. 4-14. Perform a transient analysis of the output voltage and compare the source and output noise waveforms.

4-6. Repeat Prob. 4-5 using the active bandpass filter as the test circuit. Simulate so that the output voltages from both X1 and X2 op amps can be displayed in the time domain. Study these outputs and compare them to the random noise generator source.

4-7. The circuit shown in Fig. P4-7 provides RIAA equalization for a preamplifier. Simulate this circuit and determine the noise bandwidth. Model the op amp to have an open-loop voltage gain of 2×10^6 and an input resistance of 10 MΩ. Let the op amp be ideal otherwise.

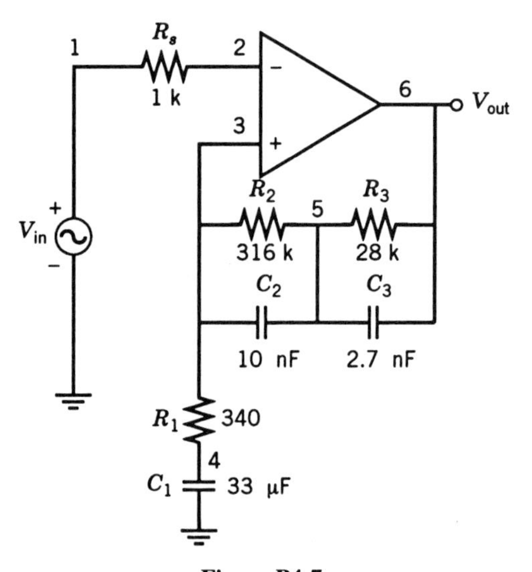

Figure P4-7

4-8. Use PSpice to determine the wideband output noise for the circuit in Fig. P4-7. Use the same op amp characteristics as given in the previous problem.

(a) First neglect all noise contributions from the op amp in finding the wideband E_{no}.

(b) Now give the op amp noise characteristics of $E_n = 20\ \text{nV}/\text{Hz}^{1/2}$, $f_{nce} = 200$ Hz, $I_n = 0.5\ \text{pA}/\text{Hz}^{1/2}$, and $f_{nci} = 1$ kHz. Place a single voltage noise source in series with the source resistor R_s. Position the current source between the inverting and noninverting inputs to the op amp.

REFERENCES

1. Connelly, J. A., and P. Choi, *Macromodeling with SPICE*, Prentice-Hall, Englewood Cliffs, NJ, 1992, pp. 6–10.
2. Byers, T. J., "SPICE—Breadboards Are Giving Way to PC-Based Circuit Simulation," *Radio Electronics* (November 1990), 63–68.
3. Smith, C. E., and Holt, I. L., "Adding SPICE to the Electronics Industry," *Printed Circuit Design* (September 1990), 39–47.
4. Rashid, M. H., *SPICE For Circuit And Electronics Using PSpice*, Prentice-Hall, Englewood Cliffs, NJ, 1990, pp. 1–3.
5. Tuinenga, P. W., *SPICE A Guide to Circuit Simulation & Analysis Using PSpice*, Prentice-Hall, Englewood Cliffs, NJ, 1988, pp. xvi–xvii.
6. Scott, G. J., and T. M. Chen, "Addition of Excess Noise in SPICE Circuit Simulations," *IEEE SOUTHEASTCON Conference Proceedings*, Tampa, FL, April 5–8, 1987, Vol. 1, pp. 186–190.
7. Skrypkowiak, S. S., and T. M. Chen, "A Method of Including Resistor Current Noise in SPICE Circuit Analysis," *IEEE SOUTHEASTCON Conference Proceedings*, Tampa, FL, April 5–8, 1987, Vol. 1, pp. 197–201.
8. Meyer, R. G., L. Nagel, and S. K. Lui, "Computer Simulation of $1/f$ Noise Performance of Electronic Circuits," *IEEE Journal of Solid State Circuits*, **SC-6**, June 1973, pp. 237–240.
9. Motchenbacher, C. D., and F. C. Fitchen, *Low Noise Electronic Design*, Wiley Interscience, New York, 1973, pp. 171–179.
10. Press, W. H., B. P. Flannery, S. A. Teukolsky, and W. T. Vetterling, *Numerical Recipes*, Cambridge University Press, Cambridge, 1986, pp. 202–203, 714–718.
11. Ortiz, J., and C. Denig, "Noise Figure Analysis Using SPICE," *Microwave Journal*, April 1992, pp. 89–93.
12. Connelly, J. A., and P. Choi, *Macromodeling with SPICE*, Prentice-Hall, Englewood Cliffs, NJ, 1992, pp. 30–35.

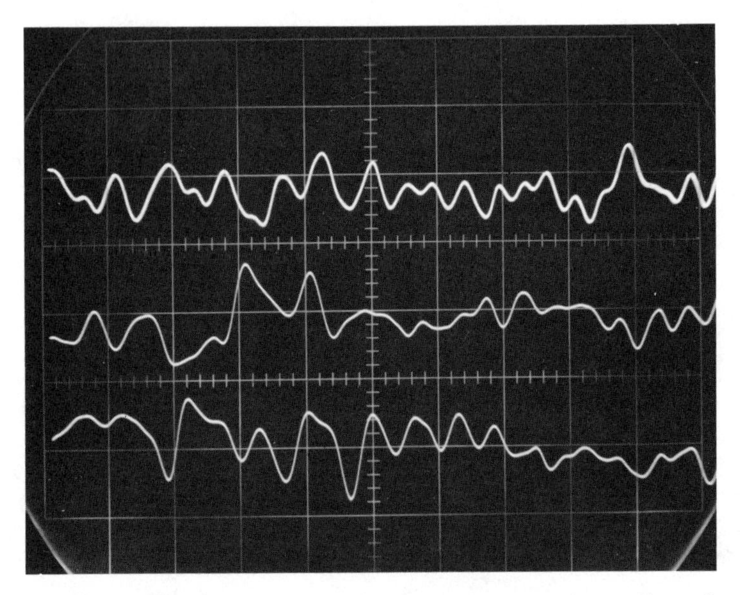

The power content of $1/f$ noise in each decade of frequency is equal, as shown in these traces. Top waveform, $\Delta f = 20$ Hz, horizontal 100 ms/cm; center waveform, $\Delta f = 200$ Hz, horizontal 10 ms/cm; bottom waveform, $\Delta f = 2$ kHz, horizontal 1 ms/cm.

CHAPTER 5

NOISE IN BIPOLAR TRANSISTORS

The objective of this chapter is to determine E_{ni}, E_n, and I_n for a bipolar junction transistor (BJT). In so doing, we will define and identify the noise sources contributing to E_n and I_n in the device noise model. Specifically, we will show that the BJT noise parameters are functions of the operating Q-point, and this operating point can be selected for reduced noise contributions.

The bipolar junction transistor (BJT) contains sources of thermal noise, $1/f$ noise, and shot noise, which are discussed in this chapter. The widely used hybrid-π small-signal equivalent circuit [1] is modified to include noise sources to represent transistor noise behavior.

From the noise circuit model, we determine the equivalent input noise parameter E_{ni} for the BJT. The resulting expression is used to predict noise versus frequency behavior and forms the basis for the derivation of expressions for E_n and I_n.

The conditions necessary for minimizing the noise figure are considered. Because noise relations are expressed in terms of operating point currents, transistor parameters, temperature, and frequency, the circuit designer can apply the results to create low-noise electronic circuits.

Comparisons of various transistor types are made and results are presented that suggest the correct type for a given application.

5-1 THE HYBRID-π MODEL

The hybrid-π transistor model represents the small-signal behavior of the bipolar junction transistor. The parameters of the hybrid-π are generally not

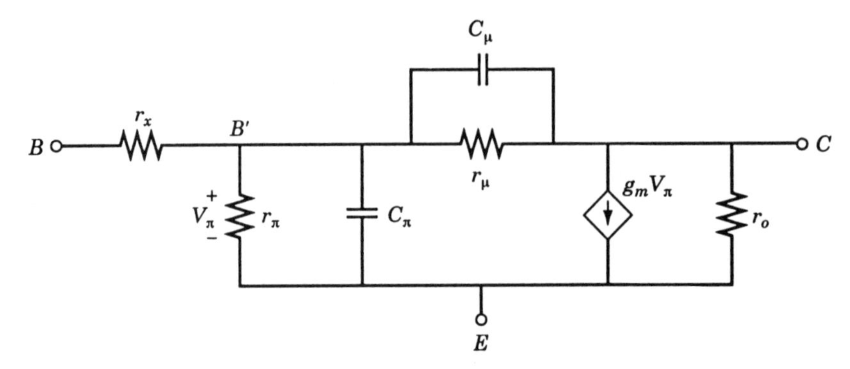

Figure 5-1 Hybrid-π bipolar transistor small-signal model.

frequency dependent, so the model can be used over a wide band of frequencies. Although the hybrid-π is best suited for the common-emitter (CE) configuration, the model is still accurate and can be rearranged for either the common-base (CB) or common-collector (CC) configurations. The basic model contains seven components shown in Fig. 5-1 and is the same whether the BJT is the *npn* or *pnp* type. Between the external base terminal B and the internal terminal B' is the base-spreading resistance r_x. The elements r_π and C_π represent the significant portion of the input impedance of the transistor.

The amplification property of the device is represented by the dependent current generator $g_m V_\pi$, where V_π is the signal potential between B' and the emitter terminal E. The elements r_μ and C_μ are caused by the base-width modulation effect and by depletion-layer capacitance. These internal feedback mechanisms within the transistor often can be omitted in developing a simplified low-frequency model. The element r_o represents the dynamic output resistance of the transistor.

The more important parameters of the hybrid-π model can be expressed in terms of easily measured quantities. The short-circuit current gain, called h_{fe} or β_o, is not a parameter of the hybrid-π. However, if we assume a short circuit between C and E in Fig. 5-1 and consider low-frequency operation where capacitances may be neglected, the low-frequency current gain β_o can be expressed as

$$\beta_o = \frac{g_m V_\pi}{V_\pi / r_\pi} = g_m r_\pi \tag{5-1}$$

Equation 5-1 assumes that r_μ is very large.

The parameter g_m can be derived from the diode equation relating I_C and V_{BE} [1]:

$$g_m = \frac{qI_C}{kT} \tag{5-2}$$

At room temperature $q/kT \cong 40/V$. Note that Eq. 5-2 relates a small-signal ac parameter to the dc collector current.

Another parameter useful for noise analysis is the so-called Shockley emitter resistance r_e which is the reciprocal of g_m. At room temperature,

$$r_e = \frac{1}{g_m} \cong \frac{0.025}{I_C} \quad \Omega \tag{5-3}$$

The base–emitter resistance can also be expressed in terms of I_C. Substituting Eq. 5-3 into Eq. 5-1 yields

$$r_\pi = \frac{\beta_o}{g_m} = \beta_o r_e \tag{5-4}$$

The gain–bandwidth product f_T is the frequency at which the short-circuit current gain equals unity. This parameter is related to the hybrid-π parameters by

$$C_\pi = \frac{g_m}{2\pi f_T} - C_\mu \tag{5-5}$$

The beta-cutoff frequency, f_{hfe} or f_β, is the frequency at which beta has declined to 0.707 of its low-frequency reference value, β_o. It can be shown that

$$f_{hfe} = f_\beta \cong \frac{f_T}{\beta_o} \tag{5-6}$$

A typical set of hybrid-π parameters is

$$r_\pi = 97\ \text{k}\Omega \qquad r_o = 1.6\ \text{M}\Omega$$

$$r_x = 278\ \Omega \qquad r_\mu = 15\ \text{M}\Omega$$

$$g_m = 0.0036\ \text{S} \qquad C_\mu = 4\ \text{pF}$$

$$\beta_o = 350 \qquad C_\pi = 25\ \text{pF}$$

These values are typical for a small-signal, high-frequency discrete *pnp* transistor similar to the 2N4250 transistor at an operating point of $I_C = 0.1$ mA and $V_{CE} = -5$ V.

The basic hybrid-π model is slightly modified for monolithic integrated circuit transistors. Because of the greater distance from the collector junction to the external contact, a small resistance may be added in series with the collector. Adding a buried layer between the substrate and the epitaxial collector region can dramatically reduce this series collector resistance. An *npn* IC transistor is fabricated on a substrate of *p*-type semiconductor. The collector-to-substrate junction is reverse-biased. The capacitance associated with that junction can be added in the equivalent circuit between the collector and a separate substrate terminal. The substrate terminal is almost always connected to the lowest dc potential which makes the substrate ac ground.

5-2 NOISE MODEL

The noise mechanisms of bipolar transistors have been widely investigated and the results are well defined [2]. Let us start by looking at the cross section of a diffused *npn* transistor as shown in Fig. 5-2 to visualize the location of the transistor noise mechanisms. In Chap. 1 we defined the three noise mechanisms: thermal, shot, and $1/f$. From the transistor model, the real resistances generate thermal noise and the diode junction currents give rise to shot noise. Current flow generates $1/f$ noise.

The base-spreading resistance r_x is the ohmic resistance of the lightly doped base region between the external base contact and the active base region. This is a true resistance, and therefore exhibits thermal noise. The base current I_B and the collector current I_C generate shot noise at their respective junctions. The flow of base current I_B through the base–emitter depletion region gives rise to $1/f$ noise [3]. Therefore, four noise generators and the thermal noise of the source resistance are shown in the hybrid-π noise model of Fig. 5-3. Dynamic resistances r_π and r_o do not generate noise since they are nonohmic, ac resistances.

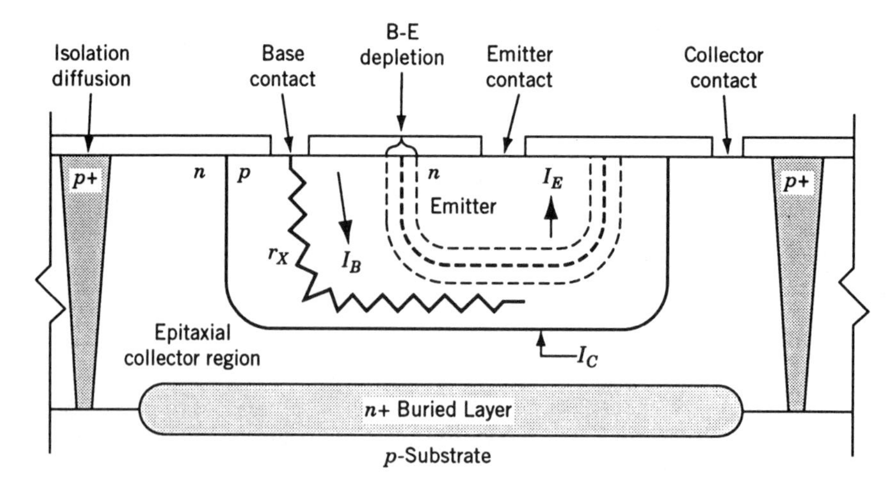

Figure 5-2 Cross-section diffused *npn* transistor.

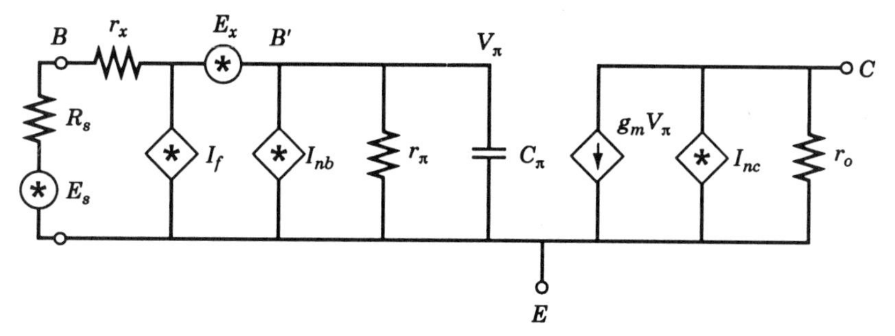

Figure 5-3 Hybrid-π bipolar transistor noise model.

Feedback elements C_μ and r_μ have been removed for clarity, which limits application of this equivalent circuit to frequencies less than $f_T/\beta_o^{1/2}$. At frequencies above f_β, the noise mechanisms are partially correlated and the total noise is slightly larger than predicted by this model. Generally, a transistor will be operated at lower frequencies to have better gain.

The E_x noise voltage generator is caused by the thermal noise of the base-spreading resistance. The noise current generator I_{nb} is the shot noise of the total base current, and I_{nc} is the shot noise of the collector current. The noise spectral density of these generators may be predicted theoretically from the discussions given in Chap. 1:

$$E_x^2 = 4kTr_x \tag{5-7}$$

$$I_{nb}^2 = 2qI_B \tag{5-8}$$

$$I_{nc}^2 = 2qI_C \tag{5-9}$$

The thermal noise voltage of the source resistance R_s is $E_s^2 = 4kTR_s$.

The $1/f$ noise contribution is represented by a single noise current generator I_f connected as shown in Fig. 5-3. It has been observed experimentally that the $1/f$ noise current flows through the entire base resistance in alloy-junction germanium transistors, but a reduced value of r_x is necessary to model correctly silicon planar transistors.

Following the discussion given in Sec. 1-10, the spectral density of $1/f$ noise current can be given by

$$I_f^2 = \frac{KI_B^\gamma}{f^\alpha} \tag{5-10}$$

The exponent γ ranges between 1 and 2, but often can be taken as unity. The experimental constant K takes on values from 1.2×10^{-15} to 2.2×10^{-12}. This constant can be replaced by $2qf_L$, where q is the electronic charge,

1.6×10^{-19} C, and f_L is a constant having values from 3.7 kHz to 7 MHz. The value for f_L is a representation of the noise corner frequency. Its value does not match the corner frequency, but there is gross correlation. The exponent α is usually unity. Thus the form for $1/f$ noise used in this chapter is

$$I_f^2 = \frac{2qf_L I_B^\gamma}{f} \tag{5-11}$$

The $1/f$ noise voltage generator is the product of the noise current given by Eq. 5-11 and the net resistance shunting I_f^2. Because the r_x contribution to $1/f$ noise is less than theory predicts for planar transistors, it is necessary to define a new r'_x, smaller than the value of r_x, to match the experimental data. Unless this change is made, calculations predict excessive $1/f$ noise. The effective r'_x is approximately one-half of r_x. An expression for the $1/f$ noise voltage generator is

$$E_f^2 = \frac{2qf_L I_B^\gamma r_x'^2}{f} \tag{5-12}$$

where $r'_x \cong r_x/2$

The $1/f$ noise results from the trapping and detrapping of carriers in surface and bulk energy states. This is a process-dependent noise mechanism. It can be caused by defects such as impurities and dislocations. Transistors that have a high beta at very low collector currents seem to have little $1/f$ noise perhaps because these trapping centers are also recombination centers. The $1/f$ noise is strongly affected by surface properties. Planar transistors are passivated so their surfaces are fairly well protected. There are epoxy-encapsulated devices with a low level of $1/f$ noise, but for the most critical applications it is desirable to use a hermetically sealed package.

5-3 EQUIVALENT INPUT NOISE

To determine an overall signal-to-noise ratio, all transistor noise mechanisms are referred to the input port. To derive the equivalent input noise E_{ni}, our method is to *calculate the total noise at the transistor output, the gain from source to output, and then divide the output noise by the gain.*

If the output is shorted in Fig. 5-3, then the output noise current is

$$I_{no}^2 = I_{nc}^2 + (g_m E_\pi)^2 \tag{5-13}$$

or

$$I_{no}^2 = I_{nc}^2 + g_m^2 \left[\frac{(E_x^2 + E_s^2) Z_\pi^2}{(r_x + R_s + Z_\pi)^2} + \frac{(I_{nb}^2 + I_f^2) Z_\pi^2 (r_x + R_s)^2}{(r_x + R_s + Z_\pi)^2} \right] \tag{5-14}$$

For an input signal V_s, *the output short-circuit signal current* is

$$I_o = g_m V_\pi = \frac{g_m V_s Z_\pi}{r_x + R_s + Z_\pi} \tag{5-15}$$

A transfer gain is

$$K_t = \frac{I_o}{V_s} = \frac{g_m Z_\pi}{r_x + R_s + Z_\pi} \tag{5-16}$$

We can now calculate the equivalent input noise E_{ni} as the ratio of Eqs. 5-14 to 5-16 according to

$$E_{ni}^2 = \frac{I_{no}^2}{K_t^2} \tag{5-17}$$

In terms of impedances and noise generators, E_{ni} is

$$E_{ni}^2 = E_x^2 + E_s^2 + \left(I_{nb}^2 + I_f^2\right)(r_x + R_s)^2 + \frac{I_{nc}^2(r_x + R_s + Z_\pi)^2}{g_m^2 Z_\pi^2} \tag{5-18}$$

Substituting the values for the noise generators ($\Delta f = 1$ Hz) into Eq. 5-18 gives the equivalent input noise:

$$E_{ni}^2 = 4kT(r_x + R_s) + 2qI_B(r_x + R_s)^2 + \frac{2qf_L I_B^\gamma(r_x' + R_s)^2}{f} + \frac{2qI_C(r_x + R_s + Z_\pi)^2}{g_m^2 Z_\pi^2} \tag{5-19}$$

The effective value of the base resistance r_x' is used in the $1/f$ term. Note that the last term contains Z_π which is frequency dependent. At low frequencies this term can be simplified to

$$\frac{2qI_C(r_x + R_s + r_\pi)^2}{\beta_o^2} \tag{5-20}$$

At high frequencies, up to about $f_T/\beta_o^{1/2}$, this term can be simplified to

$$\frac{2qI_C\left(r_x + R_s + \dfrac{1}{\omega C_\pi}\right)^2}{\dfrac{g_m^2}{\omega^2 C_\pi^2}} \cong 2qI_C(r_x + R_s)^2\left(\frac{f}{f_T}\right)^2 \tag{5-21}$$

The final expression for equivalent input noise in terms of transistor parameters, temperature, operating-point currents, frequency, and source resistance is

$$E_{ni}^2 = 4kT(r_x + R_s) + 2qI_B(r_x + R_s)^2 + \frac{2qI_C(r_x + R_s + r_\pi)^2}{\beta_o^2} + \frac{2qf_L I_B^\gamma (r_x' + R_s)^2}{f} + 2qI_C(r_x + R_s)^2\left(\frac{f}{f_T}\right)^2 \quad (5\text{-}22)$$

This form for E_{ni}^2 is an approximation. In the model of Fig. 5-3 the feedback capacitance C_μ has been omitted. This assumption may result in the actual noise at high frequencies being greater than the noise predicted by the equation. Nevertheless, Eq. 5-22 is a good engineering approximation to actual behavior. The errors are primarily at high frequencies, beyond transistor cutoff.

The first three terms in Eq. 5-22 are not frequency dependent and therefore form the limiting noise of any transistor. The first term $4kTr_x$ is the thermal noise voltage of the base resistance. The term $2qI_B(r_x)^2$ is the shot noise voltage associated with the base current and can usually be neglected since the source resistance is usually larger than the base resistance and E_s will dominate. The term $2qI_C(r_\pi)^2/\beta_o^2$ is the collector current shot noise; it can be rearranged to the more useful form $2qI_C r_e^2$. Another form for this voltage is $2kTr_e$ or half thermal noise. It should be emphasized that r_e is a dynamic resistance; therefore, it does not produce thermal noise in the BJT.

The transistor parameters for 12 types of transistors are given in Table 5-1. These values result from independent testing and do not necessarily agree with those available from manufacturers' specification sheets. Of specific interest is the variation of f_T with I_C.

5-4 NOISE VOLTAGE AND NOISE CURRENT MODEL

We now modify Eq. 5-22 to write the total expression for E_{ni}^2 of the transistor with a zero source resistance. By definition, this yields the noise voltage E_n^2 of the transistor noise model

$$E_n^2 = 4kTr_x + 2qI_B r_x^2 + \frac{2qI_C r_\pi^2}{\beta_o^2} + \frac{2qf_L I_B^\gamma r_x'^2}{f} + 2qI_C r_x^2\left(\frac{f}{f_T}\right)^2 \quad (5\text{-}23)$$

Since $r_\pi = \beta_o r_e$ and since r_x^2 is usually $\ll \beta_o r_e^2$,

$$E_n^2 = 4kTr_x + 2qI_C r_e^2 + \frac{2qf_L I_B^\gamma r_x'^2}{f} + 2qI_C r_x^2\left(\frac{f}{f_T}\right)^2 \quad (5\text{-}24)$$

TABLE 5-1 Transistor Parameters and Noise Constants

Transistor Type	β_o at I_C = 10 mA	1 mA	100 μA	10 μA	1 μA	f_T in MHz at I_C = 10 mA	1 mA	100 μA	10 μA	1 μA	C_μ (pF)	r_x (Ω)	r'_x (Ω)	f_L (MHz)	γ	I_{CBO} (max) (nA)
2N930	355	292	200	125	77	129	107	32.5	5.4	0.495	8.0	750	350	1.7	1.3	1
2N2484	400	353	322	277	180	132	101	32.2	4.88	0.475	6.0	380	200	1.6	1.5	1
2N3117	714	662	535	417	130	163	141	37.5	5.64	0.566	4.5	1000	130	7.0	1.5	1
2N3391A	704	510	345	196	111	122	67.4	13.2	1.61	0.162	7.0	800	200	0.2	1.2	10
2N3964	389	373	347	290	150	160	79.2	16.4	2.64	0.194	4.0	150	65	0.005	1.2	1
2N4058	201	250	235	192	120	106	106	43.2	7.02	0.759	5.0	280	100	0.18	1.4	10
2N4124	418	299	200	114	60	360	172	39.7	4.9	0.495	4.0	110	40	0.011	1.1	5
2N4125	170	144	118	87	45	255	101	22.0	3.63	0.302	4.5	50	48	0.004	1.2	5
2N4250	389	373	347	290	150	160	79.2	16.4	2.64	0.194	4.0	150	65	0.005	1.2	1
2N4403	278	200	136	80	48	182	68.6	11.9	1.34	0.133	8.5	40	8	0.010	1.1	10
2N5086	357	279	183	105	45	224	133	27.3	3.81	0.348	4.0	180	30	0.016	1.1	1
2N5138	215	154	77	55	29	131	64.2	14.2	2.14	0.169	5.0	100	20	1.25	1.3	1

The I_n noise current parameter of the model is obtained from Eq. 5-22 by assuming that R_s is very large valued, specifically $2qI_B > 4kTR_s$ as defined in Sec. 15-2-1. After dividing each term by R_s^2 and taking the limit as $R_s \to \infty$, we obtain the I_n^2 parameter

$$I_n^2 = 2qI_B + \frac{2qI_C}{\beta_o^2} + \frac{2qf_L I_B^\gamma}{f} + 2qI_C\left(\frac{f}{f_T}\right)^2 \tag{5-25}$$

Since $I_C/\beta_o^2 \ll I_B$, the second term in Eq. 5-25 is usually negligible. The noise current I_n becomes

$$\boxed{I_n^2 = 2qI_B + \frac{2qf_L I_B^\gamma}{f} + 2qI_C\left(\frac{f}{f_T}\right)^2} \tag{5-26}$$

Example 5-1 Determine the mean square equivalent input noise voltage, E_{ni}^2, for a 2N4250 BJT operating at a collector current of 1.0 mA, with a source resistance of $R_s = 10\ \text{k}\Omega$ in a $\Delta f = 10$ Hz band of frequencies centered at 1 kHz. Noise data for the 2N4250 transistor are shown in Fig. 5-9.

Solution There are two calculation methods possible. We will do both to illustrate the approaches and compare results. The first and easiest approach is to read the E_n and I_n from Fig. 5-9 as $E_n = 2\ \text{nV/Hz}^{1/2}$ and $I_n = 1\ \text{pA/Hz}^{1/2}$ both read at 1 kHz. Then we calculate E_{ni}^2 using the equation

$$E_{ni}^2 = \left(E_t^2 + E_n^2 + I_n^2 R_s^2\right) \Delta f$$

$$E_{ni}^2 = \left[1.6 \times 10^{-16} + (2 \times 10^{-9})^2 + (10^{-12})^2 (10^4)^2\right](10)$$

$$E_{ni}^2 = [1.6 \times 10^{-16} + 4 \times 10^{-18} + 10^{-16}](10)$$

$$E_{ni}^2 = 2.64 \times 10^{-15}\ \text{V}^2$$

$$E_{ni} = 51.4\ \text{nV}$$

In this example, the dominant noise contributions are from the thermal noise of the source resistance and the $I_n^2 R_s^2$ term. The E_n^2 term is negligible here.

The second calculation method is to use the 2N4250 transistor data in Table 5-1 plus Eqs. 5-24 and 5-26 to determine the E_n^2 and I_n^2 values directly. This is instructive because it reveals which noise mechanisms in the 2N4250 transistor are the dominant sources of noise. At $I_C = 1.0$ mA, $r_e = 25\ \Omega$. Using Eq. 5-24, we obtain $E_n^2/\Delta f = 2.63 \times 10^{-18}\ \text{V}^2/\text{Hz}$. The first term in this equation, due to the base resistance of 150 Ω, dominates. Equation 5-26 gives a numerical value for $I_n^2/\Delta f = 1.18 \times 10^{-24}$ A²/Hz. The shot noise of the base current is the largest contributor to I_n^2 with the $1/f$ term adding a significant amount of noise as well. Using these values of E_n^2 and I_n^2 in the defining E_{ni}^2 expression with a 10-Hz bandwidth produces

$$E_{ni}^2 = 2.79 \times 10^{-15} \text{ V}^2$$

$$E_{ni} = 52.8 \text{ nV}$$

The importance of this example is to show that close agreement between these two methods is possible even though it is somewhat difficult to read accurate data from the typical E_n and I_n curves.

5-5 LIMITING CASE FOR MIDBAND NOISE

Removing all frequency-dependent terms and the $1/f$ contribution from the expressions for E_n and I_n gives the limiting noise condition for the midband region. This is termed the "shot noise region" because the shot noise current sources dominate in a well-designed circuit. Equations 5-24 and 5-26 become

$$\boxed{E_n^2 = 4kTr_x + 2qI_C r_e^2} \qquad (5\text{-}27)$$

and

$$\boxed{I_n^2 = 2qI_B} \qquad (5\text{-}28)$$

Two mechanisms limit E_n, the thermal noise of the base resistance and the shot noise of the collector current times the emitter resistance. For low values of source resistance, where E_n dominates, it is desirable that the transistor have a low base resistance, and that the collector current be high. The I_n noise current generator is determined by a single noise mechanism, the shot noise of the base current. For high values of source resistance, where I_n dominates, it is desirable to operate with a low-leakage, high-beta transistor at a low level of collector current.

Plots of E_n and I_n versus collector current and transistor parameters are shown in Fig. 5-4. The E_n curve has a minimum value at high collector currents determined by the thermal noise of the base resistance. The noise voltage rises steadily with decreasing collector current. The magnitude of the noise current generator I_n decreases with decreasing collector current and with increasing β_o. This illustrates the minor effect of beta and the strong dependence on collector current.

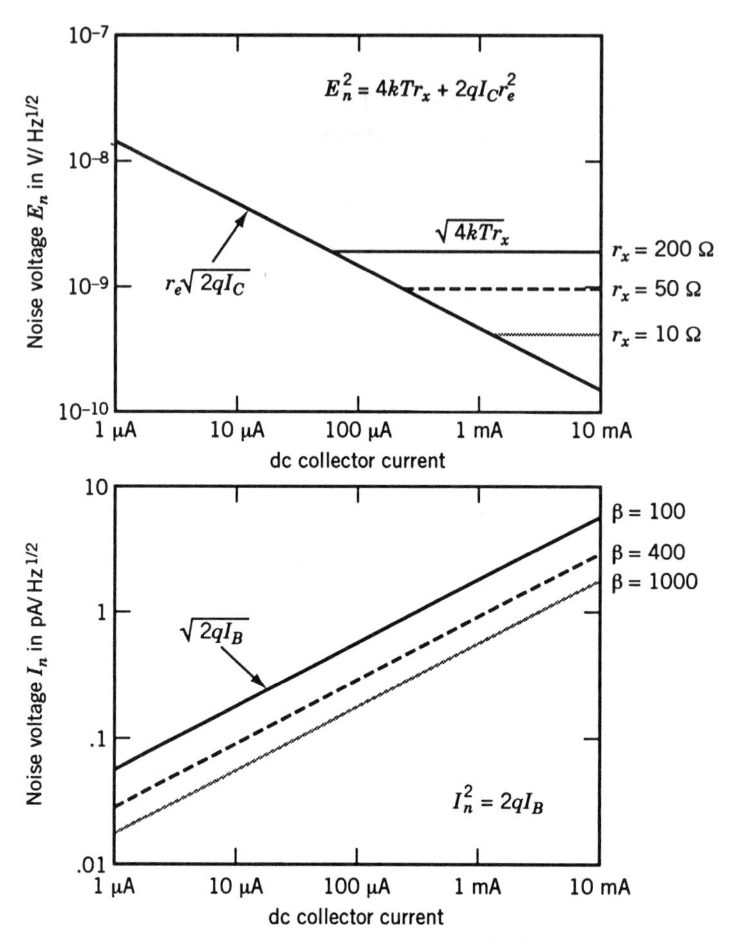

Figure 5-4 Limiting noise voltage and noise current.

5-6 MINIMIZING THE NOISE FACTOR

The optimum noise factor was defined in Chap. 2. The discussion resulted in the following expression:

$$F_{\text{opt}} = 1 + \frac{E_n I_n}{2kT\,\Delta f} \qquad (2\text{-}14)$$

F_{opt} can be obtained only when $R_s = R_o = E_n/I_n$. The lowest noise figure is obtained when the $E_n I_n$ product is low. F_{opt} is a useful number because it defines the best performance obtainable when the source resistance can be selected to match R_o. The values for E_n and I_n from Eqs. 5-27 and 5-28 can be substituted into Eq. 2-14, and the result is an expression for the optimum noise factor of a BJT:

$$F_{\text{opt}} = 1 + \sqrt{\frac{2r_x}{\beta_o r_e} + \frac{1}{\beta_o}} \tag{5-29}$$

From this expression for the optimum noise factor in terms of the base resistance, emitter resistance, and transistor beta, we observe that increasing beta reduces the minimum noise factor; similarly, reducing the base resistance and/or collector current reduces the noise factor. The lowest noise is obtained at low collector currents because the level is limited by the shot noise of the collector current and not by the thermal noise associated with the base current. Typical numerical calculations from Eq. 5-29 indicate that the best noise performance is usually obtained when the input transistor is operated at a collector current of less than 100 μA.

To design for the lowest noise figure, we require that the first transistor operate in the shot noise limited region. Minimum noise is highly dependent on a large transistor beta. When the collector current is small, the limit on F_{opt} is

$$F_{\text{opt}} = 1 + \frac{1}{\sqrt{\beta_o}} \tag{5-30}$$

The optimum value of the noise factor is only obtained when the source resistance is optimum

$$R_o = \frac{E_n}{I_n} \tag{5-31}$$

Substituting the values from Eqs. 5-27 and 5-28 gives

$$R_o = \sqrt{\frac{0.05\beta_o r_x}{I_C} + \frac{(0.025)^2\beta_o}{I_C^2}} \tag{5-32}$$

A plot of this equation versus I_C for various values of β_o and r_x is given in Fig. 5-5. It is clear from the figure that decreasing the collector current increases the required optimum source resistance. Transistor beta and base resistance have only a slight effect on R_o. In the limiting case when the base resistance is negligible, the optimum source resistance is linearly related to the collector current. Decreasing the collector current linearly increases the optimum source resistance according to

$$R_o \cong \frac{0.025\sqrt{\beta_o}}{I_C} \tag{5-33}$$

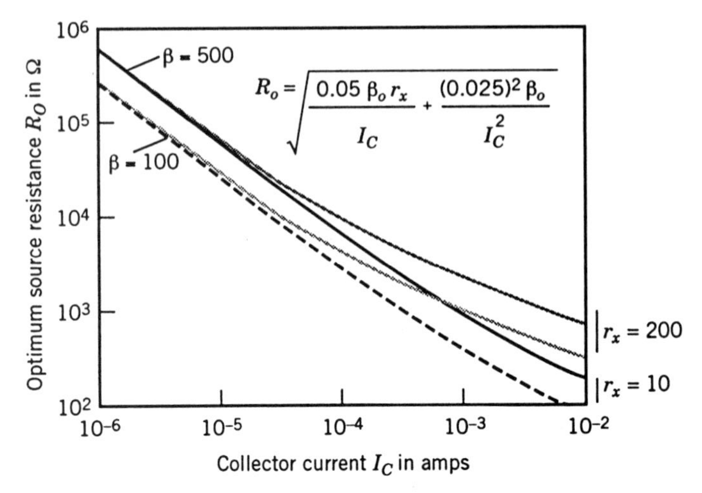

Figure 5-5 Graph of optimum source resistance versus I_C.

The preceding discussions indicate that the lowest noise figure is obtained at low collector currents and source resistances of a few thousand ohms to a few hundred thousand ohms. The lower impedance limit is set by the base resistance. A low-noise transistor should have a small base-spreading resistance and a high β_o.

5-7 THE 1/f NOISE REGION

Special consideration must be given to operation in the $1/f$ noise region. The same general design criteria apply, but different conditions exist for optimization. From the theoretical equations for E_n and I_n, the $1/f$ components can be extracted. They are

$$E_n^2 = \frac{2qf_L I_B^{\gamma} r_x'^2}{f^{\alpha}} \tag{5-34}$$

$$I_n^2 = \frac{2qf_L I_B^{\gamma}}{f^{\alpha}} \tag{5-35}$$

These two expressions differ only in the $1/f$ base resistance r_x'. The optimum source resistance is E_n/I_n. In this case, R_o *equals the base resistance and is not dependent on the operation point.*

The minimum noise factor F_{opt} obtained at source resistance R_o is

$$F_{\text{opt}} = 1 + \frac{qf_L I_B^{\gamma} r_x'}{kTf^{\alpha}} \tag{5-36}$$

The minimum noise factor is directly related to the base current and base resistance. To minimize the noise factor in the $1/f$ region, we select a good low-noise transistor with a small base resistance, operating at the lowest possible collector current. As we have seen, this criterion also gives low-noise performance in the midband region; however, it does not assure good high-frequency performance.

5-8 NOISE VARIATION WITH OPERATING CONDITIONS

It has been shown in Chap. 2 that equivalent input noise is dependent on three components

$$E_{ni}^2 = E_t^2 + E_n^2 + I_n^2 R_s^2 \tag{2-7}$$

We now wish to investigate the effect of static collector current on this noise equivalence. In Fig. 5-4 plots are shown for E_n and I_n versus I_C. The relative effect of I_n is dependent on R_s. Thermal noise E_s is constant with I_C. When these three terms are squared and added to form a plot of E_{ni}^2 versus I_C, it is clear that the total noise will be high at both extremes of I_C, and that for intermediate values of I_C the curve will perhaps experience a minimum.

The effect noted in the preceding paragraph is apparent in the plot of NF versus I_C shown in Fig. 5-6. It is seen that minima exist for all values of R_s. There is only one value of collector current and source resistance for the

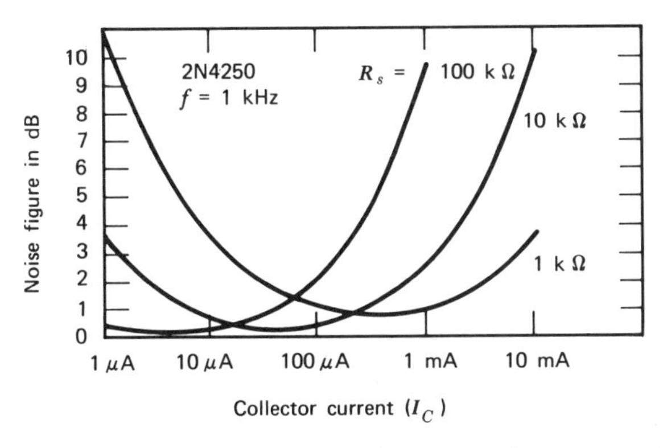

Figure 5-6 Effect of collector current and source resistance on noise figure.

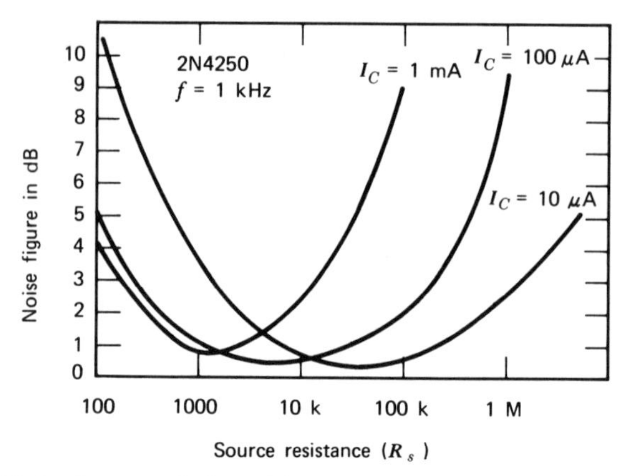

Figure 5-7 Noise figure variation with source resistance versus collector current.

lowest noise figure. Since published noise data are often given for one source resistance only, it is well to consider this curve when designing an amplifier for a different source resistance.

Test data on a 2N4250 transistor form the basis for the plot of noise figure versus source resistance shown in Fig. 5-7. Observe that the optimum source resistance decreases with increasing collector current. At a fixed value of collector current, the noise is higher for any other source resistance. A minimum value of noise figure results when the E_n noise is equal to the $I_n R_s$ noise. Below the optimum source resistance the amplifier noise E_n is constant even though the source noise is decreasing. Above the optimum value the $I_n R_s$ noise is increasing faster than the thermal noise. Both of these conditions cause the ratio of total noise to thermal noise to increase, and result in the minimum in the noise figure curve.

An alternate method of presenting noise figure data is shown in Fig. 5-8. These plots are referred to as contours of constant noise figure. Each of the graphs in the figure shows data taken at a different frequency. These data pertain to a 2N4250 silicon transistor at 10 and 100 Hz, 1, 10, and 100 kHz, and 1 MHz.

These contours of constant noise figure are convenient for analyzing noise performance. The best transistor selection will be the device with the best noise performance over the widest range of operating point. Select the transistor type with the largest noise contour below 1 dB.

The major limitation in noise contour use arises in broadband applications. For examination of frequency effects, plots such as those shown in Fig. 5-9 are more useful. We see E_n and I_n behavior for the frequency range from 10 Hz to 1 MHz. In both curves there is a flat frequency-independent region with noise increasing at both low and high frequencies. At low frequencies the slope is proportional to $1/f^{1/2}$; at high frequencies the slope is proportional to the frequency.

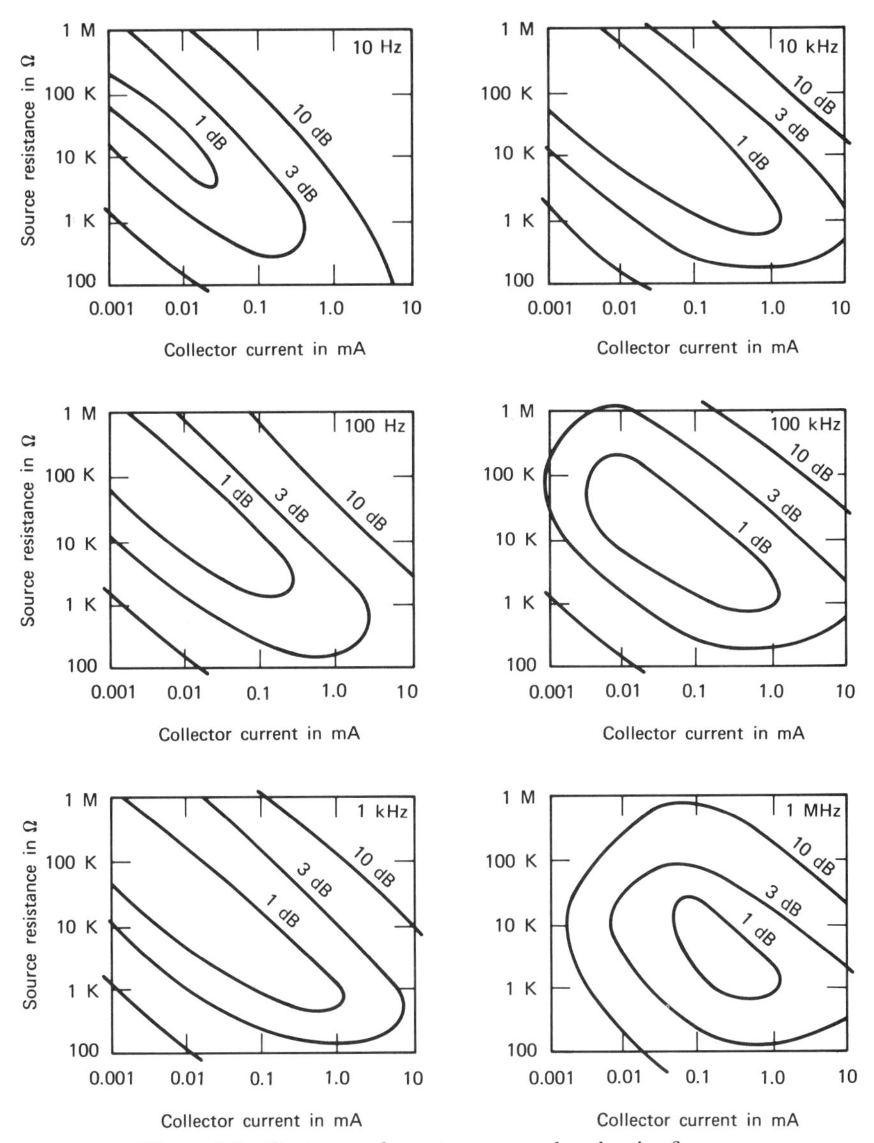

Figure 5-8 Contours of constant narrowband noise figure.

There is a small amount of $1/f$ noise at the highest collector currents in the E_n curve. The high-frequency break is located at f_T: therefore, the best performance can be obtained at high collector currents where f_T is greatest.

The I_n curves show midband noise increasing with increased collector current. The $1/f$ noise component is more prominent. For low-frequency operation it is clear that a low collector current is desirable. The high-frequency corner is approximately f_T/β_o. The high-frequency portion of the

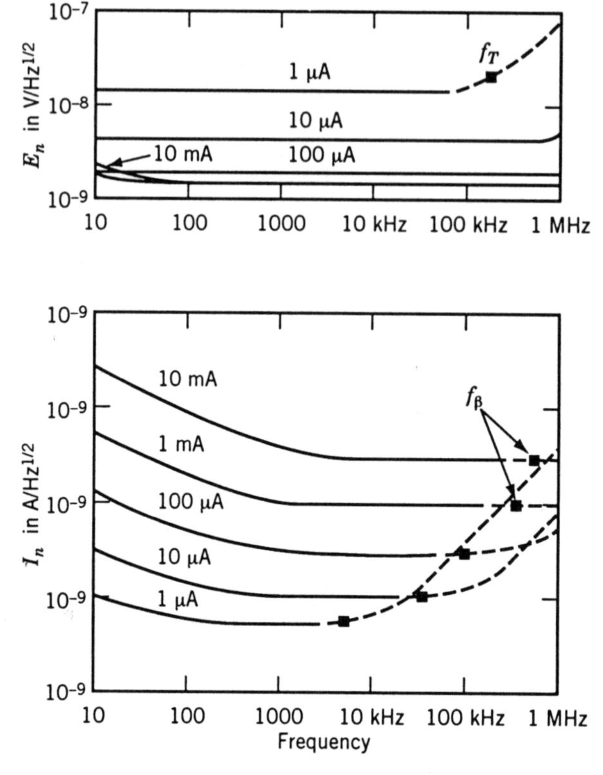

Figure 5-9 E_n and I_n performance of a 2N4250 transistor.

plot is based on calculations using the simplified hybrid-π version of Fig. 5-3. If the feedback capacitor C_μ were included in the model, the noise may be larger than indicated.

5-9 BURST OR POPCORN NOISE

Popcorn noise, also referred to as burst noise, was first observed in point contact diodes. It has since been found in tunnel diodes, junction diodes, film resistors, junction transistors, and integrated circuits. The name "popcorn" originated when the source was connected to a loudspeaker, the result sounded like corn popping.

Popcorn noise waveforms are shown in Fig. 5-11. Obviously, this is not white noise. The power spectral density of this noise is a $1/f^\alpha$ function with $1 < \alpha < 2$. It is often found to vary as $1/f^2$. The noise–frequency plot has a roller-coaster effect with one or more plateaus. The noise is masked by other mechanisms such as shot noise, unless frequency-selective networks are used to filter out higher-frequency noise.

For a *pn*-junction diode, the amplitude of popcorn pulses was found to be no higher than a few tenths of a microampere. The pulse widths are a few microseconds or longer [4, 5].

A typical noise burst of 10^{-8} A with a duration of 1 ms represents some 10^8 charge carriers. It is doubtful that such a large number of carriers would be directly involved in the process; it seems more likely that a modulation mechanism is present, and therefore small numbers of carriers are controlling the larger carrier flow. A physical model for burst noise in semiconductor junctions has been proposed that is based on the presence of two types of defects. Suppose that a metallic precipitate were to exist within a metallurgical *pn* junction, and in the space charge region adjacent to that defect a generation–recombination center (G–R) or a trap were located. It can be argued that if the occupancy of the generation–recombination center changes, the current flowing through the defect is modulated. Thus the G–R center controls the current across the potential barrier and therefore the generation of burst noise.

If we consider that the noise power spectral density of popcorn noise is inversely proportional to f^2, the equivalent noise current generator has the form

$$I_{bb}^2 = \frac{K'}{f^2} \tag{5-37}$$

where K' is a dimensional constant having units of amperes^2. Studies have refined Eq. 5-37 to

$$I_{bb}^2 = \frac{KI_B}{1 + \pi^2 f^2/4a^2} \tag{5-38}$$

where K is a constant having units of amperes per hertz and the constant a represents the number of bursts per second. Equation 5-38 predicts a leveling-off of popcorn noise at very low frequencies.

The total base resistance r_x of a BJT can be considered to be composed of two sections: the relatively small resistance from the base contact to the edge of the emitter junction and a larger resistance lying beneath the emitter. These have been called inactive (r_i) and active (r_a) base resistances, respectively. The total resistance $r_x = r_i + r_a$. It has been reported that $1/f$ noise attributable to surface recombination effects should be affected by r_i, whereas $1/f$ noise caused by recombination in the active base region should be affected by the entire base resistance [6]. It is also found that popcorn noise is associated with the value of inactive base resistance.

A $1/f$ noise equivalent circuit for a common-emitter-connected transistor that has six noise sources, including popcorn noise, two $1/f$ noise sources,

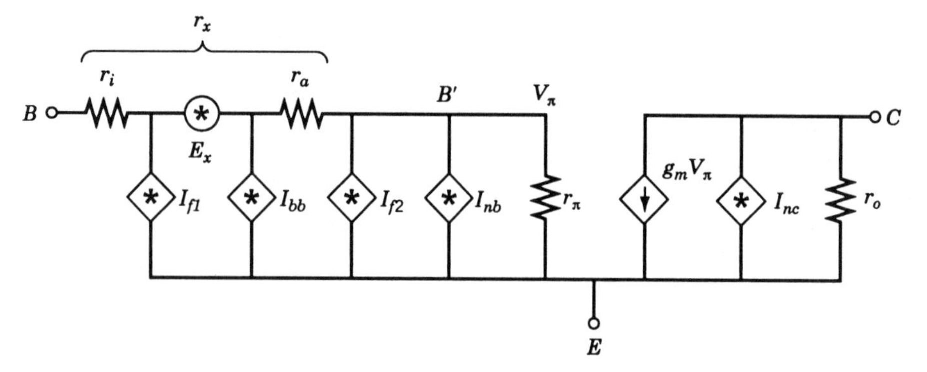

Figure 5-10 Expanded hybrid-π bipolar model with excess noise mechanisms.

two shot noise sources, and thermal noise, is shown in Fig. 5-10. The noise power spectral density of the noise generators are

Shot: $$I_{nb}^2 = 2qI_B$$

Shot: $$I_{nc}^2 = 2qI_C$$

Thermal: $$E_x^2 = 4kTr_x$$

Burst: $$I_{bb}^2 = \frac{KI_B}{1 + \pi^2 f^2/4a^2}$$

$1/f$: $$I_{f1}^2 = \frac{K_1 I_B^{\gamma_1}}{f}$$

$1/f$: $$I_{f2}^2 = \frac{K_2 I_B^{\gamma_2}}{f}$$

where K_1 and K_2 are dimensional constants having units of amperes.

Broadband noise is obtained by integrating each of these terms over frequency. For the shot and thermal noise terms, which are white noise, simply multiply by noise bandwidth Δf to obtain I_{nb}^2, I_{nc}^2 and E_x^2. Intergration of $1/f$ noise is described in section 12-1.

This noise equivalent circuit can be used for noise analysis. The constants present in the expression for $1/f$ and burst noise sources must be determined experimentally for each transistor.

Microplasma noise has been reported to be present in IC operational amplifiers [7]. The amplitude of microplasma noise can be several orders of magnitude larger than popcorn noise, whereas the pulse width can be

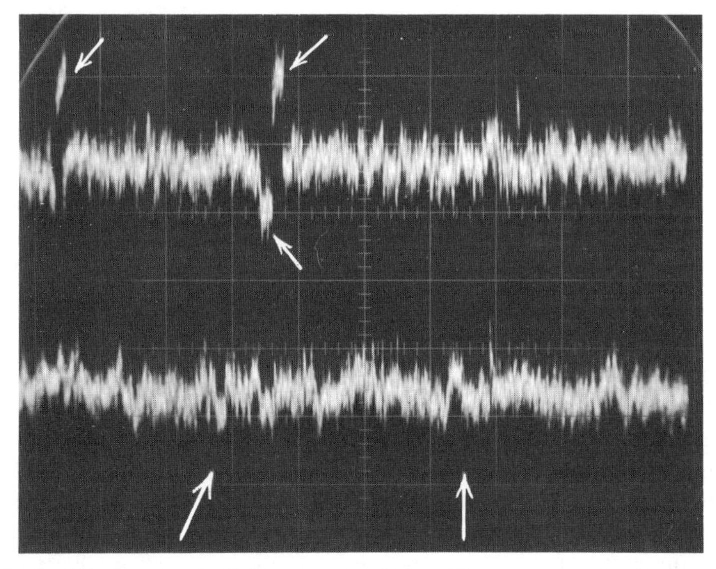

Figure 5-11 "Popcorn noise" is shown in the oscilloscope traces. The top trace is considered to represent a moderate level of this noise. The bottom trace is a low level. Some devices exhibit popcorn noise with five times the amplitude shown in the top trace. Horizontal sensitivity is 2 ms/cm.

extremely small. Microplasma noise is a local surface high-field breakdown phenomenon at the collector–base junction. It often can be traced to an input stage, but if originating in a later amplifying stage, it will still seriously degrade performance.

5-10 POPCORN NOISE MEASUREMENT

Popcorn noise becomes important in high-gain op amps which work in the $1/f$ noise frequency spectrum covering the audio frequency range (20 Hz to 20 kHz). Typical tests used to determine either narrowband or wideband noise provide only indications of the average noise power at the measurement frequency and do not reveal the "burst" noise. Analog or digital type metering circuits usually cannot respond fast enough to measure these effects. Typically, popcorn noise demonstrates a random abrupt output noise change that lasts from 0.5 ms to as long as several seconds [8–10] as shown in Fig. 5-11. Also the random rate at which the bursts occur ranges from approximately several hundred per second to less than one per minute. These rates are not repetitive or predictable. Therefore, conventional testing techniques will not work and a new procedure must be developed which counts individual bursts in periods from approximately 10 s to 1 min.

When considering how to characterize popcorn noise, one must consider these major questions:

What characteristics of popcorn noise should be measured?
What are reasonable "pass–fail" criteria?
What type of test configuration should be used?

Absolute measurement of the burst duration is not considered as important as the burst amplitude and frequency. However, the rate of occurrence as monitored by the number of counts per unit time is significant. For example, one hundred occurrences per second would be objectionable in any low-level, low-frequency application, whereas one occurrence in one minute might be perfectly acceptable. Therefore, we need a counter, timer, and threshold detector arrangement.

The bandwidth of the test system is very important. If we have excessive bandwidth to "capture" these sudden perturbations, the white noise of the various resistors and amplifier can obscure the burst noise occurrences and does not simulate realistically the low-frequency applications where popcorn noise is particularly objectionable. On the other hand, a test circuit with an excessively narrow bandwidth prevents detection of short duration bursts (approximately 0.5 ms) even if the amplitude is relatively large. Therefore, a compromise is realized where the test system's rise time permits a burst of minimum duration to reach essentially full amplitude. For a single pole system,

$$BW = \frac{0.35}{t_{\text{rise}}} \tag{5-39}$$

Assume that we require being able to count short bursts having a rise time of 0.5 ms means that we need a -3-dB bandwidth of 700 Hz, producing a noise bandwidth for the system of 1100 Hz.

Burst noise causes effects which are similar to a spurious noise current. Therefore, we should select a measurement configuration where the equivalent input noise is dominated by the $I_{\text{pop}} R_s$ product. Thus we will make R_s as large as possible without causing the thermal noise of the source resistance to be too large. Popcorn noise is most present in BJTs, and so the input offset current multiplied by a large source resistance may produce too large an offset voltage. With $R_s = 100\ \text{k}\Omega$ and an input offset current of 0.1 mA, the input offset voltage is 10 mV which is probably acceptable. With $R_s = 1\ \text{M}\Omega$, an offset of 0.1 V is probably too large.

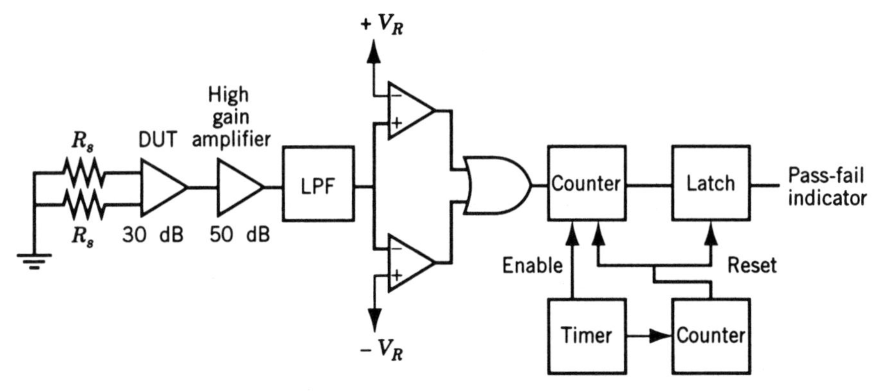

Figure 5-12 Block diagram of system for measuring popcorn noise.

Consider the popcorn noise test measurement system shown in Fig. 5-12. The two source resistors help to reduce the input offset voltage due to bias current. Together they increase the thermal noise by a factor of $\sqrt{2}$. The thermal noise in a single 100-kΩ resistor with $\Delta f = 1100$ Hz is 1.33 μV. The total E_t due to both source resistors is 1.88 μV.

Now suppose that the DUT amplifier has a wideband noise figure of 4 dB so that E_{ni} can be found from

$$\text{NF} = 10 \log \frac{E_{ni}^2}{E_t^2} = 4 \text{ dB}$$

$$E_{ni}^2 = 10^{0.4}(1.88\ \mu\text{V})^2 = 8.88 \times 10^{-12}\ \text{V}^2$$

$$E_{ni} = 2.98\ \mu\text{V} \tag{5-40}$$

If we assume a $\pm 3\sigma$ Gaussian distribution, the peaks of this background thermal noise will be approximately 3(2.98 μV) = 9 μV. This represents a lower limit of the burst amplitude which can be detected. A reasonable threshold for burst detection is 50% to 100% larger than this 9 μV. Suppose we set the $\pm V_R$ threshold to detect popcorn noise amplitudes of ± 15 μV. With 80 dB of voltage gain split between the DUT and postamplifier, we obtain $V_R = 150$ mV.

Achieving the 80-dB voltage gain from the DUT and high-gain amplifier can be accomplished using the circuit shown in Fig. 5-13. Gain is distributed. We need sufficient gain in the DUT stage to eliminate any noise contributions from the second stage while simultaneously allowing adequate loop gain in each stage so that accurate gain setting is possible with the external

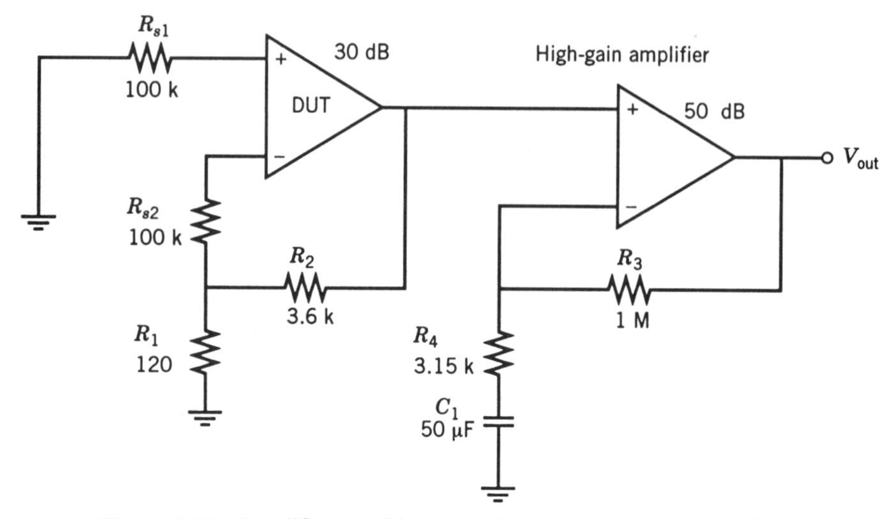

Figure 5-13 Amplifiers used in system for measuring popcorn noise.

resistors. Low-pass filtering can be done by paralleling a 230-pF capacitor with R_3 to set the -3-dB corner frequency to 700 Hz.

5-11 RELIABILITY AND NOISE

The noise mechanisms within a transistor have been attributed to thermal and shot noise effects, and, in addition, we have noted the presence of $1/f$ noise. It has been shown that a correlation exists between high values of $1/f$ noise and poor reliability. This correlation follows from the fact that $1/f$ noise is very sensitive to changes in a transistor's surface conditions and defects and these changes, if significant, result in device failure.

Two types of reliability testing programs can be performed using noise data. Routine noise tests on each transistor constitute one approach. Another means of study involves tests on a device over a period of time to determine changes in behavior. Noise tests require that some definition of normal noise levels be made. A normal level of $1/f$ noise varies among transistor types. A number of units must be tested to establish the norm.

Surface contamination can be detected by testing for $1/f$ noise current, I_n, at a few hundred microamperes of collector current and gradually increasing the collector voltage. If the noise current increases in proportion to the applied voltage, surface contamination is most probably present and is acting as a semiconductor resistor across the collector–base junction.

Faults such as defective contacts, fractures, or irregularities at the emitter–base junction can be detected by measuring the noise voltage, E_n, with the transistor operating under high-current conditions. These conditions are indicated by an abnormally high level of $1/f$ noise. A test of $1/f$ noise current I_n made at an intermediate value of collector current can indicate base region defects.

5-12 AVALANCHE BREAKDOWN AND NOISE

Caution! A circuit designer, evaluation engineer, or inspector making routine tests or adjustments can inadvertently ruin a perfectly good low-noise, high-gain transistor. This can happen when the base–emitter junction of a BJT (discrete or IC) is reverse-biased beyond its avalanche knee. The low-frequency noise of the device can be increased by 10 times because of a circuit turn-on transient, an overload signal transient, or the testing of V_{EBO} [11].

When a base–emitter injecting junction is reverse-biased, the junction will be damaged. This damage will decrease β_o slightly and increase $1/f$ noise drastically [12]. The damaging effect is observed to be proportional to the total charge flow and the logarithm of time [13]. During an avalanche breakdown there is sufficient energy in the carriers to create dislocations in the lattice structure. These crystal defects are minority carrier trapping centers and cause recombination in the base region and reduced current gain β_o. A small increase in recombination centers in the base region caused by the avalanching decreases gain slightly but significantly increases noise. Noise is probably the most sensitive measure of the quality of a transistor.

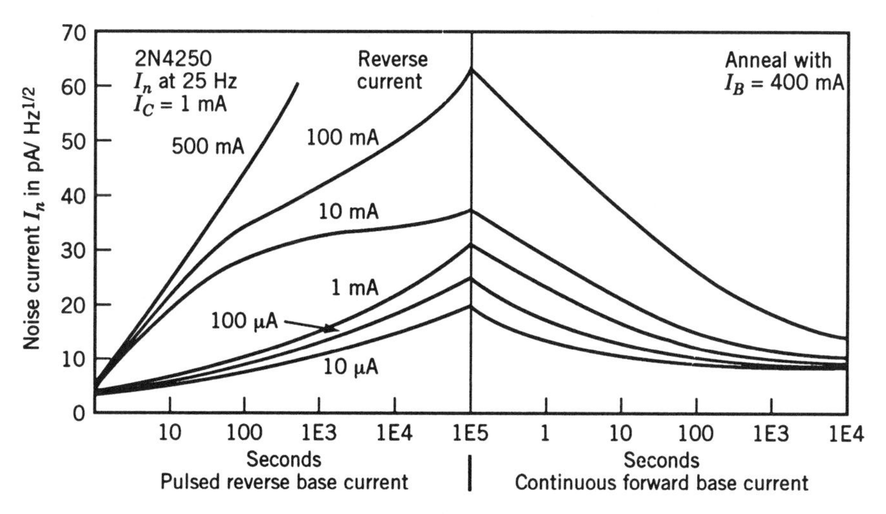

Figure 5-14 Increase in noise current with avalanching and the decrease resulting from current annealing.

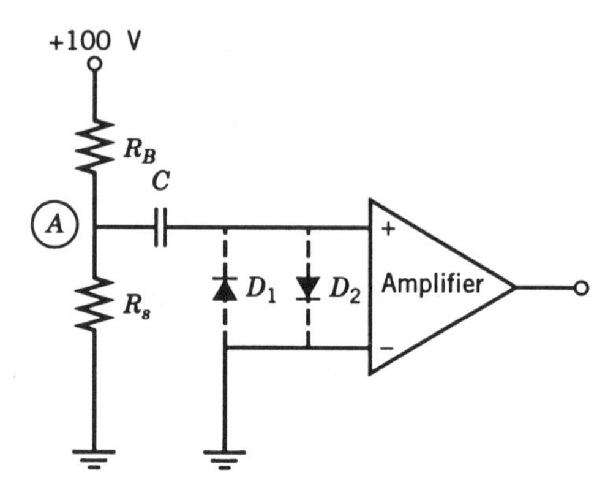

Figure 5-15 Demonstration of avalanching problem.

Although the noise observed is mainly excess $1/f$ noise, the h_{FE} and noise degradation can be a concern at RF [12]. Frequently, it includes "popcorn" noise. The equivalent noise voltage E_n can double in value, whereas the noise current I_n increases perhaps 10 times.

Figure 5-14 shows the progressive increase in noise current I_n caused by pulses of reverse-bias current applied to the emitter–base diode of a type 2N4250 silicon BJT. In normal operation the base–emitter diode is forward-biased, the pulses are avalanching the junction. To minimize heating, a 10% duty cycle was used. On the right side of Fig. 5-14 is shown the result of passing a large annealing current through the base–emitter junction. It was found that the avalanche-induced damage can be largely repaired. McDonald reports that heating the chip to approximately 300°C returns the noise to near original values [14]. The annealing was accomplished on the units in Fig. 5-14 by passing a continuous 400-mA forward current through the emitter–base junction. Although this means of repair is of interest, it is not a practical way to overcome the avalanche-caused damage. The solution is to design the circuit so that the damage cannot occur.

Avalanching can occur when operating from a biased source as shown in Fig. 5-15. When the 100-V supply is turned on, capacitor C is charged to a high voltage through R_B. Now if point A is grounded or the circuit is turned off, the voltage on C can avalanche the input transistors of the amplifier. Diodes D_1 and D_2 are added to protect the base junctions from this problem.

When an ac-coupled amplifier is turned on, it may experience wild voltage variations as the coupling and bypass capacitors charge, and there may be avalanching. To examine the conditions that spawn a turn-on transient, we consider the common direct-coupled complementary amplifier shown in Fig. 5-16. When the circuit is first turned on by connection to V_{CC}, transistors Q_1 and Q_3 are fully on while they charge C_2 and C_3. Although insufficient to burn out Q_2, this transient can result in a significant increase in the excess

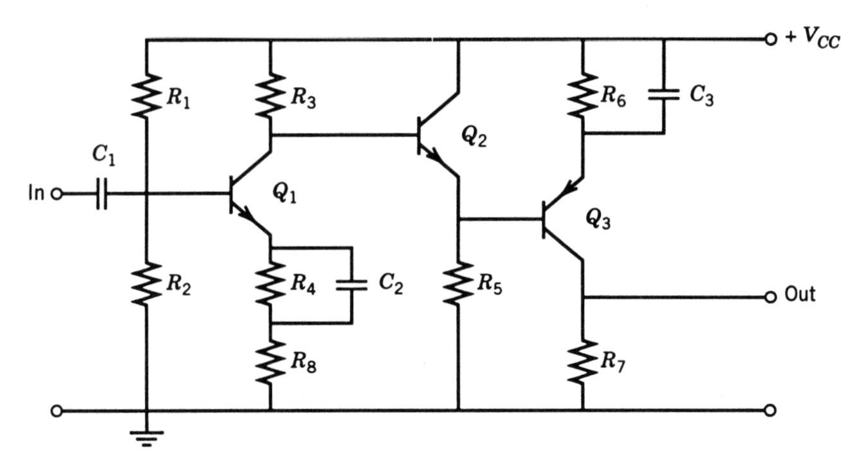

Figure 5-16 Direct-coupled complementary amplifier with single supply.

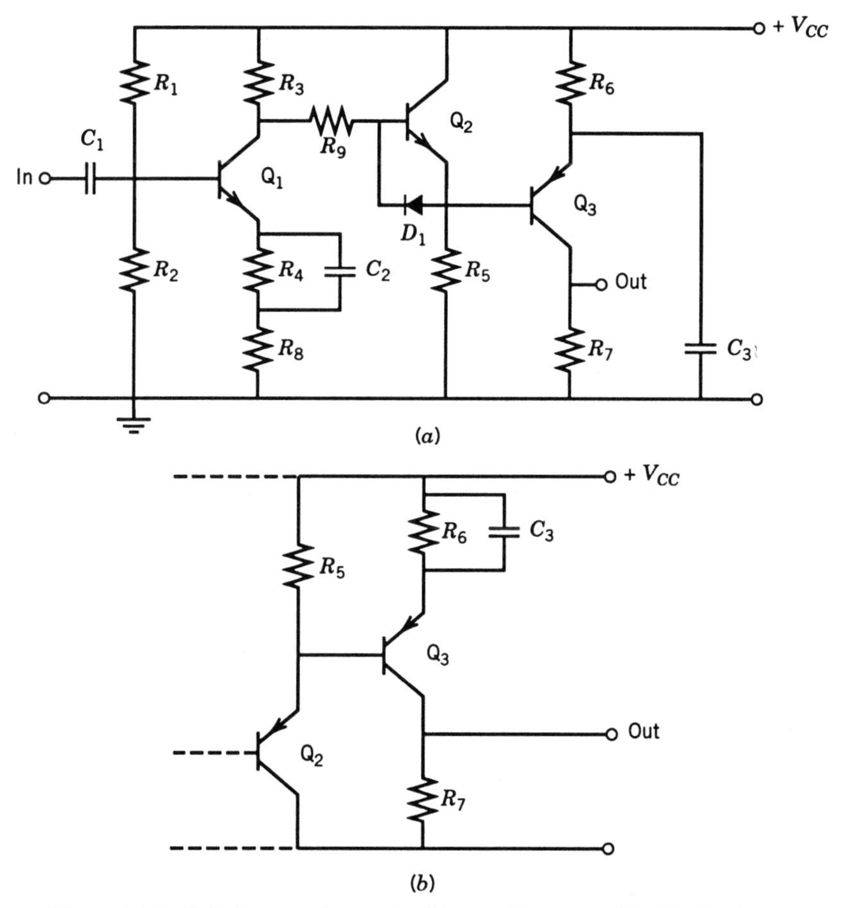

Figure 5-17 Solutions to the avalanching problem noted in Fig. 5-16.

noise current I_n. Since Q_2 usually operates from a large source resistance, R_3, the avalanching can significantly increase the total noise of the amplifier. A similar effect can result when a signal transient overloads an amplifier.

Three solutions to the turn-on problem are shown in Fig. 5-17. In Fig. 5-17*a*, C_3 is connected to bypass R_6 to ground instead of V_{CC}. A second solution is to place a reverse diode D_1 across the base–emitter junction of Q_2 and add a series limiting resistor R_9. The charging transient passes through D_1. In normal operation, D_1 has no effect on the gain or noise. In Fig. 5-17*b* the *npn* Q_2 is replaced by its *pnp* complement. The transient simple tries to turn on Q_2 harder, and the avalanching problem is therefore circumvented.

Measurement of V_{EBO}, the emitter–base breakdown voltage of a transistor, is generally considered nondestructive as long as the current is limited. In actual fact, damage may go unnoticed since a 1-s, 1-mA pulse only reduces h_{FE} by about 20%. The increased noise, however, is a significant problem in low-level circuits. There are two ways to measure V_{EBO}. One is to avalanche the base–emitter junction and measure the voltage drop. A better method is to specify a minimum acceptable V_{EBO}, then test for current flow (or breakdown) at that voltage. Frequently, it is not necessary to measure V_{EBO} especially for low-level linear circuits.

SUMMARY

a. The hybrid-π model provides a good representation of the bipolar transistor noise.

b. The BJT contains noise sources whose spectral densities are given by

$$I_f^2 = K(I_B)^{\gamma}/f \qquad 1/f \text{ noise}$$

$$I_{nb}^2 = 2qI_B \qquad \text{shot noise}$$

$$E_x^2 = 4kTr_x \qquad \text{thermal noise}$$

$$I_{nc}^2 = 2qI_C \qquad \text{shot noise}$$

c. Relative to the midband reference, the parameters E_n and I_n increase at low frequencies because of the $1/f$ noise source and increase at high frequencies because of reduced gain.

d. Both E_n and I_n are highly dependent on the collector current. E_n decreases with increasing current, while I_n increases with collector current.

e. At midfrequencies the noise parameters of a BJT are

$$E_n^2 = 4kTr_x + 2qI_C r_e^2$$

$$I_n^2 = 2qI_B$$

f. The optimum noise factor is

$$F_{\text{opt}} = 1 + \sqrt{\frac{2r_x}{\beta_o r_e} + \frac{1}{\beta_o}}$$

g. The optimum source resistance is

$$R_o = \sqrt{2\beta_o r_x r_e + \beta_o r_e^2}$$

h. The best noise performance is obtained for 10 kΩ $< R_s <$ 100 kΩ and 1 μA $< I_C <$ 100 μA.
i. The noise figure is independent of the transistor configuration.
j. Popcorn or burst noise is found in some integrated circuits, bipolar transistors, and junction diodes. It is highly process dependent and is caused by trapping centers.
k. Integrated circuits with unusually high values of $1/f$ noise tend to be less reliable than low-noise units.
l. Avalanche damage can occur in development or inspection testing, or it may result from turn-on or turn-off transients in a circuit.
m. After avalanche breakdown the emitter–base junction exhibits a large amount of $1/f$ noise, especially I_n noise.
n. Avalanche damage reduces the current gain parameter, h_{FE}.

PROBLEMS

5-1. A 2N930 transistor is biased with a dc current of 100 μA. The source resistance is 1 kΩ. Consider a 10-Hz noise bandwidth centered around an operating frequency of 3 kHz. Determine the individual contributions to E_{ni}^2 due to the source resistance, the BJT's equivalent input noise voltage, E_n^2, and the transistor's equivalent noise current, I_n^2.

5-2. Consider the same conditions as in Prob. 5-1. A 10-kΩ load resistance is placed between the collector and emitter in the noise model of Fig. 5-3. Let r_o = 40 kΩ. Determine the contribution to E_{ni}^2 from the load resistor.

5-3. A small-signal equivalent circuit model for a BJT amplifier is shown in Fig. P5-3.

(a) In this model, determine a numerical value for each noise source. Let $\gamma = 1$ as the exponent for the noise current calculations, and assume the $1/f$ noise corner frequency for I_f to be at 1 kHz. Assume a 1-Hz bandwidth centered around an operating frequency of 100 Hz. Be sure to specify proper units on all answers. Let $g_m = 5.77$ mA/V.

(b) Determine the contributions to E_{ni}^2 and E_{no}^2 due to each of the noise sources in Fig. P5-3. Specifically identify the dominant noise sources.

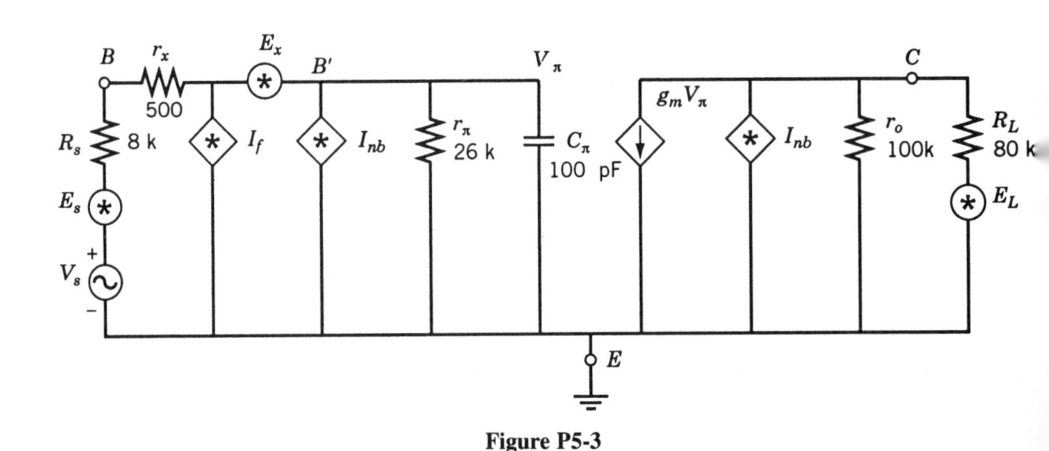

Figure P5-3

5-4. One stage of a low-noise amplifier is to be designed with a BJT operating with a collector current of 100 μA at a frequency of 1 MHz with a source resistance of 100 kΩ.

(a) From the device data in App. C, determine which device to select to produce the lowest NF.

(b) Which device would you definitely avoid specifying for the same conditions?

5-5. The noise figure of an amplifier is 5 dB with a source resistance of 10 kΩ.

(a) Determine E_{ni} as a spectral density for this amplifier.

(b) Determine the noise temperature in degrees celsius for this amplifier.

(c) Determine the noise resistance for the amplifier.

5-6. When the source resistance is equal to the optimum source resistance, then what is true? Be very specific.

5-7. Consider *npn* and *pnp* transistors with virtually the same terminal parameters. Which transistor type will more likely have the smaller equivalent input noise? Explain your reasoning.

5-8. What are the three most important factors worthy of measuring related to popcorn noise?

5-9. One stage of an amplifier has been designed with a 2N4250 BJT operating at $I_C = 100\ \mu A$ at a frequency of 100 Hz with a noise bandwidth of 3 Hz. Determine E_n, I_n, R_o, and F_{opt} for this device.

REFERENCES

1. Sedra, A. S. and K. C. Smith, *Microelectronics Circuits*, Holt, Rinehart, and Winston, New York, 1987.
2. Chennette, E. R., "Low-Noise Transistor Amplifiers," *Solid-State Design*, **5** (February 1964), 27–30.
3. Plumb, J. L., and E. R. Chenette, "Flicker Noise in Transistors," *IEEE Trans. Electron. Devices*, **ED-10**, 5 (September 1963), 304–308.
4. Hsu, S. T., and R. J. Whittier, "Characterization of Burst Noise in Silicon Devices," *Solid-State Electronics*, **12** (November 1962).
5. Hsu, S. T., R. J. Whittier, and C. A. Mead, "Physical Model for Burst Noise in Semiconductor Devices," *Solid State Electronics*, **13** (July 1970).
6. Jeager, R. C., and A. J. Brodersen, "Low-Frequency Noise Sources in Bipolar Junction Transistors," *IEEE Trans. Electron. Devices*, **ED-17**, 2 (February 1970), 128–134.
7. Hsu, S. T., "Bistable Noise in Operational Amplifiers," *IEEE J. Solid-State Circuits*, **SC-6**, 6 (December 1971), 399–403.
8. Robe, T. J., "Measurement of Burst (Popcorn) Noise in Linear Integrated Circuits," *RCA Linear Integrated Circuits*, Applicate Note SSP-202C (1975), 435–442.
9. Koji, T., "The Effect of Emitter-Current Density on Popcorn Noise in Transistors," *IEEE Trans. Electron. Devices*, **ED-22** (January 1975), 24–25.
10. Roedel, R. and C. R. Viswanathan, "Reduction of Popcorn Noise in Integrated Circuits," *IEEE Trans. Electron. Devices*, **ED-22** (October 1975), 962–964.
11. Motchenbacher, C. D., "Protect Your Transistors Against Turn-On or Testing Transient Damage," *Electronics*, **44**(25) (December 6, 1971), 92–94.
12. Schulz, D. C., "Avoid Killer Avalanches (and h_{FE} Degradation)," *RF Design* (June 1992), 35–51.
13. Collins, D. R., "h_{FE} Degradation Due to Reverse Bias Emitter–Base Junction Stress," *IEEE Trans. Electron Devices*, **ED-16**, 4 (April 1969), 403–406.
14. McDonald, B. A., "Avalanche-Induced $1/f$ Noise in Bipolar Transistors," *IEEE Trans. Electron. Devices*, **ED-17**, 2 (February 1970), 134–136.

CHAPTER 6

NOISE IN FIELD EFFECT TRANSISTORS

The basic operating principle of a field effect transistor (FET) has been known since J. E. Lilenfield's patent in 1925. In 1952, Schockley provided a theoretical description of a FET which led to the practical development of this electronic device which is routinely used to perform a wide variety of separate and useful circuit functions [1]. For example, FETs are used routinely as amplifiers, voltage-controlled resistors, switches, and capacitors.

The FET can be fabricated in two distinctly different ways. The insulated-gate FET (MOSFET) accomplishes control of current flow by a capacitor type of action. Charge in the conductive channel in the vicinity of the gate (input) electrode is attracted or repelled by the potential applied to the gate. This charge accumulation can discourage current flow through the channel.

In the junction FET (JFET), the voltage applied to a reverse-biased silicon *pn* junction modulates the conductivity of a channel below the junction by varying the size of the depletion region. A similar control mechanism occurs in gallium-arsenide FETs (GaAs FETs).

All three types, MOSFET, GaAs FET, and JFET, can be fabricated in either conductivity type depending on whether the channel is *p*- or *n*-type silicon or gallium arsenide. Most IC GaAs FET processes produce only *n*-channel type devices. The output characteristic curves for all FET types are similar to the corresponding curves for bipolar transistors. However, the important input quantity is the gate-to-source voltage instead of the base current. The input impedance of the MOSFET is due to parasitic leakage current and the capacitance associated with a silicon-dioxide dielectric. The input resistance is normally exceptionally large, on the order of TΩ. The size of the input capacitance is area dependent as set by the gate width and the length of the lateral moat diffusion under the gate. The input impedance of the JFET is due to a reverse-biased *pn* junction of a diode. The leakage current is considerably larger giving rise to a much smaller input resistance on the order of GΩ. The input capacitance is both area and voltage dependent. Additional information about these structures and the operation of FETs is available in the literature [2, 3].

In this chapter we will concentrate on the important noise characteristics of FETs and their use as a gain stage in low-noise amplifier applications.

6-1 FET NOISE MECHANISMS

The noise equivalent circuit for the common-source operation shown in Fig. 6-1 applies to all field effect type devices. The noise current generator I_g is the result of two physical processes: shot noise of the leakage current flowing through the gate and thermal fluctuations in the drain circuit that are coupled into the gate circuit. The noise generator I_{nd} is the result of thermal excitation of carriers in the channel of the device and excess or $1/f$ noise caused by trapping and detrapping of carriers. Generators I_{ng} and I_{nd} are found to be correlated at high frequencies since they both result from the thermal noise of the channel.

The E_n–I_n representation is applicable to all FET types. The curves shown in Fig. 6-2 are typical. Figure 6-2a shows the characteristic low E_n noise at high frequencies (region 2) and an increase in noise at low frequencies from excess or $1/f$ noise (region 1). The behavior of the open-circuit noise current I_n is given in Fig. 6-2b. The noise current is low in region 3 and increases linearly with frequency (region 4).

The observed behavior in region 2 is due to the thermal noise of the channel. This has been shown by van der Ziel [4, 5] to be dependent on the resistance of the channel. This thermal noise current in the channel noise can be reflected to the gate input as an equivalent input noise voltage produced by a resistance having a value given by

$$R_n \approx \frac{2}{3g_m} \tag{6-1}$$

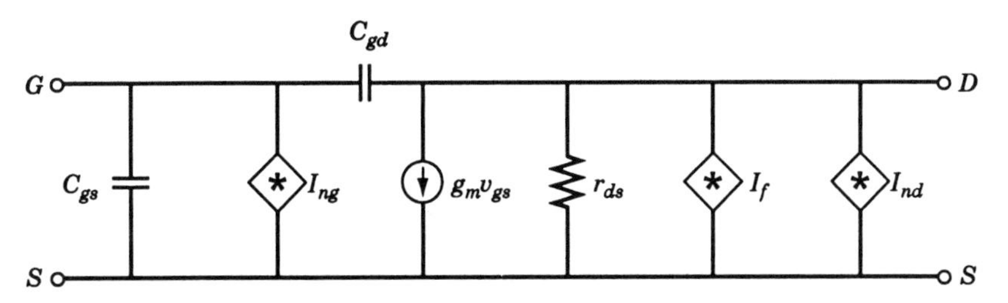

Figure 6-1 Small-signal noise equivalent circuit for a FET.

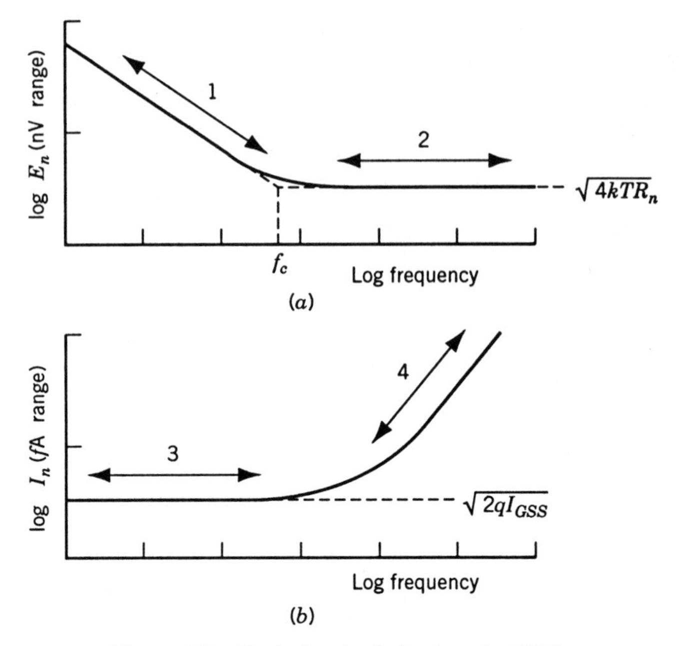

Figure 6-2 Typical noise behavior of a FET.

This value of the noise resistance R_n is related to E_n according to the thermal noise equation

$$E_n = \sqrt{4kTR_n} \tag{6-2}$$

To minimize the region 2 noise, it is necessary to operate the FET where its g_m is large. The largest g_m values can be found at high values of static drain current. Usually g_m is highest in the vicinity of I_{DSS}, the value of I_D where the gate-to-source bias voltage V_{GS} is zero. Using a large-geometry device with a large W/L ratio will give lower E_n, as discussed in the next section.

At the lower frequencies in Fig. 6-2a (region 1), there is excess $1/f$ noise. This noise arises from the trapping of carriers in trapping centers in the gate region [6]. For the MOSFET, these centers are at the silicon/silicon-dioxide interface and in the oxide layer. p-Channel enhancement-mode devices will have lower noise because carriers do not flow through this region. JFETs have trapping centers in the depletion region. Trapping centers alternately emit a hole or an electron, and simultaneously fluctuate between a charged and neutral state. This fluctuating charge looks like a change in gate voltage or a true input signal. Thus the channel current varies or "flickers." These fluctuations are the principal source of $1/f$ noise in the FET.

We now refer to the curve of the equivalent open-circuit noise current I_n in Fig. 6-2b. At low frequencies (region 3), the noise current is attributed to

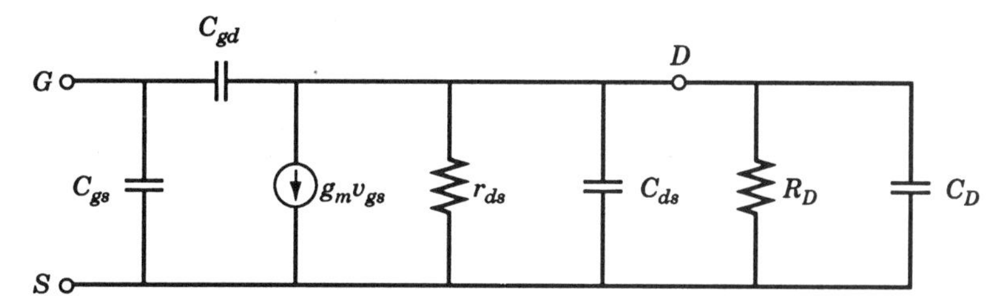

Figure 6-3 Circuit to determine high-frequency I_n noise of a FET.

the shot noise of the dc reverse saturation gate current for JFETs and to parasitic shot noise leakage current in MOSFETs. This noise is "white" and usually does not have $1/f$ noise. It is calculated as

$$I_n = \sqrt{2qI_{GSS}} \tag{6-3}$$

The higher-frequency I_n in region 4 of Fig. 6-2*b* is caused by an interesting mechanism. This is the thermal noise of g_{11}, the real part of the input admittance. In FETs, the noise of the channel is conducted to the input at high frequencies by the gate-to-drain capacitance. This appears as a shunt resistance in parallel with the input capacitance, C_{gs}, in the equivalent electrical circuit. To illustrate this, we use the small-signal model given in Fig. 6-3 with load quantities R_L to represent the parallel combination of r_{ds} and R_D and C_L to represent the sum of C_{ds} and C_D.

The capacitance C_{gd} is a feedback capacitance whose effect can be readily explained by making use of Miller's theorem [7]. Referring to Fig. 6-4, if the amplifier has a voltage gain of A_v, the feedback element Z can be separated into the two impedances Z_1 and Z_2 in the lower circuit and given by

$$Z_1 = \frac{Z}{1 - A_v} \quad \text{and} \quad Z_2 = \frac{Z}{1 - 1/A_v} \tag{6-4}$$

Making the comparison between Figs. 6-3 and 6-4, we see that the following relationships hold:

$$A_v = -g_m Z_L \tag{6-5}$$

where

$$Z_L = R_L \parallel (1/j\omega C_L) \quad \text{and} \quad Z = 1/j\omega C_{gd} \tag{6-6}$$

Using Eq. 6-4, it can be shown that

$$Z_1 = \frac{1 + j\omega R_L C_L}{-\omega^2 R_L C_L C_{gd} + j\omega C_{gd}(1 + g_m R_L)} \tag{6-7}$$

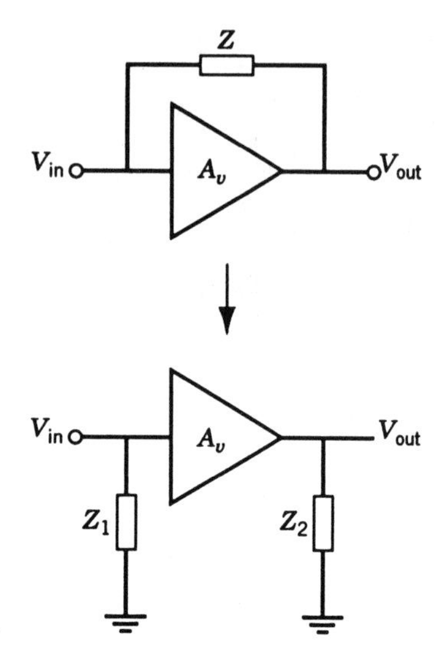

Figure 6-4 Applying Miller's theorem to Fig. 6-3.

The input admittance is the reciprocal of Z_1, or

$$Y_1 = \frac{-\omega^2 R_L C_L C_{gd} + \omega^2 C_{gd}(1 + g_m R_L) R_L C_L + j\omega C_{gd}[1 + g_m R_L + \omega^2 R_L^2 C_L^2]}{1 + \omega^2 R_L^2 C_L^2} \tag{6-8}$$

The real portion of Y_1 is

$$\text{Re}\, Y_1 = \frac{\omega^2 g_m R_L^2 C_{gd} C_L}{1 + \omega^2 R_L^2 C_L^2} \tag{6-9}$$

Thus there is an equivalent "real" reflected resistance present between the gate and source whose value is

$$R_{\text{eq}} = \frac{1}{\text{Re}\, Y_1} = \frac{1 + \omega^2 R_L^2 C_L^2}{\omega^2 g_m R_L^2 C_{gd} C_L} \tag{6-10}$$

Since $\omega^2 R_L^2 C_L^2 \ll 1$ at most frequencies of interest, this reduces to

$$\boxed{R_{\text{eq}} \cong \frac{1}{\omega^2 g_m R_L^2 C_{gd} C_L}} \tag{6-11}$$

The imaginary portion of Y_1 is

$$\operatorname{Im} Y_1 = \frac{j\omega\left[C_{gd}(1 + g_m R_L) + R_L^2 C_L^2 C_{gd}\omega^2\right]}{1 + \omega^2 R_L^2 C_L^2} \tag{6-12}$$

This forms an equivalent shunting capacitance of

$$C_{\text{eq}} = \frac{C_{gd}(1 + g_m R_L) + R_L^2 C_L^2 C_{gd}\omega^2}{1 + \omega^2 R_L^2 C_L^2} \tag{6-13}$$

Again, since $\omega^2 R_L^2 C_L^2 \ll 1$ at most frequencies of interest, this reduces to

$$C_{\text{eq}} \cong C_{gd}(1 + g_m R_L) + R_L^2 C_L^2 C_{gd}\omega^2 \tag{6-14}$$

The real "reflected" resistance in Eq. 6-11 decreases with the square of increasing frequency. The resistance R_{eq} is a real resistance and will contribute full thermal noise current $(4kT/R_{\text{eq}})^{1/2}$ in parallel with the input. The amplifier noise current I_n increases proportionally to frequency as illustrated by the noise of region 4.

The noises of E_n and I_n at high frequencies in regions 2 and 4 are correlated since they both originate from the thermal noise of the channel resistance. This results in an additional noise correlation term as shown in Eq. 1-31.

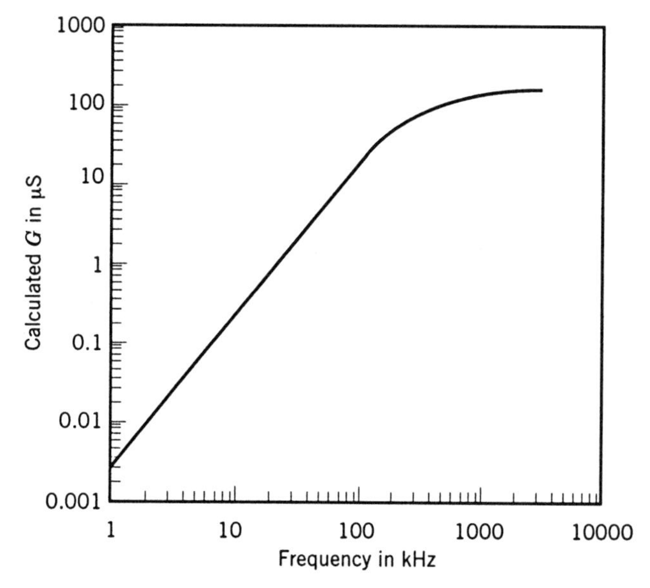

Figure 6-5 Calculated input conductance of a FET amplifier.

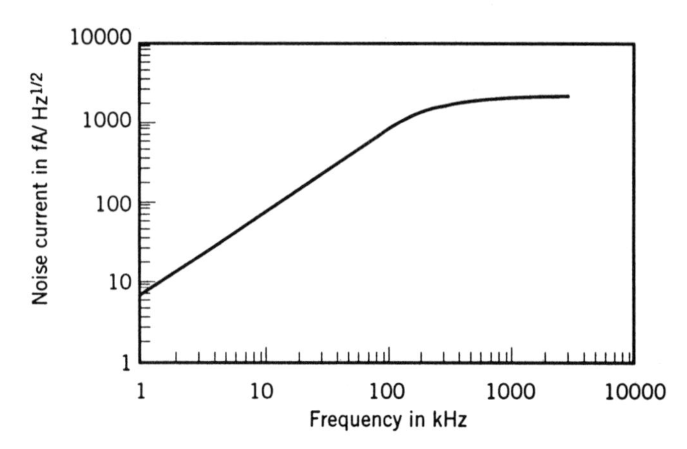

Figure 6-6 Calculated equivalent input noise current of a FET amplifier.

Example 6-1 What would the input conductance G_i and the noise current I_n be for an integrated FET amplifier whose input stage has $g_m = 500\ \mu\text{S}$, $C_{gd} = 1$ pF, $C_L = 3$ pF, and $R_L = 200$ kΩ

Solution The input conductance G_i versus frequency can be calculated using Eq. 6-9. A plot of the conductance is shown in Fig. 6-5. The equivalent input noise I_n can be calculated as the thermal noise current $(4kTG_i)^{1/2}$ of the input conductance G_i from the plot of Fig. 6-5 and plotted as shown in Fig. 6-6.

6-2 NOISE IN MOSFETs

The physical operating principles of the MOSFET are well understood and described extensively elsewhere [8, 9]. They will not be further developed here. Instead we will build upon the existing dc and ac MOSFET models to include the sources of noise. To begin, consider using the "metered" test circuit of Fig. 6-7*a* used to produce the typical output characteristics of an *n*-channel MOSFET as shown in Fig. 6-7*b*.

When the MOSFET operates in its *saturation* region, the dc drain current is related to the dc gate and drain voltages according to

$$I_D = K_p\left(\frac{W}{L}\right)(V_{GS} - V_T)^2(1 + \lambda V_{DS}) \tag{6-15}$$

where λ is the channel length modulation parameter, V_T is the threshold voltage, and W and L are the gate width and length dimensions. The

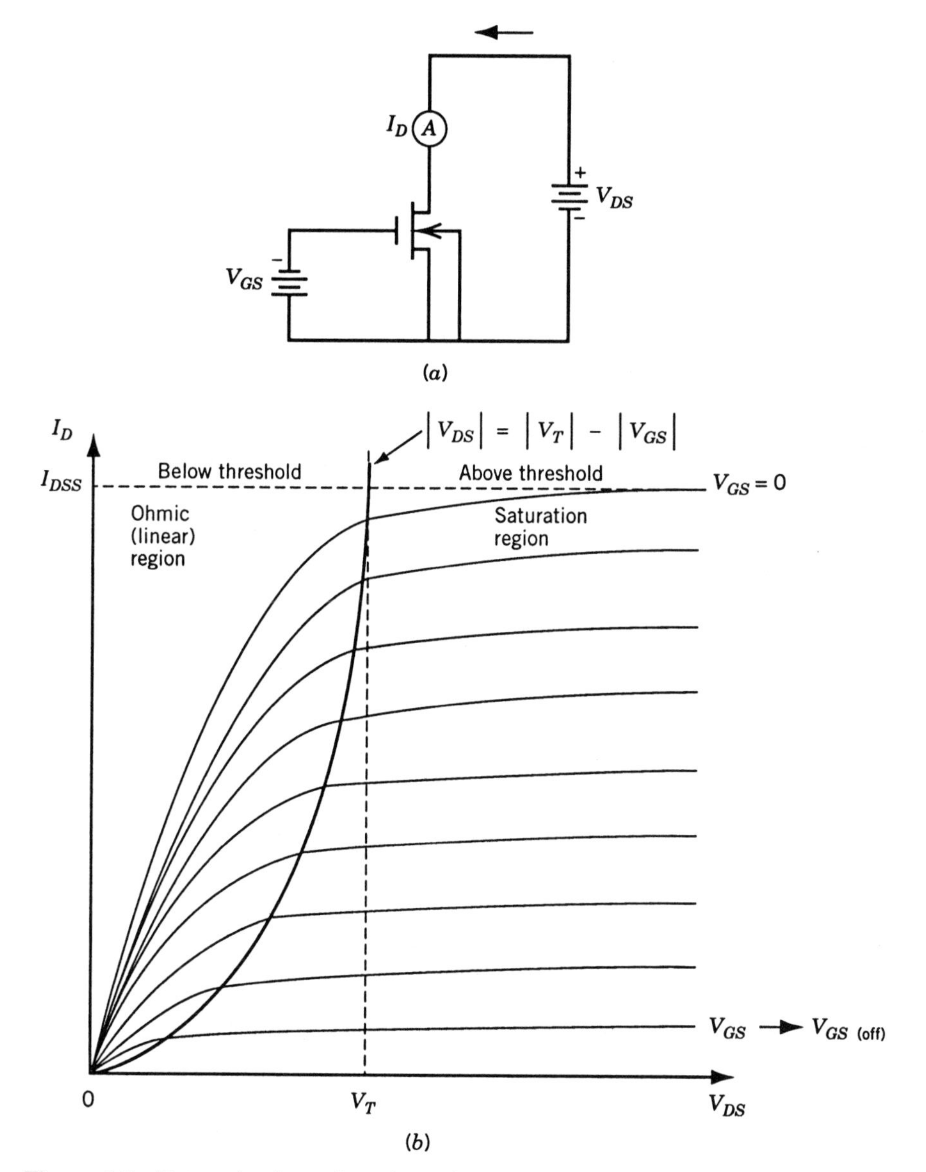

Figure 6-7 Determination of *n*-channel MOSFET *V–I* characteristics: (*a*) test circuit and (*b*) output characteristics.

transconductance parameter, K_p, is defined as

$$K_p = \frac{\mu_o C_{ox}}{2} \tag{6-16}$$

where μ_o is the mobility of the n-channel region in units of square centimeters per volt-second and C_{ox} is the capacitance per unit area of the gate oxide in units of farads per centimeter^2. The units for K_p are amperes per volt^2.

To give us some typical working values, let us assume some typical MOSFET model parameters as follows:

	n-Channel	p-Channel	Units
K_p	41.8	15.5	$\mu A/V^2$
V_T	0.79	−0.93	V
λ	0.01	0.01	V^{-1}

The ac, small-signal equivalent circuit model for a MOSFET was previously shown in Fig. 6-1. The conventional transconductance and output conductance in the model are found by differentiation of Eq. 6-15 and evaluation at the operating point. Stated mathematically,

$$g_m = \left.\frac{\partial I_D}{\partial V_{GS}}\right|_{@Q\text{-point}} \tag{6-17}$$

$$g_{ds} = \frac{\partial I_D}{\partial V_{DS}} = \left.\frac{1}{r_{ds}}\right|_{@Q\text{-point}} \tag{6-18}$$

The two model capacitances are calculated according to the following formulas:

	Region		
	Cutoff	Ohmic	Saturation
C_{gd}	$C_{ox}WL_D$	$C_{ox}WL_D + (1/2)WLC_{ox}$	$C_{ox}WL_D$
C_{gs}	$C_{ox}WL_D$	$C_{ox}WL_D + (1/2)WLC_{ox}$	$C_{ox}WL_D + (2/3)WLC_{ox}$

C_{ox} is the gate oxide capacitance per unit area in units of farads per micrometer^2. Alternatively, C_{ox} can be defined as

$$C_{ox} = \frac{\epsilon_0 \epsilon_{SiO_2}}{t_{ox}} \tag{6-19}$$

where ϵ_0 and ϵ_{SiO_2} are the permittivities of free space and the silicon-dioxide dielectric, respectively, and t_{ox} is the oxide thickness. L_D is the distance of the lateral moat diffusion under the gate.

Example 6-2 Let $W/L = 50$ for an n-channel MOSFET. At an operating point of $V_{GS} = 1$ V and $V_{DS} = 5$ V, determine the dc drain current and the ac small-signal model conductance parameters in Fig. 6-1.

Solution First we utilize the typical parameters in the table and check to make sure the MOSFET is operating in its saturated region.

Since $V_{DS} = 5 \text{ V} \geq V_{GS} - V_T = 1 - 0.79 = 0.21$ V, the FET is saturated. Next use Eq. 6-15 to calculate I_D according to

$$I_D = (41.8\ \mu\text{A/V}^2)(50)(1 - 0.79)^2\ \text{V}^2(1.05) = 97.2\ \mu\text{A} \tag{6-20}$$

Equations 6-17 and 6-18 are used to calculate g_m and r_{ds} according to

$$\begin{aligned} g_m &= 2K_p(W/L)(V_{GS} - V_T)(1 + 0.01V_{DS}) \\ &= 2(41.8\ \mu\text{A/V}^2)(50)(0.21\ \text{V})(1.05) = 926\ \mu\text{A/V} \end{aligned} \tag{6-21}$$

$$\begin{aligned} g_{ds} &= \lambda I_D = (0.01\ \text{V}^{-1})(97.2\ \mu\text{A}) = 0.972\ \mu\text{S} \\ r_{ds} &= 1/g_{ds} = 1/0.972\ \mu\text{S} = 1.03\ \text{M}\Omega \end{aligned} \tag{6-22}$$

There are three principal sources of noise in a MOSFET identified as I_{ng}, I_{nd}, and I_f in Fig. 6-1. The I_{ng} source (low-frequency I_n as shown in region 3 of Fig. 6-2) is shot noise due to leakage current through the SiO_2 gate to the source and is calculated according to

$$I_{ng}^2 = 2qI_{dc} \tag{6-23}$$

The noise current I_n at high frequencies (region 4) is caused by the Miller effect coupling of the channel resistance noise to the gate through C_{gd}. The thermal noise in the drain–source channel materializes as I_{nd}^2 and is calculated according to

$$I_{nd}^2 = \frac{8kTg_m}{3} \tag{6-24}$$

This is the same amount of noise as would be produced by a simple ohmic resistor located in the drain–source channel and having a value given by

$$R_{\text{DRAIN}} = \frac{3}{2g_m} \tag{6-25}$$

The third noise source is a flicker or $1/f$ noise current also in the drain–source channel and is given by

$$I_f^2 = \frac{K_F I_{DQ}^{A_F}}{f C_{ox} W L_{eff}} \tag{6-26}$$

where K_F is the flicker noise coefficient, I_{DQ} is the quiescent drain current, A_F is a constant, f is the frequency of operation in hertz, W is the channel width, and L_{eff} is the effective channel length.

In the MOSFET, $1/f$ noise is caused by trapping centers in the gate oxide or at the boundary between the silicon and the silicon-dioxide interface. These can trap and release electrons from the channel and introduce $1/f$ noise [10]. The mean square noise increases with temperature and the density of the surface states. It decreases with the gate area $W \times L$ and the gate oxide capacitance per unit area C_{ox}. Trapping centers in the oxide cause popcorn noise because of their longer trapping time constant.

The total noise current at the output drain–source channel is

$$I_{no}^2 = I_{nd}^2 + I_f^2 \tag{6-27}$$

We reflect this noise current to the gate as an equivalent input noise voltage using the K_t reflection coefficient defined as

$$K_{tr} = \frac{i_{d(\text{signal})}}{v_{gs(\text{signal})}} = -g_m \tag{6-28}$$

$$E_{ni}^2 = \frac{I_{nd}^2}{g_m^2} + \frac{I_f^2}{g_m^2} = E_n^2 \tag{6-29}$$

$$E_{ni}^2 = \frac{8kT}{3g_m} + \frac{K_F I_{DQ}^{A_F}}{g_m^2 f C_{ox} W L_{eff}} = E_n^2 \tag{6-30}$$

Note that the first term of Eq. 6-30 is equivalent to a single resistor of value

$$R_n = \frac{2}{3g_m} \tag{6-31}$$

connected between the gate and source producing an equivalent white-noise voltage, E_n.

Example 6-3 Calculate the noise model parameters shown in Fig. 6-2. Use the same MOSFET parameters and Q-point given previously in Example 6-1. Evaluate these noise parameters in a 1-Hz bandwidth centered at 1 kHz.

Assume the gate leakage current is 100 fA, $C_{ox} = 0.7\ \text{fF}/\mu\text{m}^2$, $K_F = 2 \times 10^{-28}$ s/V, $A_F = 1$, and

$$L_{eff} = L_{drawn} - L_{delta} = 3(1.2\ \mu\text{m}) - 0.365\ \mu\text{m} = 3.235\ \mu\text{m}$$

Solution The gate noise current is calculated as

$$I_{ng}^2 = 2qI_{dc} = 2(1.602 \times 10^{-19})(10^{-13}) = 3.2 \times 10^{-32}\ \text{A}^2/\text{Hz}$$

$$I_{ng} = 0.18\ \text{fA}/\text{Hz}^{1/2}$$

The drain–source thermal noise current is found as

$$I_{nd}^2 = \frac{8kTg_m}{3} = 2(1.6 \times 10^{-20})(926 \times 10^{-6})/3 = 9.88 \times 10^{-24}\ \text{A}^2/\text{Hz}$$

$$I_{nd} = 3.14\ \text{pA}/\text{Hz}^{1/2}$$

The equivalent drain resistance is

$$R_{DRAIN} = \frac{3}{2g_m} = \frac{3}{2 \times 926 \times 10^{-6}} = 1.62\ \text{k}\Omega$$

The gate width is $W = 50(3.6) = 180\ \mu\text{m}$, and the flicker noise current is found by

$$I_f^2 = \frac{K_F I_{DQ}^{A_F}}{fC_{ox}WL_{eff}}$$

$$I_f^2 = \frac{(2 \times 10^{-28})(97.2 \times 10^{-6})}{(1000)(0.7 \times 10^{-15})(180 \times 3.235)} = 4.77 \times 10^{-23}\ \text{A}^2/\text{Hz}$$

$$I_f = 6.9\ \text{pA}/\text{Hz}^{1/2}$$

The total noise current at the output drain–source channel is

$$I_{no}^2 = I_{nd}^2 + I_f^2$$

$$I_{no}^2 = (9.88 \times 10^{-24} + 4.77 \times 10^{-23})\ \text{A}^2/\text{Hz}$$

$$I_{no}^2 = 5.76 \times 10^{-23}\ \text{A}^2/\text{Hz}$$

$$I_{no} = 7.59\ \text{pA}/\text{Hz}^{1/2}$$

Reflecting these noise currents to the input gate as an equivalent noise voltage gives

$$E_{ni}^2 = \frac{5.76 \times 10^{-23}}{(926 \times 10^{-6})^2} = 6.71 \times 10^{-17}\ \text{V}^2/\text{Hz}$$

$$E_{ni} = 8.19\ \text{nV}/\text{Hz}^{1/2}$$

Note that the flicker contribution at 1 kHz is the dominant source of noise. The portion of the E_{ni}^2 noise due to I_{nd} is equivalent to placing a resistor at the gate of value calculated by

$$R_n = \frac{2}{3g_m} = \frac{2}{3(926 \times 10^{-6})} = 720\ \Omega$$

Equation 6-30 shows that low-noise performance in a MOSFET requires a large value of transconductance, g_m, which, in turn, means that the transistor should have a large W/L ratio and be operated at a large quiescent current level. Making the transistor physically very large with a large perimeter reduces the contribution of the $1/f$ flicker noise. This also reduces the noise corner frequency, f_c, the frequency where there are equal mean square noise contributions from the two terms in Eq. 6-30.

Example 6-4 Determine the noise corner frequency for the same MOSFET used in the previous two examples.

Solution At the noise corner frequency,

$$E_f^2 = E_n^2$$

$$f_c = \frac{3K_F I_{DQ}^{A_F}}{8kTg_m C_{ox} WL_{eff}} \tag{6-32}$$

Substituting in numerical values gives

$$f_c = 4.83\ \text{kHz}$$

6-3 NOISE IN JFETs

A noise equivalent circuit for the common-source operation of a JFET (junction field effect transistor) device was developed in Sec. 6-1 and shown in Fig. 6-1. The curves of Fig. 6-2 are typical for a JFET. Fig. 6-2*a* shows the typical characteristic of low E_n noise voltage at high frequencies (region 2) with increasing $1/f$ noise at low frequencies (region 1). Behavior of open-circuit noise current I_n is given in Fig. 6-2*b*. Noise current I_n is minimum in the low frequencies of region 3 and increases with frequency. Noise current at high frequencies (region 4) is caused by Miller effect coupling of the channel resistance noise to the gate through C_{gd}.

For the JFET, the open-circuit noise current generator I_n results from two physical processes: shot noise I_{ng} of the reverse leakage current of the gate, and thermal noise fluctuations of the drain circuit conductance that are capacitively coupled into the gate circuit. Noise current I_n at low frequencies (region 3 of Fig. 6-2*b*) is caused by the shot noise of the dc reverse saturation gate current I_{GSS}:

$$I_{ng} = \sqrt{2qI_{GSS}} \tag{6-33}$$

This noise current is white and does not usually have $1/f$ noise.

Noise current I_n at high frequencies (region 4 of Fig. 6-2*b*) is caused by the thermal noise of the input conductance as derived in Eq. 6-11. In effect, the thermal noise of the drain channel resistance is capacitively conducted to the gate at the input. Note that noise generators E_n and I_n are correlated at high frequencies because they originate from the same noise source, the thermal noise of the channel resistance.

The noise voltage E_n at midfrequencies (region 2 of Fig. 6-2*a*) is caused by the thermal noise of the channel resistance. As shown by van der Ziel [4, 5], E_n is dependent on the resistance of the channel. The channel resistance noise is reflected to the gate input as an equivalent input noise voltage E_n by the expression

$$E_n = \sqrt{4kTR_n} \tag{6-34}$$

where the noise resistance R_n is experimentally determined in Eq. 6-1,

$$R_n \approx \frac{2}{3g_m} \tag{6-1}$$

As with the MOSFET, to minimize the midband E_n noise operate the FET where its g_m is large. Since g_m increases with static drain current, g_m is highest in the vicinity of I_{DSS}. This is the value of I_D where the gate-to-source bias voltage V_{GS} is zero. A large g_m can be obtained for a FET designed with a large gate W/L ratio, but this also results in a large geometry device.

At the lower frequencies of E_n in Fig. 6-2*a* (region 1), there is excess $1/f$ noise. This noise arises from the trapping of carriers in the so-called Shockley–Read–Hall (SRH) generation–recombination centers in the junction depletion region [6]. These centers are the main sources of reverse-bias leakage current in silicon diodes. A diagrammatic cross section of a JFET is shown in Fig. 6-8.

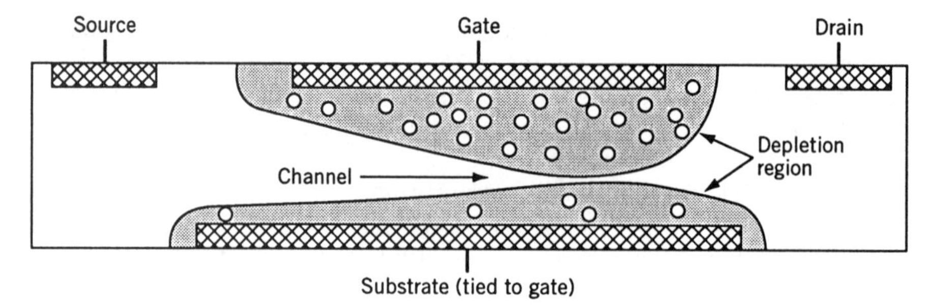

Figure 6-8 JFET cross section.

Generation centers, represented by the symbol ○, alternately emit a hole and an electron, and simultaneously fluctuate between a charged and neutral state. This fluctuating charge looks like a change in gate voltage or a true input signal. Thus the channel current varies or "flickers." These fluctuations are the principal source of $1/f$ noise in the JFET.

The generation centers are due to crystal defects or impurities. Since these centers are one of the sources of reverse-bias current, it is observed that devices with exceptionally low gate leakage current (I_{GSS}) have low $1/f$ noise. High gate leakage does not always imply a high-noise device since the currents can be of surface origin. Charge fluctuations at the surface do not modulate the width of the channel nor do they necessarily give the type of noise being discussed here.

Low-frequency E_n has bumps or a roller-coaster effect in some older devices. This is due to a distribution of generation–recombination centers with characteristic time constants.

There does not appear to be any theoretical way of reducing this $1/f$ noise. Several methods that increase $1/f$ noise, such as gold doping, are known. The noise is quite process dependent and varies from run to run and supplier to supplier. To select a FET for a $1/f$ application, it is necessary to measure the characteristics of a sample of devices.

6-4 NOISE IN GaAs FETs*

FETs can be fabricated using other semiconductors besides silicon. In particular, III–V compounds such as gallium arsenide (GaAs), indium phosphide (InP), and indium gallium arsenide (InGaAs) are becoming popular for fabricating very high performance FETs. These materials offer more desirable electronic properties, for example, higher carrier mobility and velocity,

*This section on noise in GaAs FETs was contributed by Steve Baier of Honeywell Inc., System and Research Center, Bloomington, Minnesota 55420.

lower saturation voltage, wider bandgap, and a semiinsulating substrate. Compared with identical silicon devices, these properties enable GaAs FETs to switch faster, amplify signals at higher frequencies, and operate at lower voltages and in harsh thermal or radiation environments. Consequently, GaAs FETs are utilized for specialized applications such as microwave communication, military weapons, supercomputers, and high-speed instrumentation. GaAs technology is also spurring the development of optoelectronic systems (e.g., fiberoptic communications and optical computing) due to the ability of GaAs to perform both electronic and optical functions on the same chip.

The noise characteristics of GaAs FETs are important for a variety of electronic circuits, including low-noise gain stages, wideband op amps, analog multiplexers, high-speed voltage comparators, optical receivers, and microwave oscillators (where upconverted $1/f$ noise can generate near-carrier phase noise).

6-4-1 GaAs FET Structure and Fabrication

To understand the noise characteristics of GaAs FETs, it is important to discuss several fabrication techniques and device structures that are commonly employed. These are illustrated in the tree diagram in Fig. 6-9. Basically, two types of GaAs FETs are commercially available: bulk GaAs FETs (e.g., MESFET and JFET) and heterostructure FETs (e.g., HEMT). Each type of device can be fabricated in several different ways, depending on the intended use. These fabrication options dramatically affect the resulting noise performance and this should be carefully considered when selecting discrete GaAs devices or IC technologies for low-noise applications. The important device types and process options are described next.

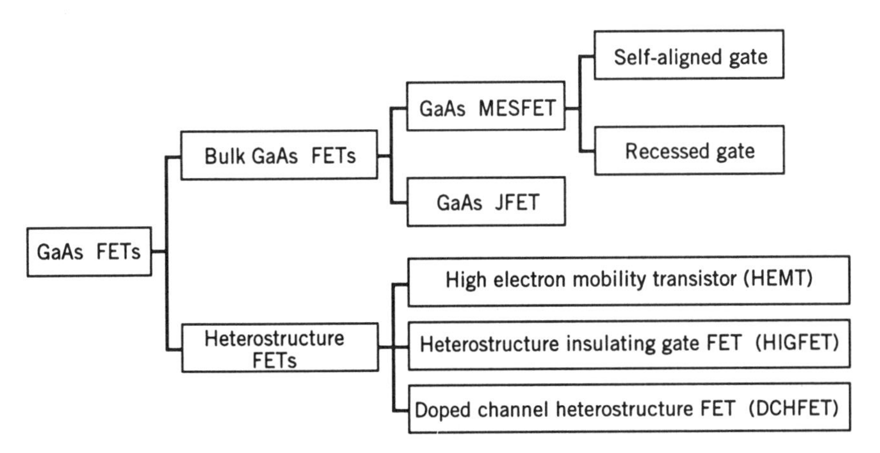

Figure 6-9 Family tree of GaAs FET processes.

6-4-1-1 GaAs MESFET The GaAs metal–semiconductor FET (MESFET) is the most common type of GaAs FET available today. It was the first GaAs FET to be invented and it is also the simplest to fabricate. The MESFET uses a Schottky metal–semiconductor contact as the gate junction. The applied gate bias controls the depth of the depletion region underneath the gate, which changes the effective channel thickness and modulates the drain current. GaAs MESFETs are only available as *n*-channel devices; the Schottky barrier height of *p*-channel GaAs MESFETs is too small to be useful. Even in the *n*-channel MESFET, however, the forward gate voltage cannot exceed ~ 0.7 V or the Schottky junction begins to conduct, resulting in loss of transconductance. Thus the useful gate swing of a GaAs MESFET is limited by forward gate turn-on at the high end and by the FET threshold voltage V_T at the low end. V_T is determined during fabrication by adjusting the channel thickness and/or doping concentration. GaAs MESFETs can be either depletion-mode (normally on) or enhancement-mode (normally off). Enhancement-mode devices are used mostly for digital logic in order to minimize power consumption. Linear and microwave circuits generally use depletion-mode devices because of their wider gate swing.

One critical difference among GaAs MESFETs is the method used to minimize the parasitic source and drain resistances. Recessed-gate MESFETs accomplish this by selectively removing the heavily doped N^+ contact layer and then depositing the gate contact as shown in Fig. 6-10. This approach permits use of low-resistivity Au-based gate metals but results in a nonplanar wafer surface. In the self-aligned process, the gate is deposited first followed by a heavy dose N^+ implant using the gate as a mask. This maintains a planar wafer surface and reduces the effect of lithographic misalignment. However, to survive the postimplant anneal ($\sim 850°C$), refractory gate metals such as tungsten–silicon are necessary and these have high resistivities (> 1 ohm per square). Consequently, recessed-gate FETs are used mainly for microwave and millimeter-wave applications where low gate resistance is mandatory for minimum noise figure. Self-aligned GaAs MESFETs are most appropriate for digital and mixed-mode circuits where high device uniformity, integration density, and circuit yield are more important.

6-4-1-2 GaAs JFET The GaAs junction FET (JFET) is very similar to its silicon counterpart, as illustrated previously in Fig. 5-8. It simply uses a *pn* gate junction in place of the MESFET's Schottky gate. The GaAs JFET's advantage over a MESFET is that it can tolerate larger forward gate biases. The effective barrier height of the *pn* junction is approximately twice that of the Schottky junction in GaAs (1.4 eV versus 0.7 eV). However, the JFET fabrication process is somewhat more complicated. Also, the threshold voltage V_T uniformity of JFETs suffers because it is determined by two dopings rather than one. Both *n*-channel and *p*-channel GaAs JFETs have been produced, although the *p*-channel JFET has much poorer transconductance and is normally used as a load device only.

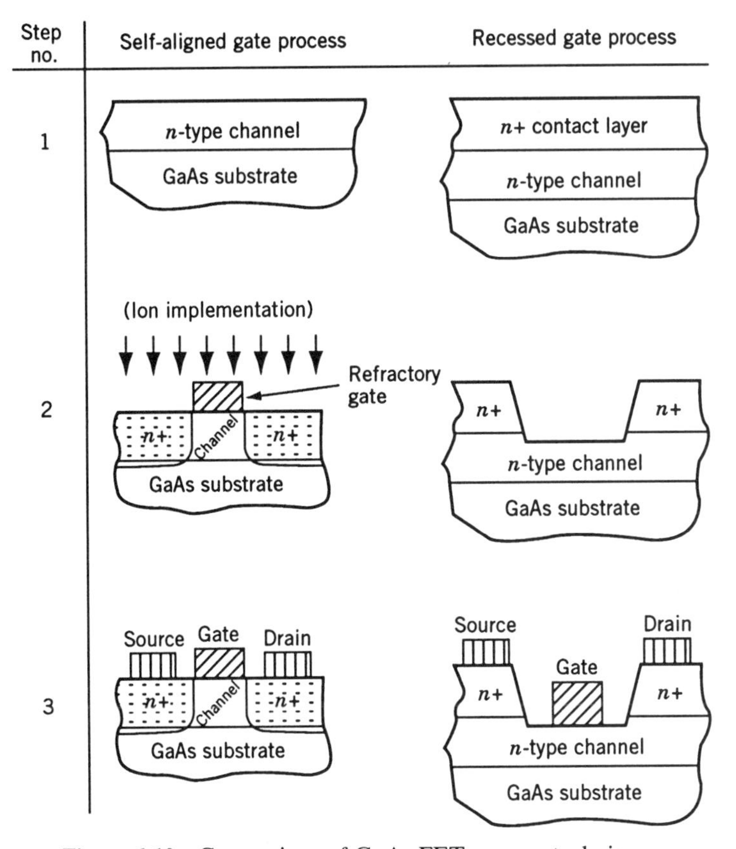

Figure 6-10 Comparison of GaAs FET process techniques.

It is important to note that the channel region of GaAs MESFETs and JFETs can be fabricated two different ways: (1) by implanting dopant atoms in the semiinsulating GaAs substrate or (2) by epitaxially growing a doped GaAs layer on top of the substrate. GaAs FETs using epi channels are generally preferred for low-noise applications because of their superior crystalline quality [11]. Ion implantation is cheaper than epi growth but creates damage that is not totally annealed out by subsequent processing. The residual damage-induced defects degrade channel mobility, which increases thermal noise. Also, the defects can act as generation–recombination (G–R) sites which randomly trap and detrap charge carriers, creating low-frequency noise. The epi doping process avoids this type of damage because the dopants are incorporated during epitaxial growth.

6-4-1-3 Heterostructure FETs Heterostructure FETs (HFETs) are actually a family of related GaAs devices. HFETs are essentially similar to the MESFET and JFET presented earlier, except that the bulk doped channel is

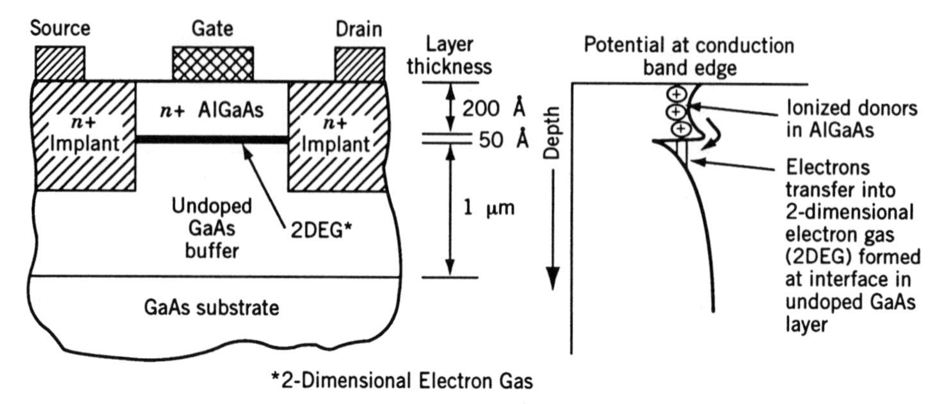

Figure 6-11 Cross section and band diagram of AlGaAs/GaAs HEMT.

replaced with a thin heterostructure sandwich containing multiple layers of various semiconductors and dopings. The heterostructure layers are carefully selected so that the energy band structure in the active transistor region is modified to enhance desirable physical effects or to suppress undesirable ones, for example, tunneling, real-space transfer, and impurity scattering. The heterostructure is deposited on a GaAs substrate using sophisticated epitaxial growth techniques such as metal-organic chemical vapor deposition (MOCVD) in which semiconductor composition and doping can be changed abruptly during deposition with atomic layer resolution. This ability to tailor the semiconductor lattice to particular device requirements is euphemistically called "bandgap engineering." It gives the device designer additional degrees of freedom in optimizing transistor performance. Research and development of these devices has been going on since approximately 1980, although GaAs HFETs have only been commercially available for the past several years.

The major type of GaAs HFET that is commercially available is the high electron mobility transistor (HEMT) in Fig. 6-11. The details of HEMT operation are quite involved. Essentially, the heterostructure is designed to spatially separate the channel charge carriers from their parent donors in the AlGaAs layer, allowing them to move from source to drain in pure GaAs where they have substantially higher mobility. Forward gate leakage is also reduced because the gate junction is formed on a wide bandgap material, typically $Al_{0.35}Ga_{0.65}As$.

For the circuit designer, GaAs HFETs offer state-of-the-art transconductance, power gain, current drive, and the lowest thermal noise and noise figure. Compared with GaAs MESFETs and JFETs, however, HFETs are generally more expensive, available from fewer vendors, and have more lot-to-lot variations. The characteristics of these devices are described next.

6-4-2 Noise Characteristics of GaAs FETs

Broadly speaking, GaAs FETs tend to be less noisy than silicon FETs at high frequencies but more noisy at low frequencies, with 10 MHz being a rough

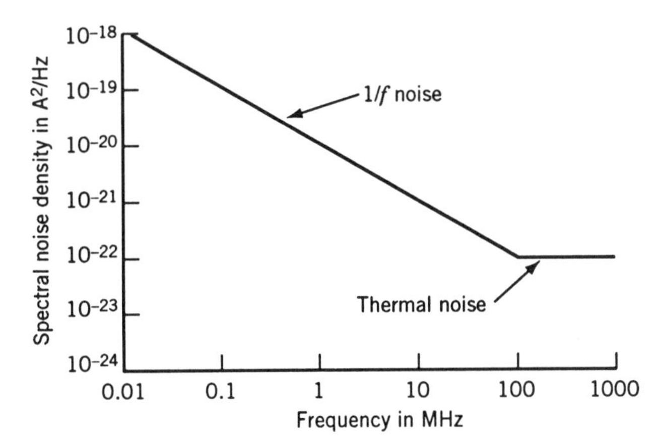

Figure 6-12 Drain current noise data for a GaAs MESFET [13]. (©1985 IEEE)

dividing line. The reasons behind this are both technological and historical. Only the major trends will be discussed here, since the understanding of noise in GaAs FETs continues to evolve. For more detailed information, see [12].

The noise equivalent circuit for GaAs FETs is somewhat more complicated than for silicon FETs because GaAs FETs are typically used over a much wider frequency range. The noise model shown previously in Fig. 6-1 can be used to model the *intrinsic* GaAs FET. In addition, two *external* noise voltage generators E_s and E_g are added in series with the source and the gate bias resistors to represent the thermal noise of these external resistances. The extrinsic noise sources are required to calculate noise parameters (e.g., noise figure, noise temperature, and noise resistance) for microwave and millimeter-wave amplifiers as defined in Chap. 2.

A typical spectrum of drain current noise I_{nd} for a GaAs MESFET is shown in Fig. 6-12. A comparison of transistor types is shown in Table 6-1. In

TABLE 6-1 Noise Comparison for GaAs FETs [13]

Device Number	Manufacturer	Technology	White Noise Current I_{nd}^2 (A^2/Hz)	Noise Coefficient P
1	Gigabit logic	Ion-implanted	1.08×10^{-22}	0.09
2	Gigabit logic	Ion-implanted	0.78×10^{-22}	0.07
3	Gigabit logic	Ion-implanted	1.3×10^{-22}	0.11
4	Gigabit logic	Ion-implanted	0.83×10^{-22}	0.07
5	AT & T	VPE	0.93×10^{-22}	0.1
6	AT & T	VPE	1.9×10^{-22}	0.06
7	NEC	VPE	1.2×10^{-22}	0.05
8	NEC	VPE	1.3×10^{-22}	0.05
9	Plessey	VPE	2.4×10^{-22}	0.08

GaAs FETs, I_{nd} is made up of thermal noise from the channel plus low-frequency noise (mainly G–R noise) from defects in the channel and the substrate buffer. For frequencies above several megahertz, the midband region, I_{nd} is independent of frequency which is indicative of white noise. There are two sources of this white noise: (1) thermal noise from the low-field "ohmic" portion of the channel and (2) thermal noise from the velocity-saturated portion of the channel, which is sometimes interpreted as diffusion noise [14]. This midband noise in GaAs FETs is usually expressed using van der Ziel's formula [15]:

$$I_{nd}^2 = 4kTG_d \, \Delta f P \tag{6-35}$$

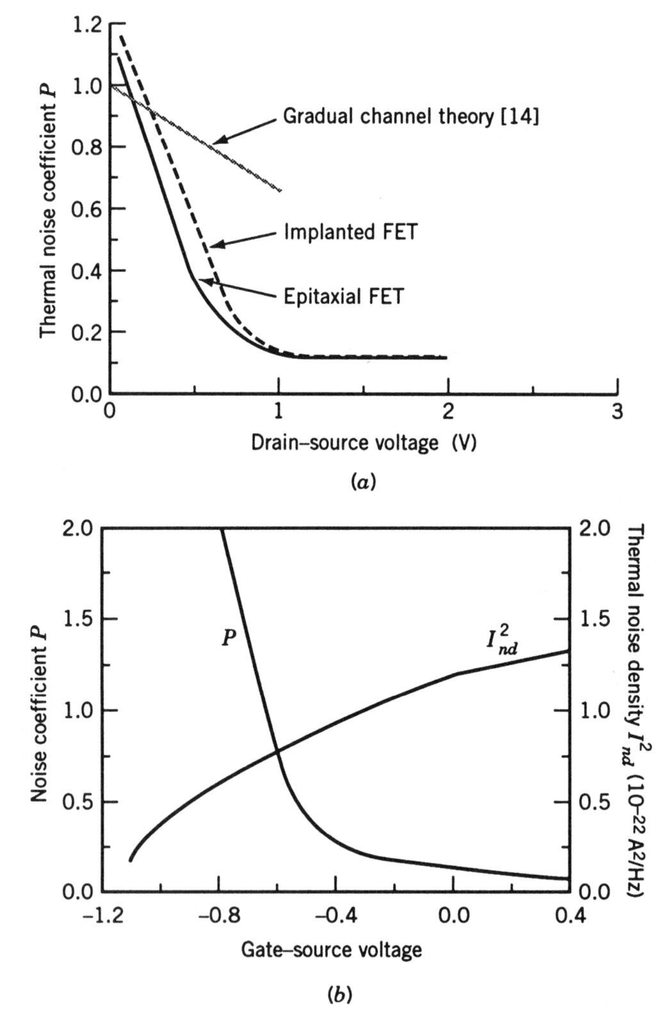

Figure 6-13 Bias dependence of the thermal noise coefficient P for 1-μm GaAs MESFET [13]. (©1985 IEEE)

where G_d is channel conductance and P is a bias-dependent noise coefficient taking into account the effects of nonuniform channel conductance. This expression has been extended [14] to include high field effects such as velocity saturation and carrier heating which occur in modern submicron gate length GaAs FETs. Thus P gives the relationship between measured conductance G_d and the generated thermal noise. The white-noise coefficient P is strongly dependent on drain and gate bias, as shown in Fig. 6-13.

GaAs FETs have extremely low noise at high frequencies and consequently are widely used for microwave and millimeter-wave circuits as illustrated in Fig. 6-14.

Their attractiveness can be understood by examining how important microwave properties such as the transistor cutoff frequency f_t and the minimum noise factor F_{min} defined by Fukui [16] are related to device parameters

$$f_t = K_1(g_m/C_{gs}) \tag{6-36}$$

$$F_{\text{min}} = 1 + K_2 f C_{gs}\left[(R_g + R_s)/g_m\right]^{0.5} \tag{6-37}$$

GaAs FETs are the best choice for these applications since their transconductance g_m is intrinsically higher than for other devices (due to the larger electron mobility and saturation velocity of GaAs), and C_{gs} is very small because the GaAs substrate is semiinsulating. However, good high-frequency operation also requires special device design to minimize the parasitic gate and source resistances R_g and R_s. As mentioned earlier, recessed-gate FETs are best suited for these requirements.

Below approximately 100 MHz, the white thermal noise in GaAs FETs is overwhelmed by frequency-dependent "$1/f$" noise. The magnitude of this

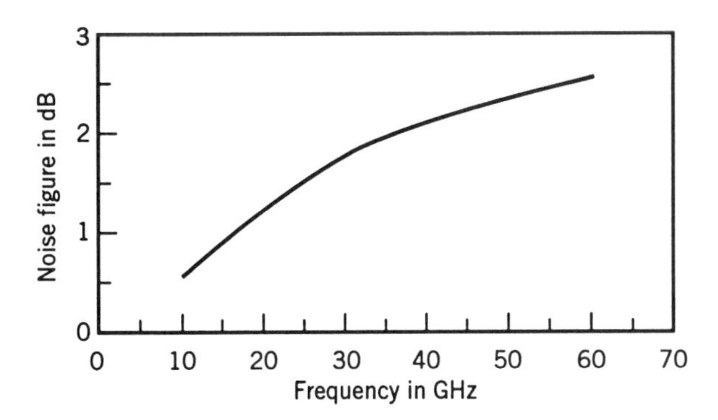

Figure 6-14 Reported HEMT noise performance at room temperature [12]. (©1988 IEEE)

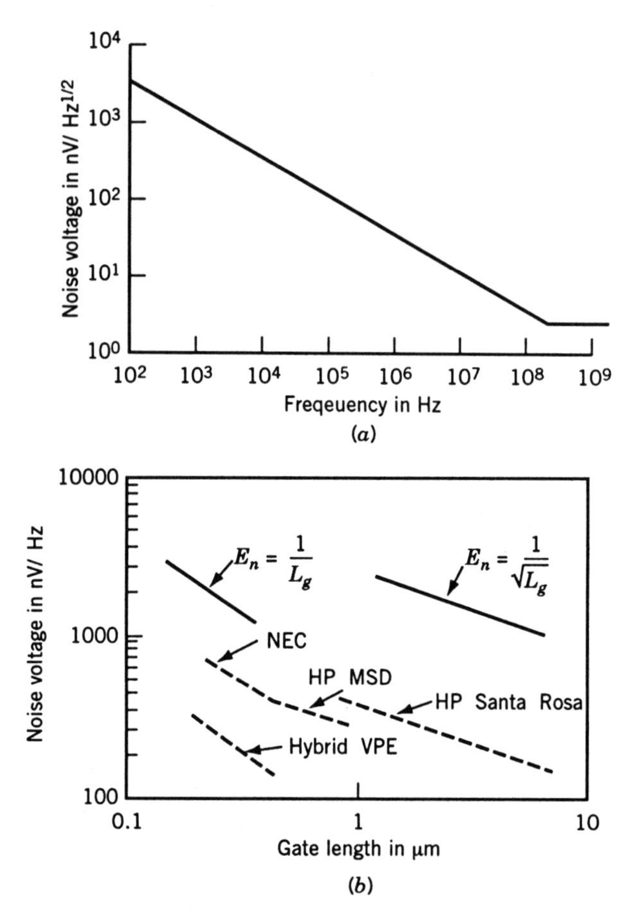

Figure 6-15 Gate-referred noise voltage characteristics for GaAs MESFETs: (*a*) noise spectrum for a 1-mm gate length device and (*b*) dependence of 1-kHz noise on gate length L_g. All measurements were made at room temperature with the drain biased into saturation [17]. (©1983 IEEE)

noise is substantially higher than in silicon FETs (see Fig. 6-15). Often, this noise is not true $1/f$ noise but rather is a summation of generation–recombination (G–R) noise from trapping centers with various activation energies. G–R noise has a Lorentzian frequency dependence given by

$$\tau/\left[1 + (2\pi\tau f)^2\right] \tag{6-38}$$

where τ is the time constant of the trap. Hence G–R noise is flat up to a characteristic corner frequency $(2\pi\tau)^{-1}$ and falls off as f^{-2} beyond. The characteristic time constant τ is strongly dependent on the activation energy

E_a of the particular trap and the local temperature T:

$$\tau = \tau_0(300/T)^2 \exp(qE_a/kT) \tag{6-39}$$

In many GaAs FETs, there appears to be a distribution of either activation energies or temperature (or both) which causes multiple G–R "bumps" to appear in the low-frequency noise spectrum. If the traps are sufficiently close in energy, the bumps can smooth together and resemble a traditional $1/f$ spectrum [18]. G–R effects can be included into Eq. 6-35 by adding the following term [19]:

$$E_n^2 = 4kT\,\Delta f\left[\rho_o\left(\frac{f_0}{f}\right) + \sum_{r=1}^{n} G_r\right] \tag{6-40}$$

where

$$G_r = \frac{\rho_r(\tau_r/\tau_o)}{1 + (2\pi f \tau_r)^2} \tag{6-41}$$

Here ρ_o is the equivalent gate resistance when $f = f_0$, ρ_r is a similar resistance for the rth component of the G–R noise, τ_r is the time constant for each active trapping center, and τ_o is a reference trapping time constant. Also, note that Eq. 6-40 is expressed as a gate-referred noise voltage as is customary for dealing with low-frequency noise. As predicted by Eqs. 6-38 and 6-39, lowering the temperature of the GaAs FET to cryogenic temperatures will dramatically reduce the $1/f$ noise as shown in Fig. 6-16.

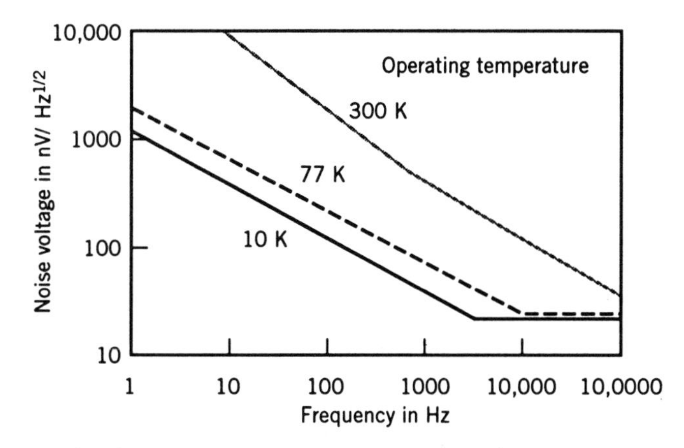

Figure 6-16 $1/f$ noise spectral density for a 20/2 (W/L) enhancement-mode GaAs MESFET [20]. (©1988 IEEE)

Two factors exacerbate the low-frequency (LF) noise problem in GaAs FETs. First, GaAs processes have to date been optimized only for digital and microwave operation, where $1/f$ noise is of little concern. Second, these same applications have driven GaAs processes to short gate lengths ($< 1\ \mu$m), which worsens $1/f$ noise as shown in Fig. 6-15*b*. Thus the circuit designer concerned with LF noise in GaAs FETs must either put up with unoptimized devices or pay for specialized device sizes and processes. In addition, there is wide variation in $1/f$ noise for each type of GaAs FET—MESFET, JFET, HEMT. None stands out as having inherently lower noise than the others.

Gate current noise I_{ng} is the other intrinsic noise generator in the GaAs FET equivalent circuit. Gate current noise in GaAs FETs is not well understood and while it continues to be of great research interest, the I_{ng} noise is normally not a significant concern for circuit designers since source resistances are generally small. Two sources in the device contribute to I_{ng}: shot noise from the gate barrier, [21], and noise capacitively coupled from the drain as described in Sec. 6-1. The biasing condition determines which mechanisms dominate as well as the degree of correlation between the intrinsic noise sources I_{nd} and I_{ng} [12].

6-5 MEASURING FET NOISE

FET noise measurements cover such wide ranges of impedance and frequency that four separate measurement methods are used. The noise voltage E_n at low and high frequencies can be measured directly as described in Chap. 15. To determine low-frequency noise current I_n, two methods are used. The I_n noise can be measured directly using a large source resistance, or the gate leakage current can be measured and the noise calculated. High-frequency I_n noise need not be measured directly. Instead, the shunt input resistance can be measured and the I_n thermal noise calculated. At midfrequencies, accurate measurement of I_n is difficult.

The noise voltage E_n of a FET can be measured using the method shown in Chap. 15, Sec. 15-2 with the noise measurement instrumentation described in Sec. 15-3. A diagram of the noise measurement instrumentation is shown in Fig. 15-2. The device under test (DUT), such as an op amp with a FET input stage, is connected as a gain stage using feedback to set the gain. Since the noise levels are low, it is a good idea to use a gain of $100\times$ in the device under test to dominate the noise of the next stage or the instrumentation.

The amplifier noise voltage E_n and noise current I_n parameters are calculated from the equivalent input noise for two source resistance values. As defined in Chap. 2, the equivalent input noise is

$$E_{ni}^2 = E_{ts}^2 + E_n^2 + I_n^2 R_s^2 + 2CE_n I_n R_s \qquad (2\text{-}8)$$

A noise measurement gives the total equivalent input noise E_{ni}. To determine each of the three quantities, E_n, I_n, and E_{ts}, make one term dominant or subtract the effects of the other two. In general, the correlation coefficient C is zero and can be neglected except for FETs operating from high source resistances at high frequencies. At high frequencies the E_n and I_n mechanisms are both generated by thermal noise of the channel so they are correlated. Measurement of C is discussed in detail in Chap. 15, Sec. 15-2.

To measure the noise voltage E_n, measure equivalent input noise with a small value of source resistance. When the source resistance is zero, the thermal noise of the source E_{ts} is zero and the noise current term $I_n R_s$ is also zero; the total equivalent input noise is the noise voltage E_n.

To measure the noise current I_n, remeasure E_{ni} with a very large source resistance. Measure the output noise and transfer voltage gain again to calculate the equivalent input noise E_{ni}. Assuming the $I_n R_s$ term to be dominant, I_n is the equivalent input noise E_{ni} divided by the source resistance R_s. If R_s is large enough, the $I_n R_s$ term dominates the E_n term, and it also dominates the thermal noise since the thermal noise voltage E_t increases as the square root of the resistance, whereas the $I_n R_s$ term increases linearly with resistance. When the $I_n R_s$ term cannot be made dominant, the thermal noise voltage $E_t = (4kTR_s\,\Delta f)^{1/2}$ can be subtracted from the equivalent input noise. Since this is an rms subtraction, a thermal noise of one-third the noise current term only adds 10% to the equivalent input noise.

The minimum value of source resistance for measuring I_n is derived in Chap. 15 as

$$R_s \geq \frac{18kT}{qI_B} = \frac{0.45}{I_B} \tag{15-6}$$

For a JFET amplifier with 2×10^{-10} A bias current, $R_s = 2.25$ GΩ.

Amplifier gain and bias requirements, however, may make it difficult or impossible to achieve these values since any amplifier with an unbalanced input will have an offset voltage of 0.45 V. From Eq. 15-6, $I_B R_s = 0.45$ V. For an amplifier with a balanced input, the same value of source resistance can be placed in each input lead to reduce the offset to about 50 mV. It may be necessary to use additional feedback to reduce the overall amplifier gain to avoid excessive offset at the amplifier output. For the amplifier with an unbalanced input, it is usually necessary to use a gain of less than 10 × to stay within linear operation. For an accurate I_n measurement, the amplifier input bias resistors must be much larger than the source resistance R_s or you will only be measuring the noise current of the shunting biasing resistors. It is best to use the measuring source resistor as the bias current path.

Another method of obtaining a high source impedance for I_n measurement is with a reactive source. A low-loss mica capacitor can be used for R_s. Now the $I_n X_c$ term is large and since the reactive impedance has no thermal noise, the equivalent input noise is

$$E_{ni}^2 = I_n^2 X_c^2 + E_n^2 \tag{6-42}$$

This method is valid only at frequencies below 100 to 200 Hz, as X_c decreases rapidly with increasing frequency.

One difficulty with a capacitive source is biasing. To provide a path for the input bias current or the offset current, it is necessary to parallel the amplifier input terminals with a large resistance. This bias resistance R_B generates thermal noise current $I_{th} = (4kT\,\Delta f/R_B)^{1/2}$ which can easily dominate the amplifier input shot noise current. A FET input amplifier often has a noise current of less than 10 $\text{fA/Hz}^{1/2}$. This is equivalent to the thermal noise of a 160-MΩ resistor. Thus the bias resistance *for all* FET I_n measurements must be in the GΩ (10^9) range.

To verify the I_n low-frequency measurement, an electrometer can be placed in series with the gate lead to measure the gate leakage current (I_{GSS}). The measured I_{GSS} is then converted to shot noise current from the formula

$$I_n^2 = 2qI_{GSS}\,\Delta f \tag{6-43}$$

The two methods of determining low-frequency I_n were in close agreement on all units measured. This agrees with the model which says that FET I_n results from the gate shot noise and there is no $1/f$ noise component of I_n.

Measurement of I_n at high frequencies is very difficult because of the shunt capacitance of the source, stray capacitances in the layout, and amplifier input capacitance. Frequency response with a large source resistance is so poor that there is little or no gain left. For the same reason that I_n is difficult to measure, it will probably not be a noise contribution in typical applications.

A more accurate method of measuring I_n at high frequencies is based on the model of the noise sources. As shown previously, the input of a FET at high frequencies is shunted by a real resistance $R_p(f)$ caused by the Miller effect. This resistance is real and dissipative so it exhibits full thermal noise. A high-frequency bridge can be used to measure this shunt resistance $R_p(f)$. I_n is determined from

$$I_n^2 = \frac{4kT\,\Delta f}{R_p(f)} \tag{6-44}$$

Noise data taken using the methods discussed are presented in Apps. A and B and summarized in Table 6-2. The values given in the table reflect the best performance obtained at the best operating point. As we may predict using

TABLE 6-2 Measured Noise Data on JFETs

FET Number	FET Channel Type	E_n at Midband ($nV/Hz^{1/2}$)	E_n at 10 Hz ($nV/Hz^{1/2}$)	I_n at Midband ($fA/Hz^{1/2}$)	I_n at 1 MHz ($fA/Hz^{1/2}$)
2N2609	n	2.8	31	13	130
2N3460	n	4	40	11	70
2N3684	n	3.5	55	8.2	34
2N3821	n	2.2	7	3.5	19
2N4221A	n	2.2	50	8.5	18
2N4416	n	2	20	13	23
2N5116	p	3	160	35	210

Eq. 6-15, operation of the JFET at drain current equal to I_{DSS} provides superior noise behavior. The measured noise data confirm that prediction.

The results of noise tests on a type 2N3821 n-channel JFET are shown in Fig. 6-17. Both E_n and I_n parameters are plotted against frequency. The roller-coaster effect and the dependence on operating point are obvious in the E_n curves. Clearly, I_n is very low valued at low frequencies.

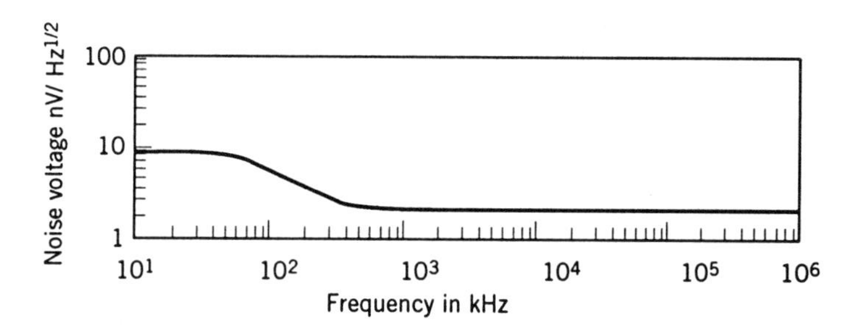

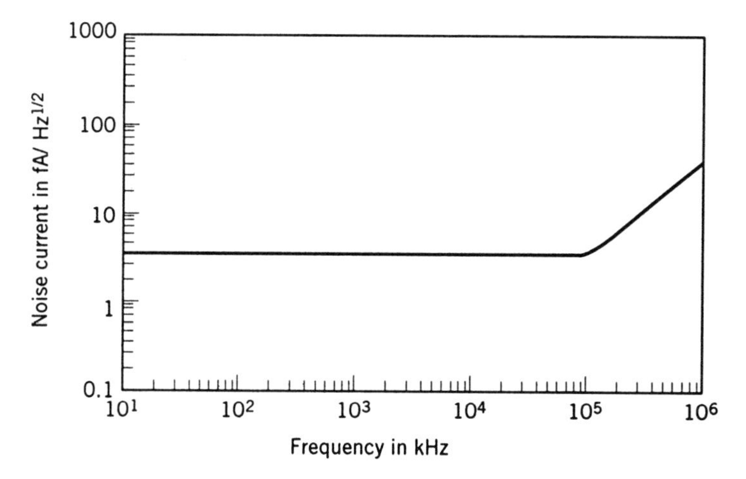

Figure 6-17 E_n and I_n for a type 2N3821 JFET.

SUMMARY

a. Noise currents in the JFET are caused by shot noise of gate leakage current, $1/f$ mechanisms, and thermal noise.

b. Because field effect devices have low I_n, they are useful in systems with sensors of high internal resistance.

c. The same small-signal equivalent circuit noise model is valid for all FET types—JFETs, MOSFETs, and GaAs FETs. Also this same model is valid for both n- and p-channel conductivity types.

d. The best noise performance for any FET occurs when the device is operated at large quiescent currents near I_{DSS}. This increases g_m, thereby reducing R_n.

e. The noise corner frequency of E_n in MOSFETs is inversely proportional to the product of the device's width and length dimension. Thus devices with large gate areas have lower f_c corner frequencies.

f. GaAs FETs offer higher speed operation at lower voltages than silicon units and permit both electronic and optical functions to be performed on the same chip.

g. MESFETs are the most common type of GaAs FET and use a Schottky metal–semiconductor contact as the gate junction.

h. GaAs JFETs can tolerate about twice the forward gate voltage of silicon JFETs. Otherwise they are very similar.

i. GaAs FETs using epi channels offer very low noise due to superior crystalline quality.

j. HFETs contain multiple layers of various semiconductors and dopings. The major type of HFET commercially available is the high electron mobility transistor, or HEMT.

k. Broadly speaking, GaAs FETs tend to be less noisy than silicon FETs at high frequencies but more noisy at low frequencies, with 10 MHz being a rough dividing line. Below about 100 MHz the white thermal noise in GaAs FETs is overwhelmed by frequency-dependent $1/f$ noise.

l. Four separate noise measurement techniques are commonly used to measure E_n and I_n noise in FETs.

m. The minimum value of source resistance for measuring I_n will cause an offset voltage of 0.45 V that can severely hamper the measurement technique.

PROBLEMS

6-1. It is desired to design a MOSFET which will have 1 nV/Hz$^{1/2}$ or less E_n noise neglecting $1/f$ contributions. Using the data from Examples

6-2, 6-3, and 6-4, determine what new W/L ratio and quiescent drain current will be required to achieve this low noise level.

6-2. Tests performed on a certain n-channel MOSFET yielded the following results: gate leakage current = 3 fA at a Q-point of $I_D = 0.5$ mA and $V_{DS} = 5$ V. The dc drain current is found from the equation

$$I_D = (50\ \mu\text{A/V}^2)(V_{GS} - 2)^2(1 + 0.001V_{DS})$$

Assume the following constants: $K_F = 10^{-27}$ s/V, $A_F = 1$, $L_{\text{eff}} = 1\ \mu$m, $W = 60\ \mu$m, and $C_{\text{ox}} = 1\ \text{fF}/\mu\text{m}^2$. A small-signal, low-frequency equivalent circuit model for this MOSFET is shown in Fig. P6-2.

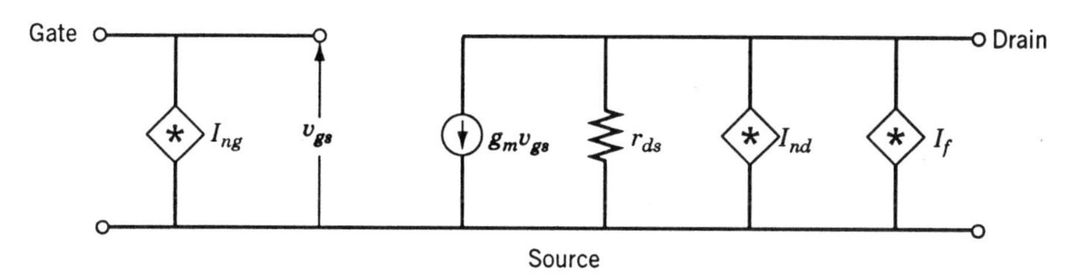

Figure P6-2

(a) In this model, determine a numerical value for each noise source and for each small-signal parameter shown. Assume a 1-Hz noise bandwidth and an operating frequency of 100 Hz. Be sure to specify proper units on all of your answers.

(b) Find the optimum source resistance for this MOSFET operating at a frequency of 100 Hz.

(c) Now suppose that you need to reduce the noise contribution due to I_D^2 by a factor of 2 without changing anything but the Q-point. Specify the parameters of the new Q-point.

6-3. Consider a type 2N3821 JFET biased at I_{DSS}. The resistance of the signal source is 10 MΩ, and gate biasing elements can be neglected.

(a) Derive an expression for the noise figure for this simple circuit in terms of R_s, E_n, I_n, and the correlation coefficient C.

(b) Using midband data from Table 6-2, calculate the noise figure. Consider that the correlation coefficient is unity.

6-4. One stage of a low-noise amplifier is to be designed with a FET operating with a drain current of $I_{DSS} = 100\ \mu A$ at a frequency of 10 kHz, and a source resistance of 1 kΩ.

(a) From the device data in App. B, determine which device to select to produce the lowest NF.

(b) Which device would you definitely avoid specifying for the same conditions given previously?

6-5. A discrete *n*-channel enhancement-mode MOSFET is biased at a drain current level of 1 mA and behaves according to the equation

$$I_D = 50\ \mu A/V^2 (V_{GS} - 2)^2$$

(a) Determine a numerical value of E_n for this MOSFET in its shot noise region.

(b) Laboratory measurements on this MOSFET yielded a value of $E_n = 10\ nV/Hz^{1/2}$ at a frequency of 500 Hz. Determine the noise corner frequency for this MOSFET.

(c) It is desired to reduce the E_n noise voltage of this discrete MOSFET in its shot noise region by a factor of 2 by changing its *Q*-point but *without* changing its geometry in any way. Explain how this can be done and find the new *Q*-point.

6-6. Determine the R_{opt} and the spectral density noise E_{ni} for the 2N4221A Motorola JFET whose characteristics are given in App. B. The source resistance is 1 MΩ, the operating frequency is 100 Hz, and the device is biased at I_{DSS}.

6-7. For the limiting case (frequency independent), what are E_n and I_n for a junction FET with a $g_m = 2$ mS, $I_D = 1.0$ mA, and $I_{GSS} = 10$ nA? Assume a noise bandwidth of 1 Hz.

REFERENCES

1. "An Introduction to FETs," Application Note AN73-7, Siliconix Inc., Santa Clara, CA.
2. Chang, Z. Y., and W. M. C. Sansen, *Low Noise Wide-Band Amplifiers in Bipolar and CMOS Technologies*, Kluwer, Boston, 1991, pp. 7–25, 41–49.
3. Alvarez, A. R. (Editor), *BiCMOS Technology and Applications*, Kluwer, Boston, 1989, pp. 72, 278.
4. van der Ziel, A., "Thermal Noise in Field Effect Transistors," *Proc. IEEE*, **50**, 8 (August 1962), 1808–1812.
5. van der Ziel, A., "Gate Noise in Field Effect Transistors at Moderately High Frequencies," *Proc. IEEE*, **51**, 3 (March 1963), 461–467.

6. Lauritzen, P. O., "Low-Frequency Generation Noise in Junction Field Effect Transistors," *Solid-State Electronics*, **8**, 1 (January 1965), pp. 41–58.
7. Sedra, A. S., and K. C. Smith, *Microelectronics Circuits*, Third Edition, Saunders College Publishing, Philadelphia, 1991, pp. 513–515.
8. Allen, P. E., and E. Sanchez-Sinencio, *Switched Capacitor Circuits*, Van Nostrand Rheinhold, New York, 1984, pp. 578–596.
9. Geiger, R. L., P. E. Allen, and N. R. Strader, *VLSI Design Techniques for Analog and Digital Circuits*, McGraw-Hill, New York, 1990, pp. 143–180.
10. Gregorian, R., and G. C. Temes, *Analog MOS Integrated Circuits for Signal Processing*, Wiley, New York, 1986, pp. 96–99.
11. Rocchi, M., "Status of the Surface and Bulk Parasitic Effects Limiting the Performance of GaAs ICs," *Physica*, **129B** (1985), 119–138.
12. Cappy, A., "Noise Modeling and Measurement Techniques," *IEEE Trans. Microwave Theory Tech.*, **36**, 1 (January 1988), 1–10.
13. Folkes, P., "Thermal Noise Measurements in GaAs MESFETs," *IEEE Trans. Electron Devices*, **6**, 12 (December 1985), 620–622.
14. Pucel, R., H. Haus, and H. Statz, "Signal and Noise Properties of GaAs FETs," *Advances in Electronics and Electron Physics*, **38** (1974), 195–265.
15. van der Ziel, A., *Proc. IRE*, **50** (1962), 1808.
16. Fukui, H., "Design of Microwave GaAs MESFETs for Broad Band, Low Noise Amplifiers," *IEEE Trans. Microwave Theory Tech.*, **27**, 7 (July 1979), 643–650.
17. Su, C., H. Rohdin, and C. Stolte, "$1/f$ Noise in GaAs MESFETs," *Proceedings of IEDM* (1983), 601–604.
18. Hughes, B., N. Fernandez, and J. Gladstone, "GaAs FETs with a Flicker Noise Corner Below 1 MHz," *IEEE Trans. Electron Devices*, **34**, 4 (April 1987), 733–741.
19. Liu, S., et al., "Low Noise Behavior of InGaAs Quantum Well Structured HEMTs from 10^{-2} to 10^8 Hz," *IEEE Trans. Electron Devices*, **33**, 5 (May 1986), 576–581.
20. Sato, R. N., et al., "Gallium-Arsenide E- and D-MESFET Device Noise Characteristics Operated at Cryogenic Temperatures with Ultralow Drain Current," *IEEE Electron Device Lett.*, **9**, 5 (May 1988), 238–240.
21. Peransin, J., P. Vignaud, and D. Rigaud, "Bias Dependence of Low Frequency Gate Current Noise in GaAs MESFETs," *Electron Lett.*, **25**, 7 (30 March 1989), 439–440.

CHAPTER 7

SYSTEM NOISE MODELING

In instrumentation, measurement, and control systems, it is often necessary to monitor some physical and/or electrical process or mechanism. The objective may be to measure a process variable, such as fluid flow, without disturbing it. The usual method is to insert a sensor or transducer that converts a small amount of the flow energy into an electrical signal. Since we do not want to affect the flow, very little power can be removed from the process, and, therefore, the resulting electrical signal output from the sensor is often weak.

The engineer's task is to amplify the weak sensor signal without masking it by noise. All sensors have internal noise generators and can be characterized by their basic signal-to-noise ratios. The methods of analysis of sensor–amplifier systems are introduced in this chapter.

7-1 NOISE MODELING

To develop the noise model of a sensor, we can start with its circuit diagram. From this we draw an ac equivalent circuit that includes all impedances and generators. To each resistance and current generator we add the appropriate noise generators to develop a noise equivalent circuit. The resistances have thermal noise and possibly excess noise. The current generators may have shot noise, $1/f$ noise, and/or excess noise. These mechanisms are described in Chap. 1. Using this equivalent circuit, an expression for gain and equivalent input noise can be derived.

It is advantageous, however, to study the entire sensor electronic system. A typical system may include a coupling device or network, as well as an amplifier. The noise equivalent circuit of the coupling network is easily obtained, and the E_n–I_n representation is valid for the amplifier. When we combine these three parts, we obtain a noise equivalent for the system.

The derivation of the equivalent input noise for the system follows three steps:

1. Determine the total output noise.
2. Calculate the system gain.
3. Divide the total output noise by the system gain to obtain the equivalent input noise.

In Chap. 8 noise models and equivalent input noise expressions are derived for six classes of sensors. These sensors are

1. Voltaic sensor.
2. Biased resistive sensor.
3. Optoelectronic detector.
4. *RLC* sensor.
5. Piezoelectric transducer.
6. Transformer model.

In this chapter we develop the method for analysis of general sensor–amplifier combinations and determine S/N ratios. The effects of shunt resistance and capacitance on equivalent input noise are determined.

7-2 A GENERAL NOISE MODEL

The diagram shown in Fig. 7-1 contains the E_n–I_n representation for the system electronics. The sensor is described by its signal voltage V_s, its internal impedance Z_s, and a noise generator E_s which represents all sources of sensor noise. To generalize the diagram, a coupling network section represented by impedance Z_C and a noise source E_C is included in shunt with the sensor. Our objective here is to combine and then reflect all noise sources to the input as shown in Fig. 7-1*b* and *c*.

A general form for the equivalent input noise voltage is

$$E_{ni}^2 = A^2E_s^2 + B^2E_n^2 + C^2I_n^2Z_s^2 + D^2E_C^2 \tag{7-1}$$

Alternatively, we can define an equivalent input noise current as

$$I_{ni}^2 = J^2I_{ns}^2 + \frac{K^2E_n^2}{Z_s^2} + L^2I_n^2 + \frac{M^2E_C^2}{Z_C^2} \tag{7-2}$$

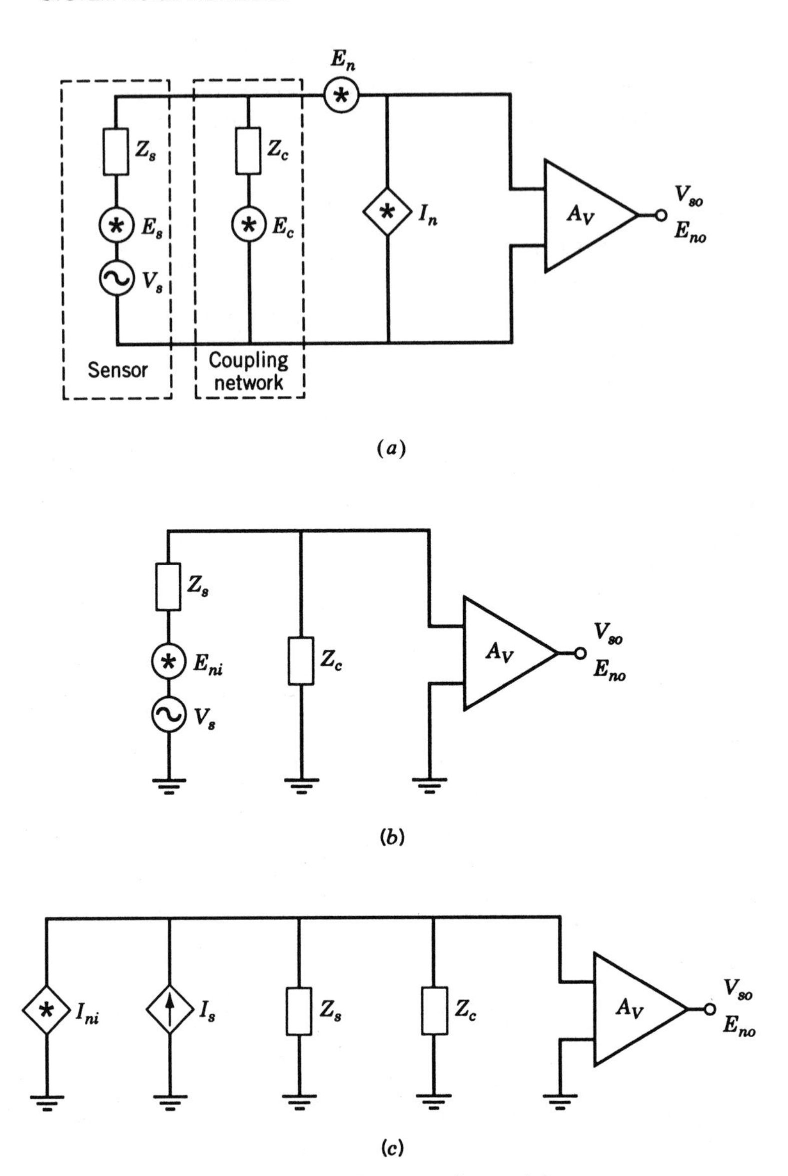

Figure 7-1 System noise model.

where $I_{ns}^2 = E_s^2/R_s^2$. To evaluate these noise equivalents, the A^2, B^2,... coefficients must be determined.

The equivalent input noise generators E_{ni} and I_{ni} are described by Eqs. 7-1 and 7-2. Either form can be used. The choice depends on the type of sensor being employed. If the signal source is a current generator, the equivalent noise current expression is more convenient. If the signal is a voltage generator, the equivalent voltage form may be more convenient.

Several cases are discussed to show the source of these constant terms. Generally, the A term is unity since it is in series with the signal source. The B term is caused by shunt impedance in the coupling network. The C term is determined by series impedance in the sensor and coupling network.

The method of calculating E_{ni} or I_{ni} is the same for any sensor. Starting from a noise equivalent diagram, the total output noise E_{no} is calculated using Kirchhoff's law or by SPICE simulation. The equivalent input noise is the output noise E_{no} divided by the system gain K_t. The system gain is either a voltage or current transfer gain as needed. As shown previously in Chap. 2, neither E_{ni} or I_{ni} depends on the input impedance or gain of the amplifier. However, E_{no} does depend on the amplifier's characteristics.

7-3 EFFECT OF PARALLEL LOAD RESISTANCE

The simplest type of sensor is represented by a resistance in series with a signal voltage generator as shown in Fig. 7-2. Also shown is a shunt network consisting of R_p and noise generator E_p. One practical purpose of this network may be to supply the sensor with bias power. For example, the sensor may be a precision potentiometer with its input being the mechanical displacement of its shaft. Unless the potentiometer is supplied with electrical bias, V_s would always be zero. Another application would be a biased photoconductive fiberoptic detector.

The signal V_s and noise E_s of the sensor are in series with the source resistance. The input signal-to-noise power ratio is simply the ratio of V_s^2 to E_s^2. When a load resistor such as R_p or other coupling network elements are added, the output signal-to-noise ratio is degraded.

Let us calculate the effect of shunt resistance on the signal-to-noise power ratio or equivalent input noise. This is done in two ways. First, we directly calculate the output noise and output signal. Next, we repeat the results by deriving an expression for the equivalent input noise E_{ni}. For $R_p \gg R_s$:

$$\frac{S}{N} = \frac{V_{so}^2}{E_{no}^2} = \frac{V_s^2}{E_s^2} \tag{7-3}$$

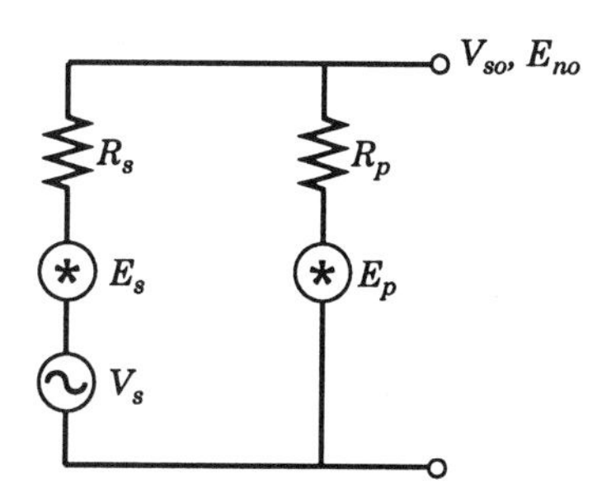

Figure 7-2 Sensor model shunted by resistance.

Example 7-1 Determine the output signal-to-noise ratio when $R_p = R_s$.

Solution When R_p is not infinite, it must be included in the expression. The output noise, output signal, and signal-to-noise ratio are determined. Since $E_s = E_p$, it follows that

$$E_{no}^2 = \left(\frac{E_p}{2}\right)^2 + \left(\frac{E_s}{2}\right)^2 = \frac{E_s^2}{2} \tag{7-4}$$

The output signal is

$$V_{so} = \frac{V_s}{2} \tag{7-5}$$

Therefore, the output signal-to-noise ratio is

$$\frac{S}{N} = \frac{V_{so}^2}{E_{no}^2} = \frac{(V_s/2)^2}{E_s^2/2} = 0.5\frac{V_s^2}{E_s^2} \tag{7-6}$$

From Eq. 7-6 we conclude that a shunt resistor decreases the signal more than the noise and the result is a decrease in the signal-to-noise ratio. For the matched condition, source resistance equal to load resistance, the S/N power ratio is reduced by 50% (3 dB) from the unloaded value. See Prob. 7-1 for other source resistance and load resistance combinations.

An analysis is now made of the more complete circuit shown in Fig. 7-3. Again, a noisy shunt resistance R_p is present. For convenience, we consider its noise to be represented by the current generator $I_{np} = \sqrt{4kT/R_p}$. Amplifier noise generators E_n and I_n are added.

To examine the effect of all noise sources, we calculate the equivalent input noise. The steps are as follows:

1. From the equivalent circuit determine the output noise E_{no}:

$$E_{no}^2 = E_s^2\left(\frac{R_p}{R_s + R_p}\right)^2 + E_n^2 + \left(I_n^2 + I_{np}^2\right)\left(\frac{R_pR_s}{R_p + R_s}\right)^2 \tag{7-7}$$

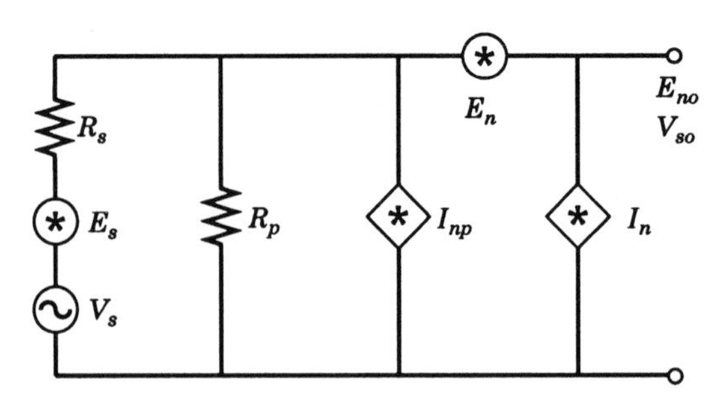

Figure 7-3 Amplifier and sensor models with shunt resistance.

2. Calculate the system gain K_t (the transfer function from sensor to output):

$$K_t = \frac{R_p}{R_s + R_p} \tag{7-8}$$

3. Divide the output noise by the system gain to obtain the equivalent input noise:

$$E_{ni}^2 = \frac{E_{no}^2}{K_t^2} = E_s^2 + \left(\frac{R_s + R_p}{R_p}\right)^2 E_n^2 + \left(I_n^2 + I_{np}^2\right)R_s^2 \tag{7-9}$$

There are two differences between Eq. 7-9 and the original equation for equivalent input noise given in Eq. 2-7. As predicted in Eq. 7-1, the E_n term has a coefficient; it is dependent on the shunt resistance R_p. If R_p were made very large, the coefficient of E_n would approach unity. The second difference is the additional thermal noise generator I_{np} of the shunt resistance R_p. Amplifier input impedance does not contribute noise (see the discussion in Sec. 2-7).

In practice, the shunt resistor should be as large as possible to reduce its noise contribution. Thermal noise of this component can be decreased by reducing temperature. An inductance has no thermal noise and can be used in certain applications in place of R_p.

7-4 EFFECT OF SHUNT CAPACITANCE

Although a capacitance is virtually noise-free, it can increase the equivalent input noise. A shunt capacitance does not affect the sensor signal-to-noise ratio because it decreases the sensor signal and noise equally, but not the following amplifier noise.

Consider the noise equivalent circuit, including shunt capacitance, shown in Fig. 7-4. Using the method of the preceding section:

1. From the equivalent circuit of Fig. 7-4 determine the output noise E_{no}:

$$E_{no}^2 = E_s^2\left(\frac{1}{1 + \omega^2 R_s^2 C_p^2}\right) + E_n^2 + I_n^2\left(\frac{R_s^2}{1 + \omega^2 R_s^2 C_p^2}\right) \tag{7-10}$$

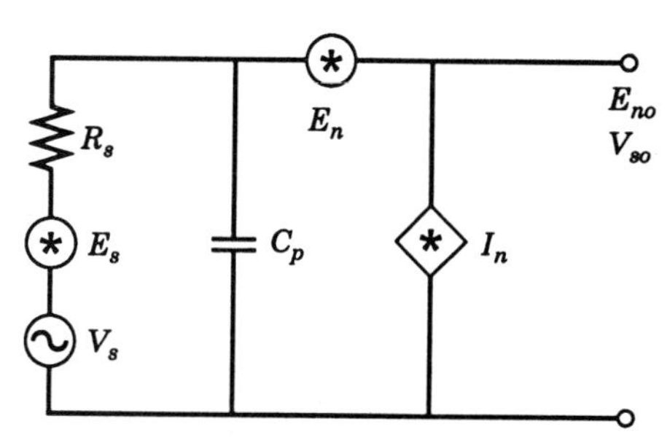

Figure 7-4 Shunt capacitance along with amplifier and sensor models.

2. Calculate the system gain K_t (the transfer function from sensor to output):

$$K_t^2 = \frac{1}{1 + \omega^2 R_s^2 C_p^2} \tag{7-11}$$

3. Divide the output noise by the system gain to obtain the equivalent input noise:

$$E_{ni}^2 = E_s^2 + \left(1 + \omega^2 R_s^2 C_p^2\right) E_n^2 + I_n^2 R_s^2 \tag{7-12}$$

Although the capacitor does not add noise, the noise voltage E_n is increased by the shunt capacitance, as is evident from the coefficient of that term in Eq. 7-12. Only the effective amplifier noise voltage contribution is increased. The capacitor is not a noise source.

The shunt capacitance C_p is not the input capacitance of the amplifier. The amplifier input capacitance drops out of the expression for equivalent input noise, but its effects are included in the values of E_n, I_n, and K_t.

7-5 NOISE OF A RESONANT CIRCUIT

Another interesting model is the resonant inductor type of sensor whose noise equivalent circuit is shown in Fig. 7-5. Following the method of this chapter, the expression for input noise is

$$E_{ni}^2 = E_s^2 + \left| 1 + \frac{R_s\left(1 - \omega^2 C_p L_p\right)}{j\omega L_p} \right|^2 E_n^2 + R_s^2 I_n^2 \tag{7-13}$$

Both the E_n and I_n terms are affected by the sensor impedance. The coefficient of the I_n term is equal to R_s at all frequencies. The coefficient of the E_n term is increased by the shunting of the LC reactance, except at

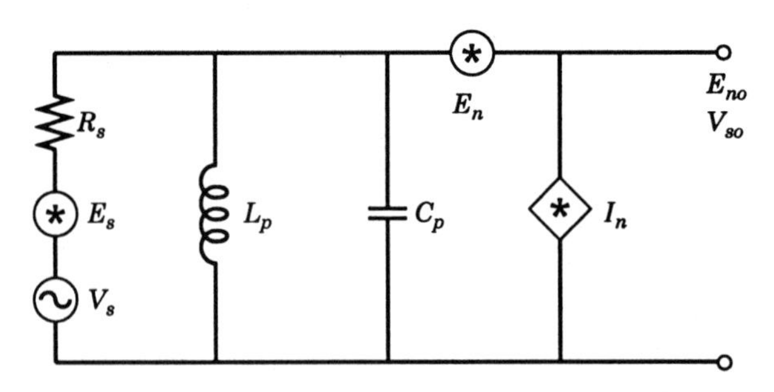

Figure 7-5 Resonant sensor equivalent circuit.

resonance. At resonance, $\omega^2 C_p L_p = 1$ so the coefficient of E_n^2 becomes one. This is logical since the complex contributions to Z_p cancel each other out so at resonance L_p and C_p effectively disappear, leaving only R_s. At resonance, Eq. 7-13 becomes

$$E_{ni}^2 = E_s^2 + E_n^2 + R_s^2 I_n^2 \tag{7-14}$$

The reactive elements do not enter into the noise expression.

7-6 PSpice EXAMPLE

To make the modeling more clear, we now take an example using the circuit in Fig. 7-5 with $R_s = 1\ \text{k}\Omega$, $L_p = 2.553\ \text{mH}$, $C_p = 100\ \text{nF}$, $E_n = 2\ \text{nV/Hz}^{1/2}$, and $I_n = 1\ \text{pA/Hz}^{1/2}$. Using the input noise calculated by PSpice, we can make a comparison with the expression for equivalent input noise given by Eqs. 7-13 and 7-14.

A PSpice equivalent circuit for the *RLC* model is shown in Fig. 7-6. The listing of the SPICE code for this example is

```
RLC Circuit
VS 1 0 AC 1
RS 1 2 1K
LP 2 0 2.553E-3
CP 2 0 1E-7
HEN 3 2 VESEN 2000
FIN 3 0 VISEN 1
RIN 3 0 1E9
E1 4 0 3 0 100
RO 4 0 1E9
RNI 5 0 16K
VISEN 5 0 0
RNE 6 0 16K
VESEN 6 0 0
.AC DEC 20 100 1E6
.NOISE V(4) VS 20
.PROBE
.END
```

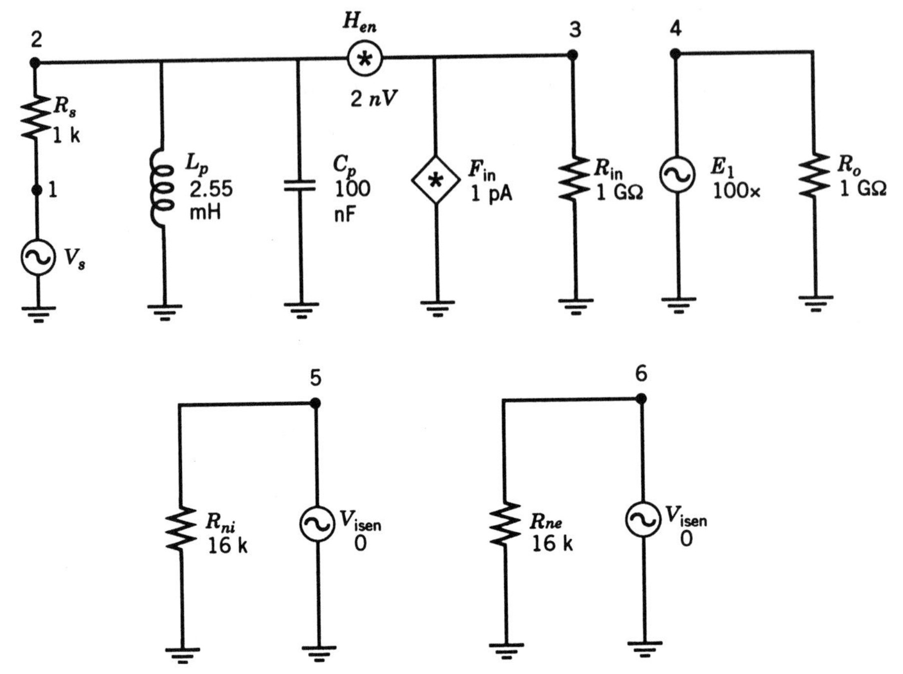

Figure 7-6 PSpice circuit for *RLC* example.

After running PSpice the output file at the resonant frequency 10 kHz is

★★★★RESISTOR SQUARED NOISE VOLTAGES (SQ V / HZ)

	RS	RIN	RO	RNI	RNE
TOTAL	1.654E-13	1.654E-19	0.000E + 00	1.033E-14	4.144E-14

★★★★ TOTAL OUTPUT NOISE VOLTAGE = 2.171E-13 SQ V/HZ
= 4.660E-07 V/RT HZ

TRANSFER FUNCTION VALUE:
V(4)/VS = 9.988E + 01
EQUIVALENT INPUT NOISE AT VS = 4.665E-09 V/RT HZ

Using the preceding values for the source thermal noise RS, noise current RNI, and noise voltage RNE and taking the square root and dividing by the gain, we obtain the components of the input noise. These are the values predicted by Eq. 7-14:

$$\mathrm{RS} = E_s = 4.067\ \mathrm{nV/Hz^{1/2}}$$

$$\mathrm{RNI} = I_n R_s = 1.016\ \mathrm{nV/Hz^{1/2}}$$

$$\mathrm{RNE} = E_n = 2.035\ \mathrm{nV/Hz^{1/2}}$$

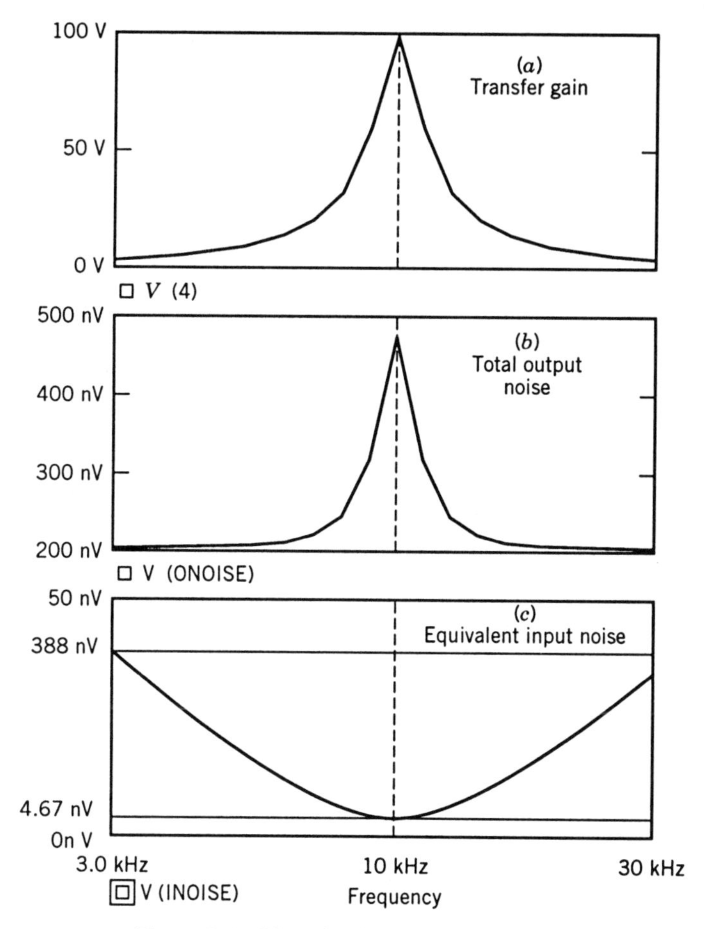

Figure 7-7 Plot of noise for *RLC* model.

Figure 7-7 illustrates the performance of the *RLC* network as a function of frequency as described in Eq. 7-13. The transfer gain has a peak at resonance as does the output noise as shown in Fig. 7-7*a* and *b*. On the other hand, the signal is amplified more than the noise at resonance so the equivalent input noise is at a minimum in Fig. 7-7*c*.

SUMMARY

a. The three major noise contributors in an electronic system are the sensor, amplifier, and coupling network.
b. Each contributor can be replaced by its noise equivalent circuit for noise analysis.

c. To determine the equivalent input noise of a system, the total output noise is divided by the system transfer gain.

d. The total noise at a location in an electronic system can be treated as the sum of the mean square values of the contributions of all sources at that location, each source acting independently.

e. E_{ni}, in its simplest form, is dependent on three noise sources: the thermal noise of the sensor resistance E_{ts} and the amplifier parameters E_n and I_n.

f. In general, the coupling network between the sensor and amplifier increase the E_n and I_n noise. Shunt components increase the E_n contribution and series elements increase I_n.

g. Although a shunt capacitance is noiseless, it will increase the E_n noise contribution at high frequencies.

h. When the sensor impedance is reactive, resonance can sometimes be employed to reduce noise.

PROBLEMS

7-1. Find the output signal-to-noise ratio for systems when (a) $R_p = 2R_s$, (b) $R_p = 0.5R_s$, (c) $R_p = 5R_s$, and (d) $R_p = 10R_s$.

7-2. A shunt capacitance of 100 pF is used as in Fig. 7-4 with an amplifier whose noise parameters are $E_n = 10\ \text{nV/Hz}^{1/2}$ and $I_n = 5\ \text{pA/Hz}^{1/2}$. The source resistance is 10 kΩ. Determine the wideband noise for E_{ni} over the frequency band from 10 Hz to 10 kHz.

7-3. Use the results of the PSpice example to compute the input and output S/N ratios and the noise figure versus frequency for the graphical data presented. Consider that the input signal is constant at 1 μV_{rms}.

7-4. Modify the resonant sensor equivalent circuit of Fig. 7-5 by adding a blocking capacitor $C_b = 1$ nF between R_s and L_p.

(a) Derive a new expression for E_{ni}.

(b) Simulate this modified circuit using SPICE.

(c) Compare your results with those of Fig. 7-7. What can you conclude about the effect of C_b?

7-5. Consider the amplifier in Fig. P7-5. The bandwidth of interest is 1 Hz centered at a frequency of 1 kHz. The amplifier's input impedance is 100 kΩ in parallel with a 100-pF capacitor. Its voltage gain is 750. This amplifier contributes noise through $E_n = 4\ \text{nV/Hz}^{1/2}$ and $I_n = 0.8\ \text{pA/Hz}^{1/2}$. E_n and I_n are not correlated.

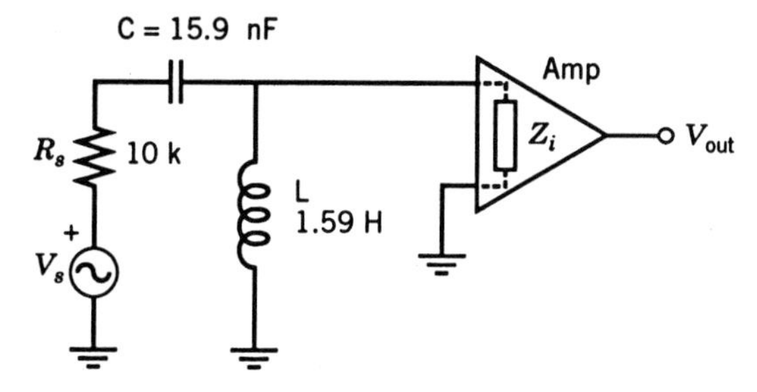

Figure P7-5

(a) Find a numerical value for K_t.

(b) Find the individual contributions to E_{ni}^2; then sum to find the total E_{ni}^2.

(c) Find the total wideband output noise which a true rms voltmeter with a -3-dB bandwidth would measure.

REFERENCE

1. *Solid State Devices*, Hewlett-Packard, Palo Alto, CA., 1968, pp. 195–213.

CHAPTER 8

SENSORS

In the previous chapter we demonstrated the method of modeling the noise of a system consisting of a sensor and an amplifier. Now let us address some practical sensors. The six models described here can be generalized to cover all types of sensors. Three models are extensions of the discussions in the preceding chapter.

These models primarily address the sensor, transducer, or detector itself. Our purpose is to address the noise sources of the sensors with respect to the signal being detected. An amplifier, coupling elements, and feedback must be added to make a noise analysis of a total system. Amplifiers are considered in more detail in the following chapters.

The terms sensor, transducer, and detector are used interchangeably in this book. There are no universally accepted definitions. A sensor or detector, by nature, senses or detects a physical parameter of some kind. A transducer converts energy of one form into another, such as radiant energy into electrical. A transducer also is defined as consisting of a sensor and actuator. In general, the term sensor is used. In the fields of optics and infrared the term detector is preferred. Piezoelectric elements are generally referred to as transducers probably because they are bidirectional. In the newly evolving field of integrated sensors, the term sensor is generally used. We have attempted to follow the commonly used terminology even though this may cause some confusion to the reader.

When using the models of this chapter in a circuit analysis program such as PSpice, enter the circuit element values but not the noise generators. The PSpice program will calculate the thermal noise and shot noise contributions, but it will be necessary to add the excess $1/f$ noise and the G–R noise generators as described in Chap. 4.

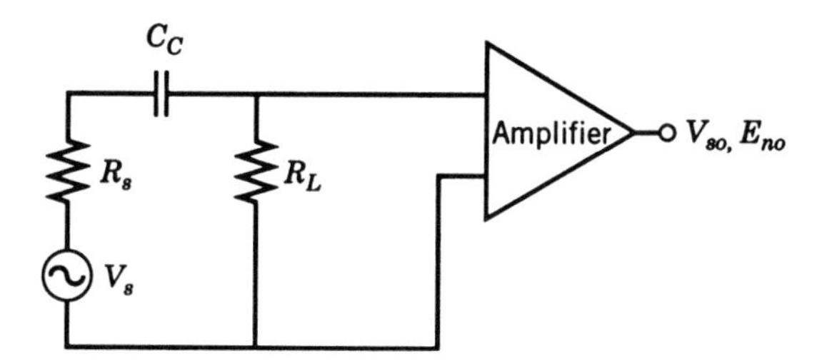

Figure 8-1 Circuit diagram for resistive sensor and amplifier.

8-1 VOLTAIC SENSOR

We consider as the first example the case of a resistive sensor that generates a voltage signal. These detectors include the thermocouple, thermopile, pyroelectric infrared cell, generators, and other sensors that are primarily resistive and generate a voltage signal. A simple circuit diagram is shown in Fig. 8-1.

The sensor is symbolized by the signal source V_s and the internal series resistance R_s. The voltage V_s is the output from the sensed physical or electrical parameter such as pressure or radiation. A coupling capacitor C_C can be used if we are interested exclusively in the time-varying output of the sensor. The element R_L may be needed for impedance matching. The noise model for the sensor–amplifier system is shown in Fig. 8-2.

The shunt capacitance C_p can be in the sensor assembly or it may represent parasitic stray capacitance between lead wires. The amplifier is now represented by the noise parameters E_n and I_n.

For low noise, the noise contribution of R_L is kept low if R_L is large. The shunt capacitance C_p should be minimized to avoid increasing E_n at high frequencies. The decoupling capacitor C_C should be very large or removed to reduce its effect on the amplifier's I_n noise at low frequencies. The amplifier input resistance R_i can often be reduced with overall negative feedback to increase the corner frequency caused by C_p. Of course, if R_s is kept small

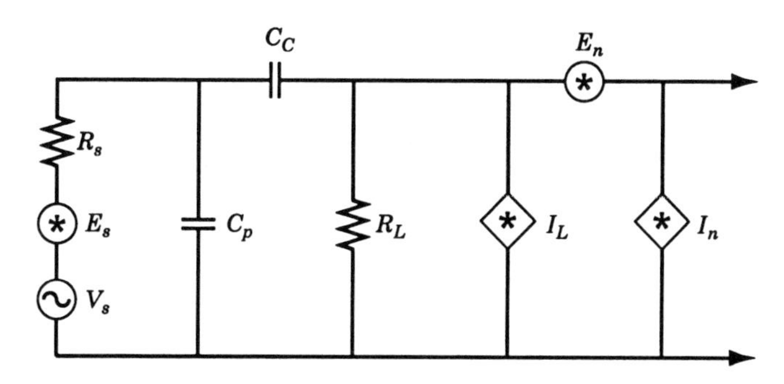

Figure 8-2 Noise equivalent circuit for voltaic sensor.

the limiting system thermal noise will be reduced. The amplifier should be chosen so that the optimum noise resistance R_o is equal to the source resistance R_s and the $E_n I_n$ product is as small as possible.

8-2 BIASED RESISTIVE SENSOR

One of the most common type of sensor is the variable resistance sensor whose resistance or conductance changes in response to a sensed input. Since this sensor does not generate a signal directly, it must be biased with a voltage or current source and coupled with a load resistance. This adds two more noise sources, the biasing signal noise and the load resistance noise. If the sensor is placed in a bridge, the other resistive elements of the bridge contribute noise.

Some examples of resistive sensors are the strain gauge, photoconductive infrared cell, bolometer radiation detector, resistance thermometers, and piezoresistive sensors. A simple circuit diagram is shown in Fig. 8-3.

The bias voltage is supplied by V_{BB}. The sensor signal is developed as a voltage drop across R_B. The variable resistance component of the sensor R_s is represented by the incremental resistance ΔR_s which will be very small with respect to R_s. The coupling capacitor C_C will remove the common-mode dc bias voltage from the amplifier input. The load resistance R_L provides a bias path for the amplifier input. A noise equivalent circuit is shown in Fig. 8-4.

The voltage generator V_s represents the detected signal. It can be either a voltage generator in series with the sensor resistance or a current generator, $I_s = V_s/R_s$, in parallel with the source. Since $\Delta R_s \lll R_s$ (or we would not be designing a low-noise amplifier), it will not be included in the expression. The signal V_s is caused by the resistance or conductivity modulation of the

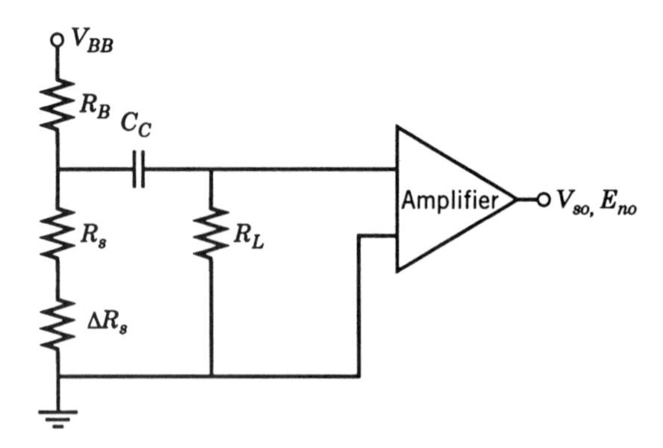

Figure 8-3 Circuit diagram for biased resistance sensor system.

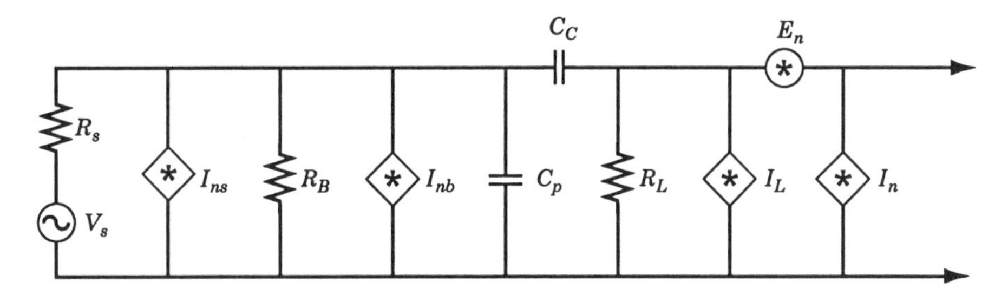

Figure 8-4 Noise equivalent circuit for biased resistive source.

sensor resistance R_s and is

$$V_s = I_B \, \Delta R_s \cong \frac{V_{BB} \, \Delta R_s}{R_s + R_B} \tag{8-1}$$

When the bias is a constant current source I_{BB}, the voltage signal is $V_s = I_{BB} \, \Delta R_s$. In this case, the bias resistor R_B is very large and the thermal noise negligible. The current noise generator I_{nb} becomes $I_{nb}(f)$ which represents the noise of the current source which is probably frequency dependent.

The noise current generator I_{ns} in Fig. 8-4 represents three noise generators in the sensor. Many sensors, especially photoconductive radiation detectors, show thermal noise, $1/f$ noise, and G–R noise. Thermal noise is caused by a fluctuation in the velocity of carriers as described in Chap. 1. Second, because current is flowing through a nonperfect medium, there is $1/f$ excess noise generated. This has a $1/f$ noise power spectrum as described in Chap. 1. The third noise mechanism, generation–recombination noise or G–R noise, is caused by fluctuations in the generation, recombination, and trapping of carriers in the semiconductor. Generation–recombination fluctuations cause a variation in the number of carriers available which causes a modulation in the conductivity of the semiconductor. Since this is caused by the random trapping and detrapping of conductors, it appears as noise. In photoconductive radiation detectors, this noise is "white," has a flat spectrum over the useful frequency range, and will usually dominate the thermal noise.

The bias resistor R_B affects both the sensor signal and its noise. The noise current generator I_{nb} is the thermal noise current and excess noise of the load resistance R_B. Since the thermal noise of I_{nb} is inversely related to the square root of the resistance, we again desire R_B to be large. It is sometimes possible to replace R_B by a noise-free inductor for biasing and load. This gives an additional low-frequency roll-off term. When a resistor is used for biasing instead of a current source, the signal is dependent on R_B as shown in Eq. 8-1. A larger resistor decreases the bias current and reduces the output signal although a large resistor also reduces the noise. The resistor R_B

requires a trade-off between gain and noise that depends on the sensor characteristics.

The coupling capacitor C_C can be used to remove the common-mode voltage from the amplifier. C_C must be large so that $I_n X_C$ will not contribute noise at the lowest frequencies. X_C is the reactance of C_C. R_L should be kept much larger than R_s so that I_L does not contribute noise.

For the lowest-noise system, R_s will be the dominant noise source. Select an amplifier E_n and I_n to provide an R_o equal to R_s and the minimum E_n I_n product.

8-3 OPTOELECTRONIC DETECTOR

An optoelectronic detector is used to detect various forms of visible and nonvisible radiation and has a wide range of applications such as infrared detection, heat measurement, light and color measurement, fiber optic detectors, sensors for compact disk players, laser detectors, and many other uses.

There are two general types of solid-state photon detectors: photoconductive and photovoltaic. In a photoconductive detector, radiation on a cell produces a current in addition to the dark current. Bias is applied to the cell to collect the current. In a photovoltaic detector, radiation on the cell produces a voltage directly. Photoconductive cells can be fabricated from bulk semiconductor material where the conductivity increases as radiant energy is absorbed. This type of photoconductive detector was modeled in Sec. 8-2. A more common form of this detector is a reverse-biased or unbiased diode. The diode photodetector model will be developed in this section. A simplified circuit diagram is shown in Fig. 8-5. Biasing voltage is provided by supply V_{BB}. This reverse bias will collect all the current generated by the radiant photon signal λ, and a voltage signal is developed across the load or bias resistor R_B.

Most often photodiodes are used with op amps employing negative feedback to produce the photoconductive detector as shown in Fig. 8-6.

The feedback resistor R_B produces a virtual ground at the anode of the photodiode which reduces the input impedance, thereby improving the frequency response. The output voltage is $V_o = -I_D R_B$ where I_D is the reverse-bias current in the photodiode, also called the dark current. Ideally,

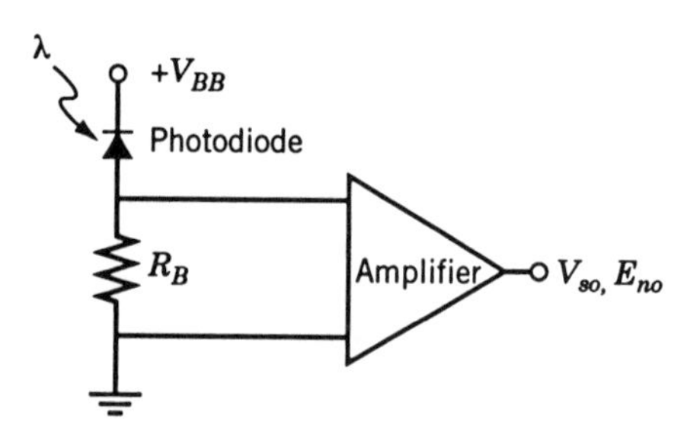

Figure 8-5 Circuit diagram for photodiode detector.

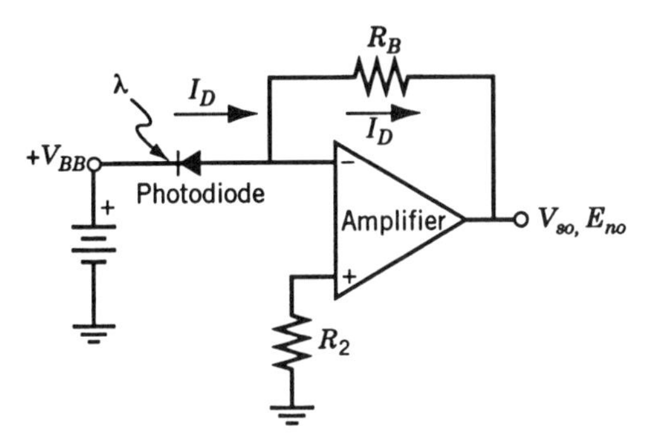

Figure 8-6 Feedback amplifier for photodiode detector.

$R_2 = R_B$ to reduce the output offset voltage caused by the input bias current. However, R_2 adds noise as can be seen from the noise equivalent circuit in Fig. 8-7. The load resistor R_B has the same effect on equivalent input noise and gain for either circuit. The noise equivalent circuit of the photodiode detector is shown in Fig. 8-7. The signal current source I_s is located at the input and

r_d = noiseless dynamic reverse-bias resistance of the photodiode
R_B = feedback resistance
R_{cell} = cell series resistance ($< 50\ \Omega$)
R_2 = bias resistor for noninverting input
E_{cell} = thermal noise of R_{cell}
E_n = amplifier noise voltage
C_d = cell capacitance
C_W = stray wiring capacitance
I_D = sensor dc photocurrent plus dark current
I_{nB} = $(4kT/R_B)^{1/2}$ = thermal noise of R_B
I_p = $(I_{\text{sh}}^2 + I_{\text{G-R}}^2 + I_{1/f}^2)^{1/2}$
I_{n1} = amplifier noise current for inverting input
I_{n2} = amplifier noise current for noninverting input
I_2 = thermal noise current of R_2

The photodiode signal I_s is a current proportional to the light intensity. The shot noise current term of I_p depends on the dc leakage and photocurrent I_D. I_p should also include the G–R noise and the excess low-frequency $1/f$ noise. Most of the circuit impedances shunt the signal source. The series resistance R_{cell} is usually less than 50 Ω so E_{cell} is negligible. The load resistance (or bias resistance) R_B is a thermal noise current generator so it should be large to minimize the noise contribution.

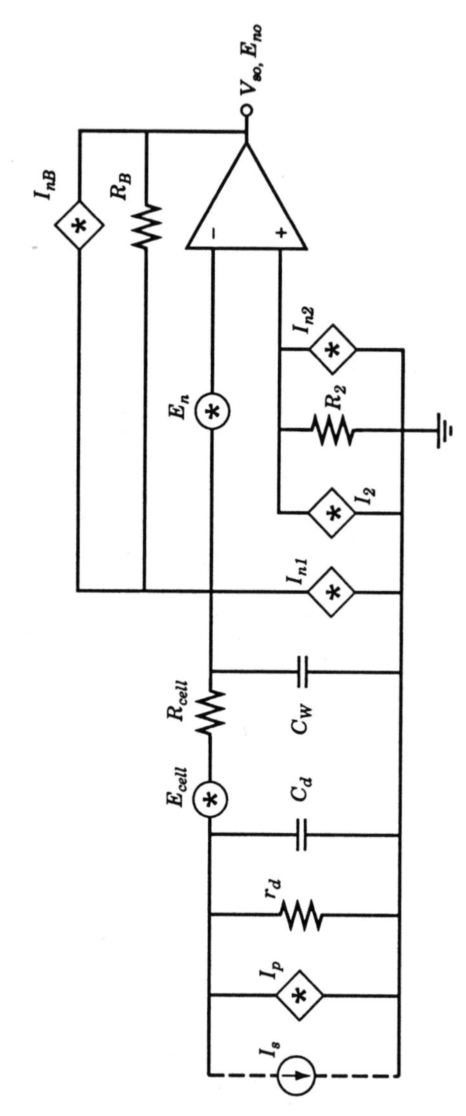

Figure 8-7 Noise equivalent circuit of photodiode detector.

The cell capacitance C_d and wiring capacitance C_W probably will be the frequency-limiting elements so they should be kept as small as possible. A "guarded" amplifier input terminal will reduce the wiring capacitance. Although the amplifier input capacitance C_i and input resistance R_i drop out of the noise expression, they do affect the amplifier gain. This gives us a mechanism for optimizing the frequency and noise responses separately.

An unbiased diode or a photovoltaic diode uses a similar model. The biasing resistor R_B may be removed. If there is dc current flowing due to the ambient light level, there will be a shot noise component. The shunt resistance of the diode can be determined from the diode equation. Diode shot noise was discussed in Sec. 1-11.

A field effect transistor (FET) input stage amplifier may be ideal for photodiodes with their high impedance and low capacitance. If the FET is connected in the common-drain configuration, the Miller effect is not present, reducing the amplifier input capacitance. The high-resistance levels in the sensor–amplifier network result in a low value of the upper cutoff frequency for the system. To make the system broadband, two approaches can be followed. A "lead" network can be connected near the amplifier output to compensate for the high-frequency loss, or negative feedback can be employed around the amplifier.

8-3-1 Photodiode Noise Mechanisms

Figure 8-7 shows the noise circuit of a photodiode amplifier system and illustrates the four types of principal noise mechanisms: $1/f$ noise, shot noise, generation–recombination noise, and thermal noise. The first three are the noise current generators included in I_p. The thermal noise comes from the diode series resistance R_{cell} and is included as E_{cell}. A plot of noise versus frequency for a photodetector is shown in Fig. 8-8.

Excess noise or $1/f$ noise is characterized by its frequency spectrum. The square of the noise current is inversely proportional to the frequency. $1/f$

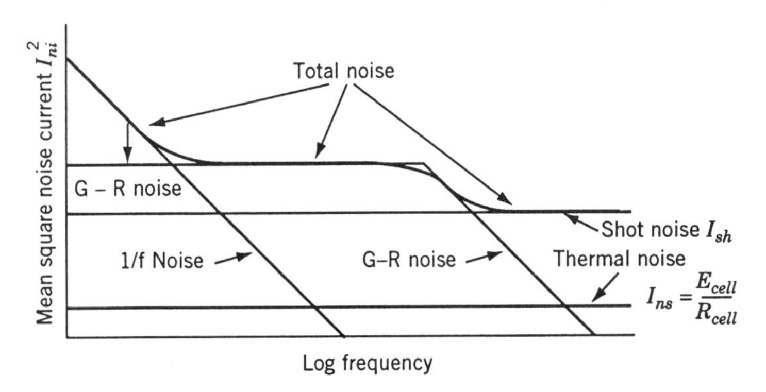

Figure 8-8 Noise sources in a photodiode.

noise is primarily caused by carrier trapping at or near the surface and is a result of lattice defects and impurities. This noise was discussed in Chap. 1.

Shot noise is a fundamental noise mechanism as is thermal noise. Shot noise is caused by the random arrival of carriers crossing a diode junction. It can be calculated as $I_{\text{sh}}^2 = 2qI_D\,\Delta f$ as derived in Chap. 1. Thermal noise is caused by the random motion of carriers in a conductor or semiconductor. It can be calculated as $E_{\text{th}}^2 = 4kTR\,\Delta f$ as derived in Chap. 1.

Generation–recombination, G–R, noise is often the dominant noise mechanism in photodetectors. Fluctuations in the generation, recombination, and trapping of carriers in semiconductors cause G–R noise. This causes the number of free carriers to fluctuate. Hence the conductivity of the material fluctuates. Although G–R noise appears similar to thermal noise, it is a separate mechanism where the conductivity of the solid fluctuates because the number of carriers changes, whereas for thermal noise the number of carriers is essentially constant but the instantaneous velocity of the carriers changes. The spectral response of G–R noise is white up to a frequency determined by the lifetime of the carriers in the photodetector [1].

Thermal noise is the only noise which occurs in a passive resistance and which is independent of current bias. All other noise sources require the existence of a current and the noise increases with current flow. The shot noise is proportional to the square root of the current. Generation–recombination and $1/f$ noises are proportional to the current. Thermal and shot noises are white, but G–R and $1/f$ noises have a characteristic frequency response.

Background photon noise is the ultimate noise limit of a photodetector. This is the noise due to the random arrival of photons and the generation of carriers in the detector, much as the random collection of electrons generates shot noise. An ideal detector is background photon noise limited and has a quantum efficiency of unity. Background-limited infrared photodetectors are said to be in the *blip* condition where blip is the acronym [2].

The threshold noise performance can be best specified in terms of a noise equivalent input signal. The *noise equivalent power*, NEP, is a figure-of-merit used for comparison of sensors. NEP *is the value of the input signal (in this case light power) that produces an output electrical signal equal to the noise output present when no input is applied*. Thus in equation form we have

$$\text{NEP} = \frac{\text{Noise current (A/Hz}^{1/2})}{\text{Current responsivity (A/W)}} \qquad \text{W/Hz}^{1/2}$$

or

$$\text{NEP} = \frac{I_p}{\text{CR}} \tag{8-2}$$

A low value for NEP corresponds to a high sensitivity.

8-3-2 PIN Photodiode Sensor

One type of sensor commonly used for visible and near-infrared detection is the PIN photodiode. The PIN photodiode is a photodiode as shown schematically in Fig. 8-6 with an intrinsic layer added to improve performance. It is easy to fabricate and gives good sensitivity. In this section several of the engineering considerations relevant to the application of this device are discussed. The device is a detector for visible and near-infrared radiation. It is contained in a TO-18 can with a glass-window cap to admit radiation. The diode is a biased detector, with the biasing voltage between 5 and 20 V dc for best performance.

The PIN photodiode is a semiconductor "sandwich" composed of an "*I*" or intrinsic layer of silicon with *p*-type silicon diffused into its upper face and *n*-type silicon forming the lower face. The *I*-layer is purposely thick; the *p*-layer is very thin. The external surfaces of the sandwich are covered with gold to prevent optical radiation from reaching the semiconductor material. A dc bias of less than 50 V is applied to the structure with its positive connected to the *n*-material and its negative connected to the *p*-material. No connection is made to the *I*-layer.

Light energy can reach the diode through an aperture in the gold coating above the *p*-layer. When a photon is absorbed by the silicon, a hole and an electron are liberated from a broken covalent bond. Because the *p*-layer is so thin, the bond that is broken is in the *I*-layer. The applied potential establishes an electric field in that layer that accelerates the hole toward the *p*-layer and the electron toward the *n*-layer. Naturally, this charge movement constitutes a current, the value of which is dependent on the intensity of light impinging on the device.

Electrically, the PIN diode's behavior is similar to a conventional junction diode. The reverse bias controls the capacitance of the structure; increasing the value of the bias to 20 V reduces that capacitance to less than 1 pF in some types. The thin *p*-layer can be represented by a low-valued resistance; the current leakage through the device under reverse bias can be modeled by a large resistance.

In the small-signal and noise model of Fig. 8-7, R_{cell} is the series resistance of the semiconductor, which is less than 50 Ω. The elements C_d and r_d are the diode's capacitance and dynamic resistance, typically 5 pF and 10 GΩ, respectively. I_s is the signal current resulting from illumination. The major noise source is shot noise, $I_{\text{sh}} = (2qI_D\,\Delta f)^{1/2}$, where I_D is the dark or leakage current which is as low as 100 pA.

Some excess $1/f$ noise is generated in the photodiode. It can be included in the I_p generator. It has a noise corner of about 20 to 30 Hz and rises at about 10 dB/decade for frequencies below that break.

The signal source I_s is capable of high current responsivity (CR); it can provide an output of some 0.5 μA/μW in the band of light wavelengths from 0.5 to 0.8 μm. This band includes part of the visible spectrum and a portion of the infrared spectrum. The quoted responsivity corresponds to 0.75 electrons/photon or a quantum efficiency of 75%. The NEP for a PIN sensor can be as low as $-110\ \mathrm{dB_m/Hz^{1/2}}$ ($\mathrm{dB_m}$ refers to dB below a 1-mW reference). The corresponding numerical value of NEP for this case is 1.1×10^{-14} $\mathrm{W/Hz^{1/2}}$. For an input light power level of 1 pW, the S/N ratio for the device, found by dividing the signal current by the noise current, is nearly 40 dB.

The threshold sensitivity of a PIN photodiode system is adversely affected if the diode is forced to feed a low-resistance load. A low shunting resistance value results in a disproportionate amount of thermal noise at the input terminals, relative to the signal and noise from the photodiode sensor. In the noise equivalent circuit shown in Fig. 8-7, sensor shot noise is I_p, shunt resistance is R_B, and the noise current associated with R_B is shown as I_{nB}. Applying the definition of NEP to this circuit, we may obtain NEP = $(I_p^2 + I_{nB}^2)^{1/2}/\mathrm{CR}$. It is clear that I_{nB}^2 must be somewhat less than I_p^2 to reach an NEP level comparable with the unloaded figure given earlier. For $I_D = 100$ pA, $I_p^2 = 3.2 \times 10^{-29}\ \mathrm{A}^2$. For I_{nB}^2 to equal I_p^2, R_B must be 500 MΩ. Whereas it is usually not necessary for R_B to be that large in a practical application, it may be recognized that $I_{nB}^2 = 4kT/R_B$ and, consequently, noise current increases for decreasing values of R_B.

The amplifier requirement can be satisfied in several ways. A bipolar transistor, selected for low noise, can be connected as an emitter–follower input stage; its operating point and load resistance are selected for low-noise and high-input impedance. A field effect transistor (FET) is ideal for this application. If the FET is connected in the common-drain configuration, the Miller effect is not present and thus the amplifier input capacitance is low.

The high-resistance levels in the sensor–amplifier network result in a low value of the upper cutoff frequency for the system. That frequency is approximately $1/2\pi(C_d + C_i)R_p$, where C_i is the capacitance of the amplifier and $R_p = R_B \| R_i$. We see that for $C_d = C_i = 5$ pF and $R_p = 10$ MΩ, the cutoff frequency is a few thousand hertz. To broadband the system, negative feedback can be employed around the amplifier to reduce the input impedance.

Good examples of the design of high-impedance amplifiers for photodiodes are given in the Analog Devices specification sheet [3] and in its 1992 Amplifier Applications Guide [4].

8-4 *RLC* SENSOR MODEL

The *RLC* sensor models magnetic tape recorder heads, coils, inductive pickups, dynamic microphones, linear variable differential transformers, and various other inductive sensors. *RLC* sensors may be resonant devices, but their principal characteristic is the inductive signal source. A general circuit diagram is shown in Fig. 8-9.

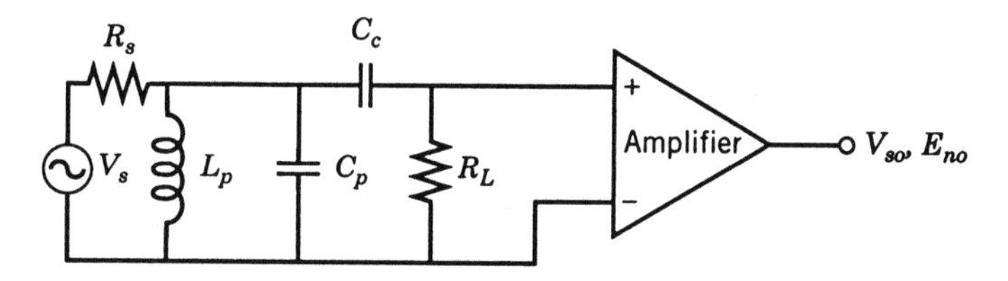

Figure 8-9 System diagram for *RLC* sensor.

The noise model shown in Fig. 8-10 contains the noise source E_s to represent the thermal noise that may be present in the real part of the sensor impedance R_s. The sensor is assumed to have shunt inductance represented by L_p. C_p represents the inductor capacitance and wiring capacitance or external capacitance added to resonate the sensor.

For the circuit configuration of Fig. 8-10:

R_s = sensor series resistance or real part of the impedance
R_L = load resistance
L_p = sensor inductance
C_p = shunt capacitance
E_s = thermal noise voltage of R_s
E_n = noise voltage of amplifier
I_L = thermal noise current of R_L
I_n = noise current of amplifier

In the *RLC* example in Chap. 7, the effect of resonance is seen in Eq. 7-14. At resonance the E_n value is minimum. The I_n term is dependent only on the impedance of the series inductance and resistance.

The signal in the *RLC* sensor is usually a voltage proportional to the rate-of-change of flux linkage. The coil L_s and resistor R_s can also be

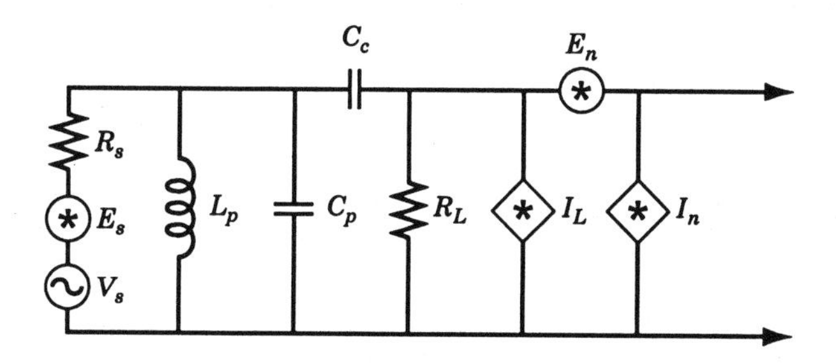

Figure 8-10 Noise equivalent circuit for *RLC* sensor.

expressed as a parallel resistance and inductance. Then the voltage generator E_s would become I_s, a thermal noise current generator.

Magnetic-core coils show a decreasing inductance and increasing series resistance loss at high frequencies due to eddy-current losses. Since the real part of the coil impedance is a thermal noise generator, it may be necessary to calculate or model the inductance and resistance at each frequency. The real and reactive parts of impedance as a function of frequency can be measured on an impedance bridge.

The design of an inductive sensor can be optimized to obtain the maximum signal-to-noise ratio. The voltage signal is proportional to the number of turns. The coil resistance is proportional to the turns for small diameters, and the noise is proportional to the square root of the turns so the signal increases faster than the noise as the turns increase. When the coil becomes large enough in diameter, the resistance increases faster than the square of turns, and the signal-to-noise ratio begins to fall off.

8-5 PIEZOELECTRIC TRANSDUCER

A piezoelectric transducer generates an electrical signal when it is mechanically stressed, as the name implies—"piezo," mechanical, and "electric" for an electrical output. This is usually a reversible process so the same ceramic element can be used as the sonar "pinger" and the signal detector. That is, a mechanical strain gives an electrical output and an electrical voltage on the element gives a mechanical deflection. Butane matches "snap" a piezoelectric element to make the spark.

A common example of a piezoelectric sensor is the crystal phonograph cartridge. Other common applications are microphones, hydrophones, sonar, seismic detectors, vibration sensors, accelerometers, and other devices where there is mechanical to electrical energy conversion.

Ferroelectric ceramic elements and quartz crystals are common examples of piezoelectric transducers. They are generally characterized as a capacitor which generates a charge when mechanically strained. The output signal is a small current flowing through a high external impedance. Often these sensors are hard such as ceramic elements but there are also plastic films such as polyvinyldiflouride, PVDF, which are also piezoelectric. Often pyroelectric

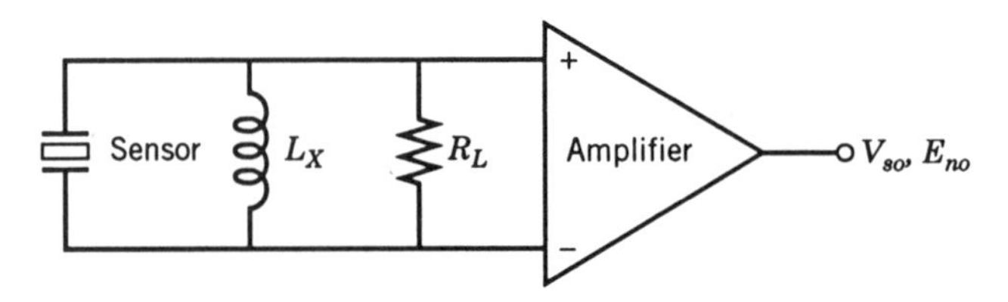

Figure 8-11 System diagram for piezoelectric sensor.

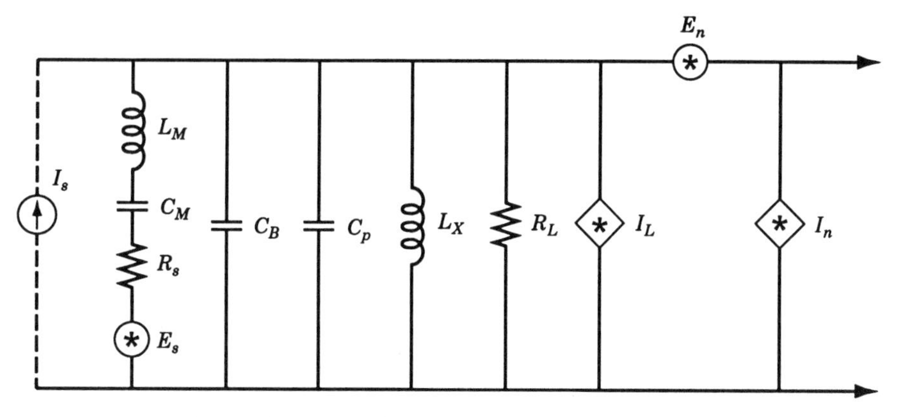

Figure 8-12 Noise equivalent circuit for piezoelectric transducer.

radiation detectors also can be modeled since they are capacitive elements that generate a charge when heated by incident radiant energy.

A system diagram is shown in Fig. 8-11 and a noise equivalent circuit for the piezoelectric system is shown in Fig. 8-12. The sensor is modeled by series resistance R_s and by so-called "mechanical" inductance L_M and capacitance and C_M. Electrical reactances are modified by the mechanical coupling to the acoustic medium. The resulting electrical parameters of the crystal are a combination of mechanical and electrical properties. The block or bulk capacitance C_B of the transducer parallels the L_M–C_M circuit.

At frequencies below the L_M–C_M resonance, C_M and C_B form a voltage divider. Since C_B may be $10C_M$, the signal to the amplifier is only $I_s/10$. The external inductance L_X can be resonated with sensor and wiring capacitances and can improve the signal transmission. These transducers have two resonances, a series resonance of L_M and C_M as well as the parallel resonance of $(C_M + C_B)$ and L_X. Because of the reactance of the transducer, it is commonly operated at its parallel resonance frequency.

For the circuit configuration of Fig. 8-12:

L_M = mechanical inductance
L_X = external inductance
C_M = mechanical capacitance
C_B = block or bulk capacitance
C_p = cable capacitance
R_s = series loss resistance
R_L = load resistance
E_s = thermal noise of $R_s = (4kTR_s)^{1/2}$
I_L = thermal noise of $R_L = (4kT/R_L)^{1/2}$
I_s = signal current source

The system equivalent input noise is

$$E_{ni}^2 = 4kTR_s + E_n^2\left(\frac{Z_s + Z_L}{Z_L}\right)^2 + \left(I_n^2 + I_L^2\right)Z_P^2 \tag{8-3}$$

Z_s is the series impedance of R_s, C_M, and L_M. Z_L is the parallel impedance of C_B, C_P, L_X, and R_L. Z_P is Z_L in parallel with Z_s.

The equivalent input noise has two noise voltage and two noise current terms. Since R_s is small, thermal noise is usually negligible. The E_n contribution is small with respect to I_n for this high-impedance device. The noise current terms dominate. At low frequencies Z_P is extremely large, primarily the impedance of C_B and C_P. To keep the noise current contribution small, R_L should be large and I_n small. This implies a bipolar transistor biased at a very low collector current (in the microampere range or less), or a FET with low gate leakage current. The $I_n Z_P$ term is most prominent at low frequencies where $1/f$ noise is likely to manifest itself. This constitutes a second vote for the FET since its noise current I_n has a very small $1/f$ component. Also, the FET requires little or no biasing so R_L can be extremely large. Resonating the sensor decreases the equivalent input noise, that is, increases the signal-to-noise ratio at the resonant frequency.

Examples of charge amplifiers are given in the Analog Devices specification sheet and applications guide [3, 4].

8-6 TRANSFORMER MODEL

There are three main reasons for using an input transformer to couple the signal source to the amplifier. The first is to transform the impedance of the source to match the amplifier noise resistance R_o of the amplifier and minimize the system noise. The second is to provide isolation between the source and amplifier. A third reason is for impedance matching to obtain maximum signal power transfer. Although the transformer can potentially reduce the equivalent input noise of the amplifier, its own noise mechanisms can contribute to the overall system noise.

A transformer has the ability to transform impedance levels. To demonstrate this statement, consider the ideal transformer shown in Fig. 8-13. It is lossless; therefore, the power levels at the primary and secondary are equal

$$\frac{V_1^2}{R_1} = \frac{V_2^2}{R_2} \tag{8-4}$$

where R_2 is the resistive load connected to the secondary terminals, and R_1 is the effective (reflected) load at the primary terminals (not the dc resistance of the transformer winding, because that resistance is assumed negligible

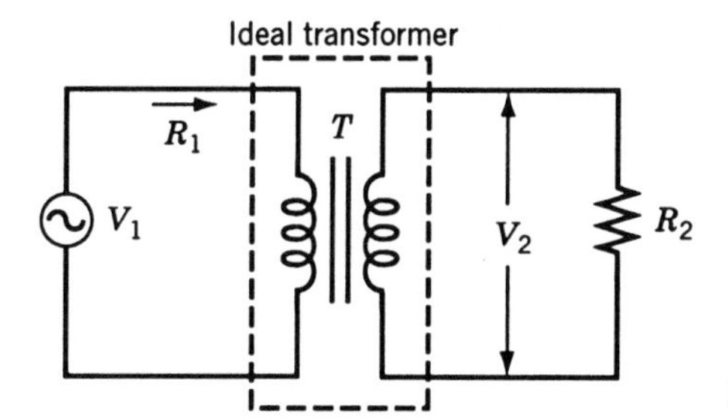

Figure 8-13 Ideal transformer.

when we state that the device is ideal). Let the ratio of primary to secondary turns be 1:T. Then the voltages are related according to

$$V_2 = TV_1 \tag{8-5}$$

Substitution of Eq. 8-5 into Eq. 8-4 gives the reflected secondary resistance at the primary

$$R_1 = \frac{R_2}{T^2} \tag{8-6}$$

By an appropriate choice of T, the desired impedance level can be realized.

The use of an input transformer between the sensor and amplifier improves the system noise performance by matching the sensor resistance with the amplifier optimum source resistance R_o. Consider the secondary circuit as shown in Fig. 8-14 containing noise generators E_n and I_n. We have defined $R_o = E_n/I_n$. When these quantities are reflected to the primary as E_n' and I_n', we obtain

$$E_n' = \frac{E_n}{T} \qquad \text{and} \qquad I_n' = TI_n \tag{8-7}$$

The ratio of the reflected noise parameters is R_o':

$$R_o' = \frac{E_n'}{I_n'} = \frac{E_n}{T^2 I_n} = \frac{R_o}{T^2} \tag{8-8}$$

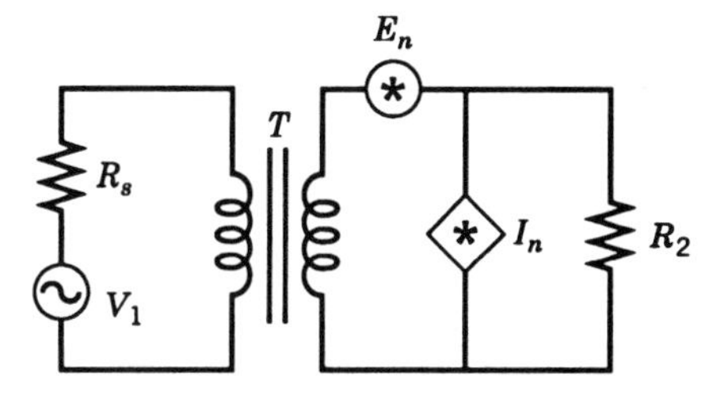

Figure 8-14 Amplifier noise generators located at secondary.

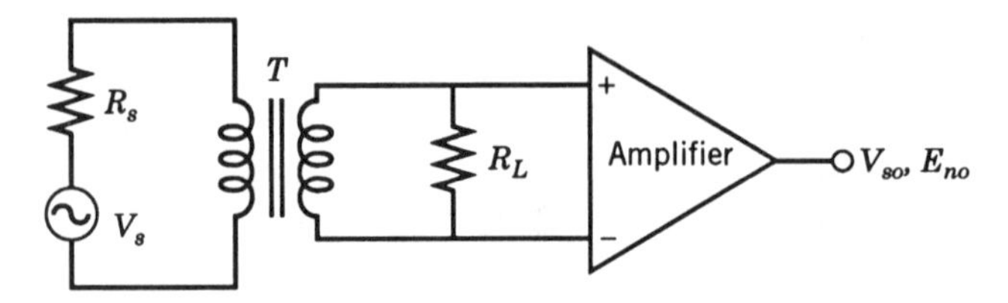

Figure 8-15 Transformer-coupled source.

Now match R'_o to the source resistance R_s. Since we are free to select T, it follows that

$$T^2 = \frac{R_o}{R_s} \tag{8-9}$$

This ensures that the amplifier sees the optimum source resistance R_o to achieve the highest signal-to-noise ratio.

A transformer-coupled input stage is shown in Fig. 8-15. The signal source is represented by a resistance R_s but it could be an impedance or any network.

The circuit diagram given in Fig. 8-16 is a small-signal ac equivalent of the transformer-coupled system. The transformer is represented by the primary winding resistance R_p, the primary inductance L_p, and the reflected secondary series resistance R'_{sec}. This "T" equivalent circuit is valid for low-frequency analysis. The load resistance R'_L and the input quantities R'_i and C'_i carry the prime designation to indicate they have been "reflected" by the transformer turns ratio. A transformer turns ratio T means it has a primary-to-secondary turns ratio of 1:T.

A noise model with a noiseless transformer can be developed from the "T" model as shown in Fig. 8-17. This model *does not* reflect the secondary network to the primary side. Transformers have ohmic resistances in both the

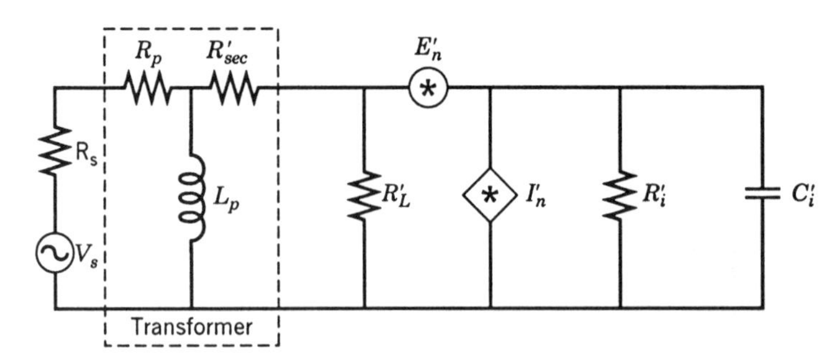

Figure 8-16 Ac equivalent circuit for transformer-coupled source.

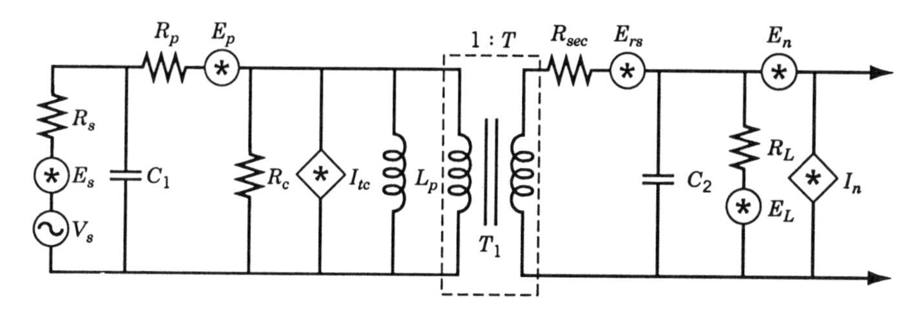

Figure 8-17 Noise equivalent circuit for transformer-coupled source.

primary and secondary windings that generate thermal noise and may result in signal power loss. The primary and secondary resistances are R_p and R_{sec} and their thermal noise generators are E_p and E_{rs}. The element R_c is a representation of the core losses caused by eddy currents in the conductive core. Since this is a real dissipative resistance, it will exhibit thermal noise I_{tc}. The primary winding inductance L_p shunts the ideal portion of the practical transformer, and with the source resistance is the cause of low-frequency roll-off. High-frequency performance is limited by the wiring capacitances C_1 and C_2. These reactive impedances are noiseless. Only dissipative elements generate thermal noise so they do not have thermal noise generators. The source and load resistances R_s and R_L also have thermal noise generators E_s and E_L. Generally $R_p = R'_{sec}$ for most transformers.

For the circuit configuration of Fig. 8-17:

R_p = resistance of the transformer primary
R_{sec} = secondary resistance of the transformer
R'_{sec} = secondary resistance reflected to primary
R_c = resistive core loss
R_s = source resistance
R_L = secondary load resistance
L_p = transformer primary inductance
T = transformer secondary to primary turns ratio
T_1 = noiseless transformer with turns ratio T
E_p = thermal noise of $R_p = (4kTR_p)^{1/2}$
E_{rs} = thermal noise of R_{sec}
E_s = thermal noise of R_s
E_L = thermal noise of R_L
I_{tc} = thermal noise of R_c
V_s = input signal voltage
C_1, C_2 = primary and secondary shunt capacitances

When the transformer equivalent circuit in Fig. 8-17 is used in a PSpice simulation, only the components are entered. Values for the thermal noise generators are automatically calculated and entered by PSpice when the .NOISE option is selected. Be sure to include the actual source and load impedances as well as the amplifier noise model that represents your circuit.

In addition to the resistive noise mechanisms, there are second-order noise effects in a transformer that must be reduced. The transformer must be magnetically shielded and wound with balanced windings to minimize pickup of external ac fields. Also, the transformer should be tightly packaged to minimize microphonic shock sensitivity. The high-permeability core in a low-level instrument transformer tends to be highly magnetostrictive and sensitive to flexing and mechanical motion.

For high common-mode rejection, interwinding electrostatic shields should be used. When testing a low-level transformer, an ohmmeter should not be used. It could magnetize the core and therefore decrease the inductance and increase the microphonics. It is difficult to degauss the windings once they are magnetized. The inductance and resistance are best measured by using a bridge.

SUMMARY

a. To design the lowest-noise system with the highest signal-to-noise ratio, it is necessary to use the sensor model as part of the noise analysis.

b. A low-noise design requires matching the amplifier's noise characteristics to the sensor noise and impedance.

c. A sensor noise model is developed from the ac equivalent circuit by adding thermal noise generators to all resistances and shot noise generators to all diode junctions. Excess $1/f$ noise and G–R noise generators are added at the appropriate points.

d. When analyzing a sensor noise model with PSpice, the thermal and shot noise generators are automatically entered by PSpice. The $1/f$ noise and G–R noise generators must be inserted as circuit elements in the circuit description.

e. Biased resistive detectors do not generate a signal directly so they must be biased with a low-noise dc generator through a load impedance. This load impedance is critical since it may contribute noise and it affects the gain of the detector.

f. Optoelectronic devices are ideally limited by photon noise rather than the internally generated G–R noise.

g. For RLC and piezoelectric sensors, circuit resonance can have a large effect on system noise.

h. Eddy current and core losses can contribute noise to the inductive sensor as well as the thermal noise of the coil resistance.

i. A transformer matches sensor impedance to an amplifier's optimum noise resistance R_o, but the transformer winding resistance and core loss will contribute thermal noise.

PROBLEMS

8-1. Using Fig. 8-2, derive expressions which show the effect that the coupling capacitance C_C has upon the four noise sources which contribute to E_{ni}^2.

8-2. A diagram of the TSL250 light-to-voltage optical sensor is shown in Fig. P8-2. The typical output voltage for this unit at a frequency of 20 Hz is specified as 0.6 $\mu V/Hz^{1/2}$. When the irradiance is zero, the dc output voltage is 3 mV. The irradiance responsitivity is 85 $mV/(\mu W/cm^2)$. Assume $R_F = 100$ kΩ and that the op amp is noiseless.

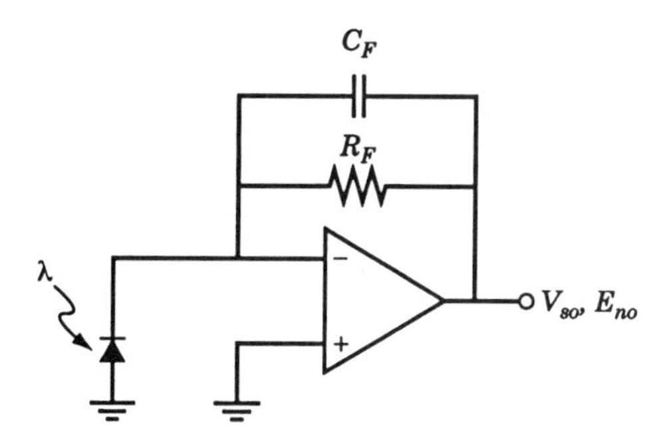

Figure P8-2

(a) Determine the shot noise and $1/f$ noise components which contribute to the total photodiode noise current, I_p. Neglect I_{G-R} noise.

(b) Find the noise corner frequency of the photodiode.

(c) Now assume that the reverse dynamic resistance of the photodiode is 100 kΩ. Also assume that all the output noise is $1/f$ noise due to the noise voltage of the op amp. Find E_n at the given frequency of 20 Hz.

(d) Determine what additional information is required in order to calculate the NEP for this TSL250 optical sensor.

8-3. The RLC sensor of Fig. 8-10 has the following component values: $R_s = 150\ \Omega$, $R_L = 10\ \text{k}\Omega$, $L_p = 1$ mH, $C_p = 35$ pF, $C_C = 1$ nF, $E_n = 10$ nV/Hz$^{1/2}$, $I_n = 3$ pA/Hz$^{1/2}$.

(a) Find the resonant frequency, f_o.

(b) Perform a SPICE simulation to determine the total noise voltage E_{ni} evaluated over the frequency range from 1 kHz to 1 GHz. Express your answer in I_{rms}.

8-4. Refer to the equivalent noise circuit for the piezoelectric transducer shown in Fig. 8-12. The components have the following numerical values: $R_s = 40\ \Omega$, $R_L = 5\ \text{k}\Omega$, $L_M = 1.0$ mH, $L_X = 0.1$ mH, $C_M = 1$ nF, $C_B = 4$ nF, $C_p = 200$ pF, $E_n = 4$ nV/Hz$^{1/2}$, $I_n = 0.02$ pA/Hz$^{1/2}$.

(a) Find the series and parallel resonant frequencies, f_{os} and f_{op}.

(b) Perform a SPICE simulation to determine the total equivalent input noise current I_{ni} evaluated over the frequency ranges from 200 kHz to 400 kHz. Express your answer in I_{rms}.

REFERENCES

1. Van Vliet, K. M., and J. R. Fassett, in *Fluctuation Phenomena in Solids*, R. E. Burgess (Editor), Academic, New York, Chap. 7, 1965.
2. Dereniak, E. L., and D. G. Crowe, *Optical Radiation Detectors*, Wiley, New York, 1984.
3. AD745 Specification Sheet, Analog Devices Inc., 1991.
4. 1992 Amplifier Applications Guide, Analog Devices Inc., 1992.

PART III

DESIGNING FOR LOW NOISE

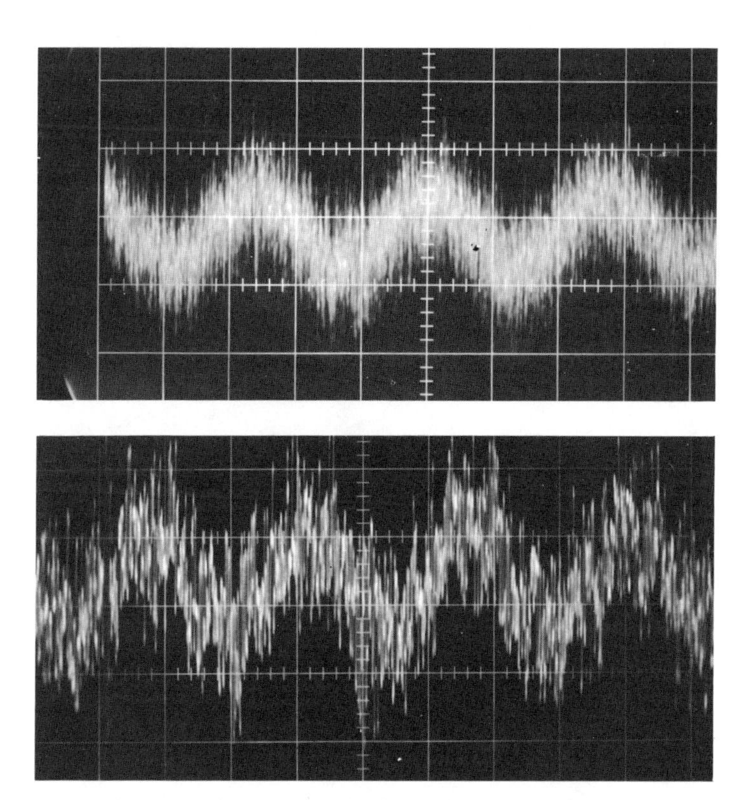

The signal-to-noise ratio is about unity in each photo. The top waveform is a sinusoidal signal in white noise; the bottom waveform is a sinusoidal signal in $1/f$ noise. White noise appears "furry" or "grassy" on an oscilloscope; $1/f$ noise appears "rough" and "jumpy."

CHAPTER 9

LOW-NOISE DESIGN METHODOLOGY

In this chapter we will deemphasize models and equations and provide an overview of low-noise *design*. We will give an overall perspective on the design process. This combines all of the material from the previous chapters plus years of design experience to provide some general guidelines on how to start the design of a low-noise system. Other chapters are very specific and mathematically rigorous, but this one generalizes to show the big picture, to give an overview.

We have discussed the noise mechanisms present in electronic devices, shown models to represent the noise behavior of those devices, and described the noise models for a wide variety of sensors. Furthermore, we have treated the methods of circuit analysis peculiar to noise problems. Now we turn our attention to system design. The information contained in the preceding chapters is used to create new low-noise systems that perform according to the design requirements.

Low-noise design from the system designer's viewpoint is concerned with the following problem: Given a sensor with known signal, noise, impedance, and response characteristics, how do we optimize the amplifier design to achieve the lowest value of equivalent input noise?

The amplifying portion of the system must be *matched* to the sensor. This matching is the *essence* of low-noise design.

9-1 CIRCUIT DESIGN

When designing an amplifier for a specific application, there are many specifications to be met and decisions to be made. These include gain, bandwidth, impedance levels, feedback, stability, dc power, cost, and, as expected, signal-to-noise requirements. Amplifier designers can elect one of two paths. Typically, they worry about the gain and bandwidth first and later in the design process they check for noise. We strongly urge the reverse approach, with initial emphasis on noise performance. Although there are many low-noise devices available, all do not perform equally for all signal sources. To obtain the optimum noise performance, it is necessary to select the proper amplifying device (FET, BJT, or IC) and operating point for the specific sensor or input source. Feedback and filtering can then be added to meet the additional design requirements. A critical noise specification can be one of the most serious limiting factors in a design; it is best to meet the most stringent requirement head-on.

To start the design, first select the input-stage device, discrete or IC, BJT, or FET. The operating point is then selected. If preliminary analysis shows that the noise specification can be met, a circuit configuration (CS, CD, or CG) can be selected and the amplifier designed to meet the remainder of the circuit requirements. Noise is essentially unaffected by circuit configuration and overall negative feedback. Therefore, the transistor and its operating point can be selected to meet the circuit noise requirements. Then the configuration or feedback can be determined to meet the gain, bandwidth, and impedance requirements. This approach allows the circuit designer to optimize for the noise and for the other circuit requirements independently.

After selecting a circuit configuration, analyzing it for nonnoise requirements may indicate that it will not meet all the specifications. If the bandwidth is too narrow, more stages and additional feedback can be added, the bias current of the input transistor can be increased, or a transistor with a larger f_T can be selected. Then noise can be recalculated to see if it is still within specifications. This iterating procedure ensures obtaining satisfactory noise performance and prevents locking in on a high-noise condition at the very start of the design. Once the design is frozen and you find that you have too much noise, it is difficult and very expensive to make minor changes that lower the noise.

Much of the work that the authors have done as consultants is to reduce the noise of a system in or near production. This is always a challenging task. It is always better and less expensive to spend the time on the initial design to obtain the desired noise than to fix it later during production. A noise specialist is very helpful during the early analysis and design phase in providing the optimum design concept.

9-2 DESIGN PROCEDURE

The ultimate limit of equivalent input noise is determined by the sensor impedance and the first stage Q_1 of the amplifier. The source impedance $Z_s(f)$ and noise generators $E_n(f)$ and $I_n(f)$ representing Q_1 are each a different function of frequency. Initial steps in the design procedure are the selection of the type of input device, such as BJT, FET, or IC, and the associated operating point to obtain the desired noise characteristic as described in Chaps. 5 and 6. In the simplest case of a resistive source, match the amplifier's optimum source resistance R_o to the resistance of the source or sensor. This minimizes the equivalent input noise at a single frequency as described in Chap. 2. If the amplifier is operated over a band of frequencies, the noise must be integrated over this interval. Since the noise mechanisms and sensor impedance are functions of frequency, the PSpice computer program is used to perform this integration. Use of the PSpice program for noise analysis is discussed in Chap. 4. The resulting analysis may indicate that there is too much noise to meet the requirements or it may suggest a different operating point. By changing devices and/or operating points, theoretical performance can approach an optimum. If the initial analysis indicates that the specifications cannot be met, then no amount of circuit design is going to reduce the equivalent input noise to an acceptable level.

The noise of the first stage *must* be low to obtain overall low system noise. The following stages cannot reduce noise no matter how good. Subsequent stages, however, can add noise so the design of these stages must be considered for a low-noise system. This is frequently a problem when designing high-frequency amplifiers. There is not enough gain–bandwidth in the first stage to provide high gain with the bandwidth needed. Then the noise of the following stages may contribute.

After selecting the input stage the circuit is designed. Set up the biasing, determine the succeeding stages (Chap. 10), the coupling networks, and the power supply (Chap. 13). Then analyze the total noise of the entire system, including the bias-network contributions to ensure that you can still meet the noise specifications. Finally, add the overall negative feedback to provide the desired impedance, gain, and frequency response. The noise models for noise in feedback amplifiers are derived in Chap. 3. Of course, these requirements should have been kept in mind while doing the initial noise design. Input impedance, frequency response, and gain are not basic to one type of transistor or to one particular operating point. Overall negative feedback can be used to increase or decrease input impedance or to broaden the frequency response and set the gain.

If it appears at this time that one or more of these circuit requirements cannot be met, then readjust the circuit to suit these needs and make another noise analysis to ensure that the design continues to meet the noise specifications. Continuing around the loop of noise analysis, circuit synthesis, and back to noise analysis will provide the best design.

The question can be asked, "In a low-level application do you always design the amplifier for the lowest possible noise?" The answer is "not necessarily." Often there is not much value in designing for a noise figure of less than 2 to 3 dB. A NF of 3 dB is equivalent to a signal-to-noise ratio of 1:1. In other words, the amplifier and sensor are each contributing equal noise, and the total system noise is $\sqrt{2}$ times either component. If, from supreme design effort, the amplifier noise is reduced to 1/10 of the source noise, the total system noise is now reduced by $1/\sqrt{2}$ or about 0.707 of the 3-dB condition. It is impressive to say that your amplifier has a NF of 0.3 dB, but really, in terms of the signal-to-noise ratio, only 30% is gained with a NF of less than 3 dB.

Applications such as cooled cryogenic sensors, reactive sources, and active systems do require very low equivalent input noise. For an active system consisting of a transmitter and receiver, a reduction of 3 dB in the receiver noise is equivalent to doubling the transmitter power. That *can* be significant.

Another difficult task is designing a system to operate from a reactive source with only a small resistive component. In this case a NF of 3 dB may not be meaningful. For example, with an inductance and small series resistance, thermal noise can be negligible, but in reality the system noise is dominated by $I_n X_L$ at the higher frequencies. *To obtain the best performance, do not design for minimum NF, but design for lowest equivalent input noise E_{ni} over the bandwidth of interest.*

9-3 SELECTION OF AN ACTIVE DEVICE

An active input device can be an IC with a bipolar or FET input stage or a discrete transistor. Selection depends primarily on the source impedance and frequency range. It is difficult to say exactly where each type of device should be used. To assist in decisions of this type, a general guide is shown in Fig. 9-1.

At the lowest values of source resistance, it is usually necessary to use transformer coupling at the input to match the source resistance to the amplifier R_o.

Bipolar transistors and bipolar input ICs are most useful at midrange impedances. Adjustment of R_o to match the source impedance is made by

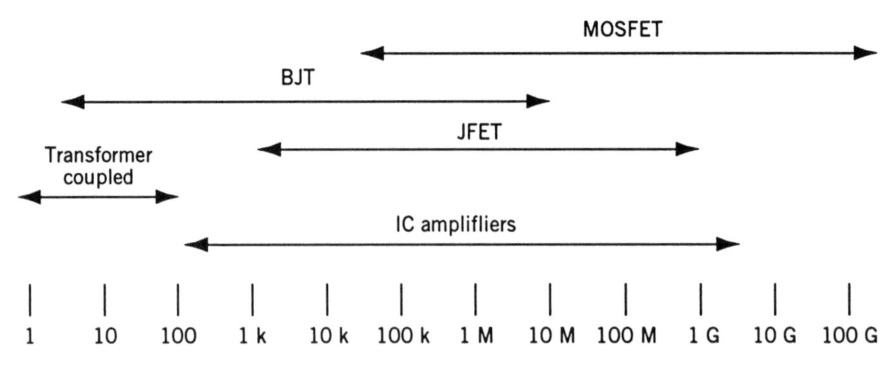

Figure 9-1 Guide for selection of active input devices.

changing the transistor collector current with higher currents for lower resistances as derived in Chap. 5. There is a slight difference between *pnp* and *npn* transistors; the *pnp* can have a lower base resistance due to the higher mobility of the *n*-type base region. The *pnp* thus has a lower thermal noise voltage and can be used with smaller source resistances. On the other hand, *npn* transistors often have a slightly larger β_o and f_T, and so are more useful at the higher end of the resistance range. This is especially true for IC *npn* transistors made in the standard bipolar process. Lateral and substrate integrated *pnp* transistors are almost always inferior to *npn* types.

At higher values of source resistance, FETs are more desirable because of their very low noise current I_n. In some instances, they are even preferred when a low E_n is desired. In fact, when operating with a very wide range of source resistances, such as in an instrumentation amplifier application, a JFET is generally preferred for the input stage. A typical JFET has an E_n slightly larger than that of a bipolar transistor, but its I_n is significantly lower. This is of particular value when operating from a reactive source over a wide frequency range because the source impedance is linearly related to frequency. Another advantage of the FET is its higher input resistance and low input capacitance; thus it is particularly useful as a voltage amplifier. FET input impedance is highly frequency sensitive because of the high input resistance.

For the highest source resistances the MOSFET with its extremely low I_n has an advantage. The MOSFET may have 10 to 100 times the $1/f$ noise voltage E_n of a JFET or BJT. As processing techniques have improved, the MOSFET is becoming more attractive as a low-noise device. The advantages of MOSFET devices include low cost and compatibility with the digital IC process. It is frequently desirable to combine a MOSFET input signal amplification stage with digital signal processing on the same chip.

Integrated amplifiers, ICs, are usually the first choice for amplifier designs because of their low cost and ease of use. There are many good low-noise IC devices available with guaranteed maximum noise specs. However, when selecting the one IC for your design, the source impedance and the transistor type must still be considered. All of the preceding statements about input stages apply when selecting the IC for your design. For lower source impedances, select a bipolar unit IC, and for higher impedances, select a FET input IC. In general, ICs will have slightly higher noise than discrete devices because of the extra processing and IC fabrication design constraints. Usually, an IC can be selected that will provide noise levels equal to a discrete transistor circuit at moderate impedances and frequencies. If state-of-the-art performance is needed, use a discrete BJT or JFET stage ahead of the IC. This is usually only required at the lowest and highest frequencies as well as very low or high impedances. In most ICs the circuit designer does not have control over the *Q*-point of the input device or devices. Therefore, the major consideration for designing with IC amplifiers is the type of device being used in the first stage.

9-4 DESIGNING WITH FEEDBACK

After determination of the input device, its operating point, and configuration, we can add overall multistage negative feedback to achieve the required input impedance, amplifier gain, and frequency response. As shown in Chap. 3, negative feedback *does not* increase or decrease the *equivalent input noise* except for the additional noise contribution of the feedback resistors themselves. Feedback *will* decrease the output noise and signal. In effect, single-stage negative feedback is utilized if the design employs either the common base–gate or common-collector–drain configurations because these stages have 100% current and voltage feedback, respectively. Improved performance is achieved by using negative feedback around an op amp and a discrete input stage.

If an amplifier with a low input resistance at the amplifier terminals is desired, use negative feedback to the inverting input of an op amp as shown in Fig. 9-2. The input resistance is reduced in proportion to the feedback factor, the ratio of the open-loop-to-closed-loop amplifier gains. This makes the amplifier inverting input a virtual ground. A low input resistance is required when an amplifier must respond to a signal current rather than voltage. Low input impedance is useful to minimize the effects of amplifier input shunting capacitance on the frequency response of the source resistance R_s.

An amplifier with high input impedance and reduced input capacitance results from the addition of negative feedback to the inverting input of an op amp as shown in Fig. 9-3*a*. A high impedance is needed when the amplifier must respond to the voltage signal from a high-impedance source without loading. This makes the best general-purpose instrumentation amplifier. FET input op amps offer the highest input resistance. The addition of a discrete common-source stage at the input can offer lower noise. Overall negative feedback to the source of a common-source input stage, as shown in Fig. 9-3*b*, raises the amplifier's input impedance.

Determining the input impedance of an amplifier with overall negative feedback can be confusing. If the input impedance is measured with the typical low-resistance impedance bridge, the measured values will agree well with the calculations. On the other hand, you can measure the input impedance by inserting a variable resistor in series with the input signal

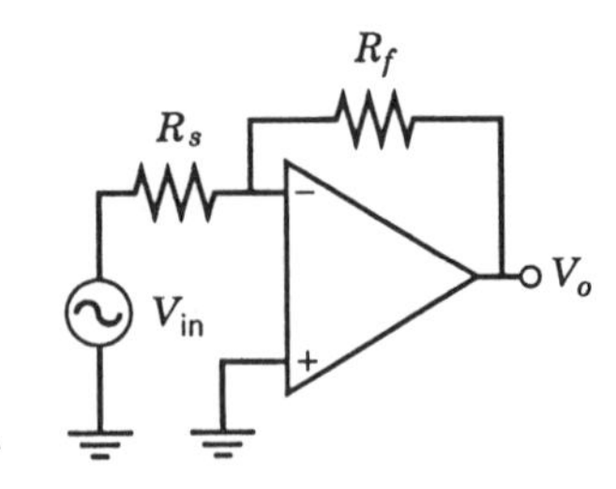

Figure 9-2 Low-input-impedance feedback amplifier.

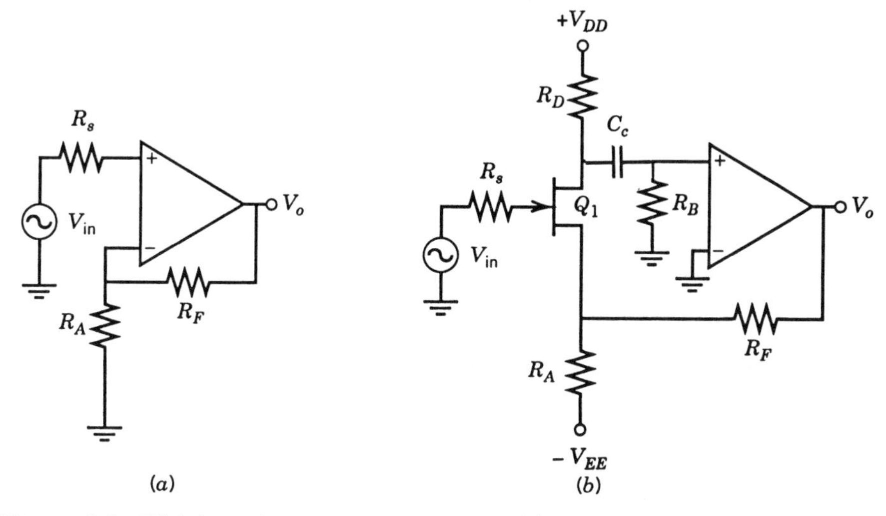

Figure 9-3 High-input-impedance amplifiers: (*a*) Noninverting amplifier and (*b*) CS stage added for higher input resistance.

source and increasing its value until the output signal halves. The input impedance then should equal the value of the series resistor. Using this method on a negative feedback amplifier often produces a lower measured value of input impedance than measured with the impedance bridge. The measuring resistor has become a part of the overall feedback network. Which value is correct? That depends on the source resistance. For low source resistances the bridge value is correct. For very large resistances, greater than the open-loop R_i, the second method is correct. If we do not simulate actual system conditions when making measurements of this type, the test results obtained can be in error.

9-5 BANDWIDTH AND SOURCE REQUIREMENTS

An important point to remember when designing low-noise amplifiers is not to overdesign for wide bandwidth. Be sure that the amplifier or system has definite low- and high-frequency roll-offs and that these are set as narrow as possible to pass only the signal spectrum required. The amplifier has a certain amount of noise in each hertz of bandwidth, and the greater the amplifier bandwidth the greater the output noise. There is no value in having an amplifier with a frequency response wider than the spectrum of the signal. In fact, it is detrimental to good low-noise design practices and principles.

The noise curves of discrete BJTs indicate that the noise spectral density is constant at the midfrequencies, but increases at both low and high frequencies. In general, increasing the transistor quiescent collector current increases the $1/f$ noise and decreases the high-frequency noise, so there is not one single collector current that always provides the best performance for all applications. To determine the best performance for a specific application, it is necessary to integrate the total noise over the frequency range of interest. Although it is possible to approximate the total noise, usually it is better to use a computer program with the exact circuit model as described in Chap. 4, especially if the source is reactive.

The impedance of an inductive source rises with frequency. This means that the amplifier E_n is important at low frequencies and I_n is important at high frequencies. For this type of application, either a BJT operating at a low value of collector current, or a FET having a low E_n at low frequencies, may be recommended. It was shown in Chap. 7 how the series inductance of the source influences the amplifier noise current I_n contribution to E_{ni}.

For a sensor that has shunt capacitance, the noise voltage E_n of the amplifier is the most critical as pointed out in Chap. 7. The best performance can be obtained with a bipolar transistor operated at a high collector current to minimize E_n. If the source resistance is large, it may be necessary to use a FET having a low-noise voltage at high frequencies.

A sensor that has both inductance and capacitance and is resonant in the passband can be a problem because of the opposite operating conditions required. In this case, it is difficult to generalize because of the wide range of impedance encountered over the frequency range. Since both low E_n and I_n are required, a good FET input would be suggested. As shown in Chap. 7, the sensor resonance decreases the effect of the amplifier noise voltage at resonance. This can be used to reduce the amplifier noise contribution at the resonant frequency, but can make signal conditioning more difficult because of the resonant peak. The best improvement is generally obtained if the resonance is placed at the high-frequency corner near cutoff. A current amplifier such as the one shown in Fig. 9-2 can be used for signal linearization. The low-input-impedance amplifier will shunt the tuned circuit input and linearize the output. Note this shunting by the negative feedback does not affect the noise model and the resonance will still minimize the equivalent input noise voltage.

Designing a low-noise amplifier to operate at high frequencies from a high source resistance requires a low value of I_n at the frequencies of interest, and, in addition, a low input capacitance to assure adequate frequency response. These requirements can be met with a FET operated in the source follower configuration or with an emitter follower operated at low collector current. The input capacitance can be further reduced by the use of overall negative feedback.

9-6 EQUIVALENT INPUT NOISE

To a system engineer, low-noise operation implies an acceptable signal-to-noise ratio at the system output terminals. To the designer, this means designing for the minimum equivalent input noise E_{ni}. In effect, E_{ni} is a normalized reciprocal of the signal-to-noise ratio referred to the point where the signal originates. The general expression for equivalent input noise E_{ni} for a resistive source was given in Eq. 2-7:

$$E_{ni}^2 = E_t^2 + E_n^2 + I_n^2 R_s^2 \tag{2-7}$$

where E_t is the thermal noise of the source resistance R_s, E_n and I_n are the equivalent input noise voltage and current generators of the amplifier without including the coupling elements.

In the case of a reactive source, we must deal with the total impedance of the source. Only the equivalent series resistive component of the impedance contributes thermal noise. Reactive components do not have thermal noise. The noise current generator I_n multiplies the absolute value of the total source impedance $|Z_s|$. Equation 2-7 thus becomes

$$E_{ni}^2 = E_t^2 + E_n^2 + I_n^2 |Z_s|^2 \tag{9-1}$$

where E_t is the thermal noise voltage of the real part of the impedance.

Graphs of the total equivalent input noise in terms of the noise voltage E_n, noise current I_n, and source resistance R_s for two IC amplifiers are plotted in Figs. 9-4 and 9-5. The values used for E_n and I_n can be either spot

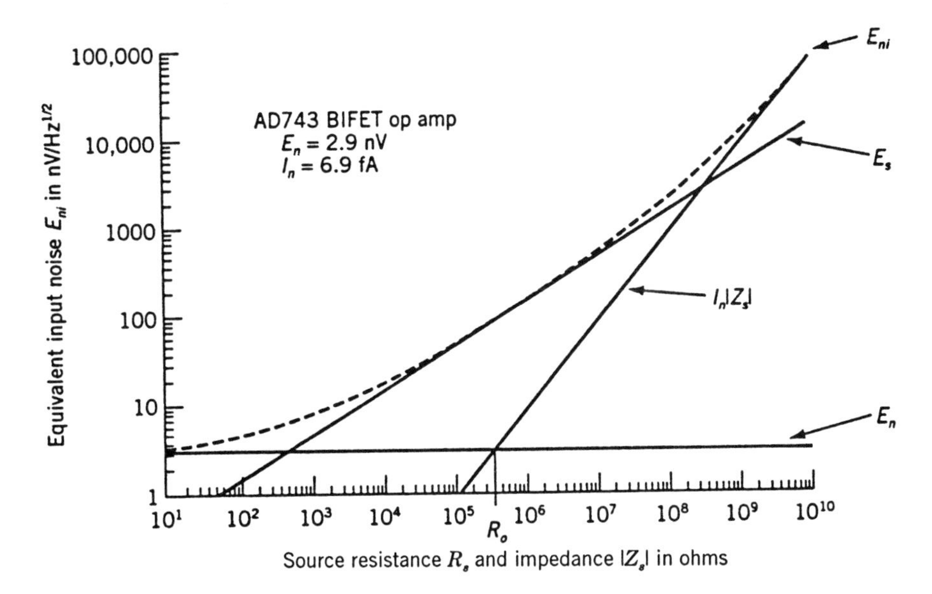

Figure 9-4 Plot of total equivalent input noise versus source resistance for an AD743 BIFET op amp.

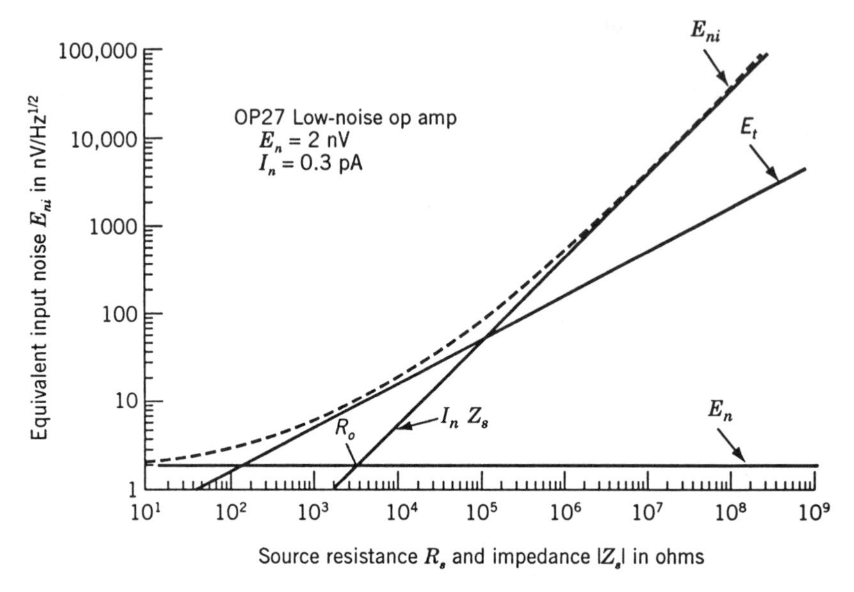

Figure 9-5 Equivalent input noise for an OP-27 low-noise bipolar op amp.

noise at a single frequency or total integrated noise over a bandwidth. From this curve the total equivalent input noise can be determined for any source impedance Z_s. Remember that the thermal noise line E_s is determined by the real part of the source impedance only. On the other hand, E_{ni} is dependent on the absolute value of the total source impedance.

From the plot of Fig. 9-4, we can see that the optimum source resistance is 420 kΩ but that good noise performance can be obtained from 600 Ω to 300 MΩ. This makes a versatile amplifier capable of operating from a wide range of source impedances. FET amplifiers generally have more $1/f$ noise voltage than bipolar amplifiers.

The OP-27 bipolar op amp is shown in Fig. 9-5. This amplifier has an optimum source resistance R_o of 6.7 k Ω and will operate well from 200 Ω to 200 kΩ.

When designing for the maximum signal-to-noise ratio and the source impedance is not specified, select a sensor having the highest internal signal-to-noise ratio with internal impedance equal to the amplifier optimum noise source resistance R_o. If the sensor resistance is much less than R_o, adding a resistor in series with the sensor does not reduce the equivalent input noise. In fact, it will increase E_{ni} because of the additional thermal noise voltage. Only when an increase in sensor resistance increases the signal proportionally should the total resistance value be adjusted, assuming you are the sensor designer also.

In the amplifier, design for maximum signal-to-noise ratio. For a sensor whose impedance is already determined, the values of E_n and $I_n Z_s$ are adjusted by active device selection so that the lowest E_{ni} can be obtained.

As an example, consider the E_{ni} curves on the two amplifiers shown in Figs. 9-4 and 9-5. With a 100-Ω source resistance, E_{ni} for the OP-27 is the lower. If the source resistance is 1 MΩ, the AD743 amplifier clearly gives the lower equivalent input noise. In between these resistances, both may work equally well. Methods for optimizing E_n and I_n were described in Chaps. 5 and 6.

9-7 TRANSFORMER COUPLING

To couple the sensor and amplifier in an electronic system, it is sometimes better to use a coupling or input transformer. When it is not possible to achieve the necessary NF using device selection, transformer coupling may be the solution. Very low resistance sources can cause this problem. According to Fig. 5-6, this source characteristic suggests that the BJT be operated at a collector current near 10 mA for low-noise operation. However, this high value of I_C is shown in Fig. 5-5 to result in relatively poor F_{opt}. If transformer coupling is employed, a lower value of I_C could be selected, and the noise performance improved.

The use of an input transformer between the sensor and amplifier improves the system noise performance by matching the sensor resistance with the amplifier's optimum source resistance R_o. Consider the secondary circuit as shown in Fig. 9-6 containing noise generators E_n and I_n. We have defined $R_o = E_n/I_n$. When these quantities are reflected to the primary as E_n' and I_n', we obtain from Eq. 8-9:

$$E_n' = \frac{E_n}{T} \quad \text{and} \quad I_n' = TI_n \tag{9-2}$$

The ratio of the reflected noise parameters is R_o':

$$R_o' = \frac{E_n'}{I_n'} = \frac{E_n}{T^2 I_n} = \frac{R_o}{T^2} \tag{9-3}$$

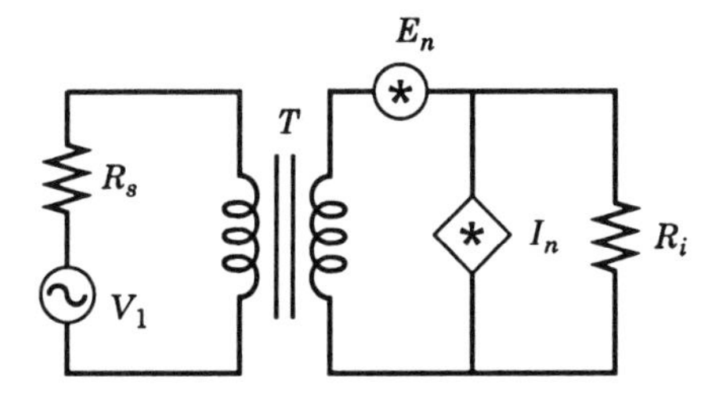

Figure 9-6 Amplifier noise generators located at secondary.

Now match R'_o to the source resistance R_s. Since we are free to select T, it follows that

$$T^2 = \frac{R_o}{R_s} \tag{9-4}$$

This ensures that the amplifier sees the optimum source resistance R_o to achieve the highest signal-to-noise ratio.

A more complete equivalent circuit of a transformer was shown in Fig. 8-13.

In addition to matching a source resistance to the amplifier's optimum source resistance, transformer coupling will also isolate portions of a system to reduce common-mode noise transfer and ground loops.

Coupling transformers are discussed in detail in Sec. 8-6, and the design of transformers is considered in Sec. 12-5.

9-8 DESIGN EXAMPLES

The design examples presented in this section apply the information presented here and in preceding chapters. The examples are not complete in all respects; however, they should prove to be of assistance in illustrating low-noise design procedures.

Example 9-1 Device Selection Procedure (Amplifier Comparisons)

Objective To select and compare first-stage active devices for an amplifier operating at 10 kHz from a resistive source of 100 kΩ.

Solution Figure 9-1 indicates that an IC, a BJT, or a JFET could be used for this application. We consider each of these devices in this example.

The μA741 op amp is a likely candidate for this job. According to the curves in App. A, $E_n = 20\ \text{nV/Hz}^{1/2}$ and $I_n = 0.4\ \text{pA/Hz}^{1/2}$ at a frequency of 10 kHz. The optimum source resistance is $R_{\text{opt}} = E_n/I_n = 50\ \text{k}\Omega$. Assuming a $\Delta f = 1$ Hz at 10 kHz, the noise figure can be obtained from

$$\text{NF} = 10 \log \frac{E_{ni}^2}{E_t^2} = 10 \log \frac{E_t^2 + E_n^2 + I_n^2 R_s^2}{E_t^2} \tag{9-5}$$

where

$$E_{ni}^2 = (1.61 \times 10^{-20} \times 10^5) + (20 \times 10^{-9})^2 + (0.4 \times 10^{-12} \times 10^5)^2$$
$$= 3.61 \times 10^{-15}\ \text{V}^2 \tag{9-6}$$

For the μA741 op amp, we find that $E_{ni} = 60$ nV and NF = 3.5 dB.

Now let us consider using a BJT. The curve of NF versus dc collector current I_C shown in Fig. 5-6 for the 2N4250 transistor, which is typical for most BJTs, indicates that a minimum noise figure occurs at a specific collector current. Fortunately, achieving low NF is possible over a relatively broad range of collector currents in the vicinity of 5 μA. From the contours of constant NF given in App. C, three transistors would be good design choices because they have large 1-dB noise contours for 100 kΩ at 10 kHz. These are the *npn* types 2N4124 and MPS-A18 and the *pnp* 2N4250 transistor. Assuming that other circuit design considerations make using an *npn* transistor easier, we select the 2N4124 device. The 10-kHz noise contour indicates that for $R_s = 100$ kΩ, the bias current, I_C, for minimum noise should be chosen to be approximately 2 μA. The resulting NF is less than 1 dB. Extrapolating between the 1- and 10-μA curves for the 2N4124 suggests that $E_n = 12$ nV/Hz$^{1/2}$ and $I_n = 0.12$ pA/Hz$^{1/2}$ for an $R_{\text{opt}} = 100$ kΩ matching the given source resistance. Repeating the calculations of Eqs. 9-5 and 9-6 gives $E_{ni} = 43.5$ nV/Hz$^{1/2}$ and a NF of 0.718 dB.

Lastly we consider this design using an *n*-channel JFET. From the noise data in App. B we select the 2N3821 as a good choice. Its noise data for I_{DSS} is $E_n = 4$ nV/Hz$^{1/2}$ and $I_n = 3$ fA/Hz$^{1/2}$ for an $R_{\text{opt}} = 1.33$ MΩ. We calculate as before, obtaining $E_{ni} = 40.32$ nV/Hz$^{1/2}$ and a NF of 0.043 dB.

The JFET gives the lowest E_{ni} and lowest NF. The 2N4124 is also an acceptable design choice since its noise figure is much less than 3 dB and the dominant noise source is the thermal noise of the source resistance. The μA741 op amp is not a good choice because its added noise exceeds that of the source resistance. As in all designs, other factors will need to be considered, such as the amount of noise tolerable and the component, manufacturing, and testing costs. These factors taken together are required to determine lowest system cost for acceptable performance.

Example 9-2 Transformer-Coupled Amplifier

Objective To design an amplifier for a sensor test instrument that measures signal and noise performance at 100 Hz for sensors with a source resistance varying between 5 and 500 Ω.

Solution From Fig. 9-1 we conclude that it is usually desirable to use an input transformer with such low values of source resistance. Then a transistor can be selected for operation at its minimum NF, and the transformer impedance designed to match the source resistance to the R_{opt} of the transistor.

First we examine the noise contours for the BJTs given in App. C. Here we see that good noise performance is easily achieved if we have an equivalent source resistance of 10 kΩ. The turns ratio of the transformer makes the sensor resistance look like R_{opt}. Since 5 $\Omega \leq R_s \leq$ 500 Ω, we design our transformer based upon the geometric mean value of 50 Ω as a compromise. The turns ratio determined from Eq. 9-4 is

$$T^2 = \frac{R_{\text{opt}}}{R_s} = \frac{10\ \text{k}\Omega}{50\ \Omega} = 200 \tag{9-7}$$

$$T = \sqrt{200} = 14.14 \tag{9-8}$$

To specify the coupling transformer completely, additional information is necessary. The primary dc resistance plus reflected secondary resistance should be much less than the smallest source resistance, 5 Ω. The primary resistance can be about 0.1 Ω and the secondary resistance about 20 Ω. To minimize the noise contribution of the shunting core loss resistance, R_c should be much greater than the largest source resistance, 500 Ω. A value of 10 kΩ would certainly be satisfactory. Both frequency response and noise dictate that the primary inductance L_p must be much larger than the 500-Ω source resistance at the lowest frequency of interest, namely, 100 Hz. This requirement is met if the primary inductance is greater than 16 H. This inductance can be achieved in shielded transformers having a volume of a few cubic inches.

Next we choose the best BJT for this application. We review and tabulate the size of the noise contours at 10 Hz over the range of reflected source resistances from 1 to 100 kΩ:

Device	Noise Contour
2N930	Small 3-dB
2N4124	Large 3-dB for high and medium R_s
2N4125	Small 1-dB
2N4250	Large 1-dB for high and medium R_s
2N4403	Large 3-dB
2N5138	Small 3-dB
MPS-A18	Medium 1-dB for high R_s

From this comparison it is easy to see why we select the type 2N4250. The NF contour at 100 Hz suggests that choosing $I_C = 30\ \mu\text{A}$ will position us near the center of the 1-dB noise contour and allow for the variation in source resistance. At the selected dc operating point, $E_n = 3\ \text{nV/Hz}^{1/2}$, $I_n = 0.3\ \text{pA/Hz}^{1/2}$, and $R_{\text{opt}} = 10\ \text{k}\Omega$.

The reflected noise voltage of the transistor is divided by the turns ratio in order to refer it to the primary, or sensor, side of the transformer. The current is multiplied. This results in $E_n' = 0.21\ \text{nV/Hz}^{1/2}$ and $I_n' = 4.2\ \text{pA/Hz}^{1/2}$.

If we assume that the transformer does not contribute noise, the following summary is valid for the three values of sensor resistance:

R_s	5 Ω	50 Ω	500 Ω
R_2	1 kΩ	10 kΩ	100 kΩ
E_t	0.284 nV/Hz$^{1/2}$	0.89 nV/Hz$^{1/2}$	2.82 nV/Hz$^{1/2}$
E_{ni}	0.355 nV/Hz$^{1/2}$	0.946 nV/Hz$^{1/2}$	3.55 nV/Hz$^{1/2}$
NF	1.94 dB	0.460 dB	1.96 dB

SUMMARY

a. Low-noise design implies attaining a low value for the equivalent input noise E_{ni}. Since

$$\frac{S_o}{N_o} = \frac{V_s^2}{E_{ni}^2}$$

minimizing E_{ni} maximizes S_o/N_o.

b. The noise specification is often the most important and may be the most difficult to meet. A NF below 3 dB may be unnecessary.

c. A recommended low-noise design procedure with sensor characteristics fixed is as follows:
   1. Select the input device and configuration (Chap. 9).
   2. Decide on an operating point (Chaps. 5 and 6).
   3. Design the bias circuit (Chap. 10).
   4. Determine the succeeding stages (Chap. 12).
   5. Consider the coupling (Chap. 9).
   6. Determine the power supply requirements (Chap. 13).
   7. Select the location of the frequency equalization (Chap. 9).
   8. Determine the overall feedback (Chap. 3).

   An iterative analysis–synthesis procedure determines the final system. Digital computer assistance is valuable.

d. The input devices considered are the BJT (conventional or SBT), the FET (JFET or MOSFET), and the IC (BJT or FET input stage).

e. Because it can transform impedance levels, a coupling transformer can be used with low-resistance sensors.

PROBLEMS

9-1. Determine the R_{opt} and the E_{ni} for the 2N4221A Motorola JFET whose characteristics are given in App. B. The source resistance is 1 MΩ, the operating frequency is 100 Hz, and the device is biased at I_{DSS}.

9-2. Find E_n and I_n for the MPS-A18 transistor at a frequency of 1 kHz, $R_s = 50\ \mathrm{k}\Omega$, and $I_C = 1$ mA. Also find R_{opt}, E_{ni}, and NF.

9-3. Find the turns ratio required to couple a 200-Ω source resistance to a transistor with $R_{\mathrm{opt}} = 5\ \mathrm{k}\Omega$. What is the signal level at the transformer input terminal if the sensor signal is 2 mV?

9-4. A given sensor has a series resistance of 30 kΩ. (a) Select an operating point for the 2N4403 BJT transistor from App. C. Then determine the wideband equivalent input noise of your design for the frequency band from 10 Hz to 100 kHz. Find the wideband NF over this band and then determine the NF in 1-Hz bandwidths centered at 10 Hz, 100 Hz, 1 kHz, 10 kHz, and 100 kHz. (b) Repeat for the MPS-A18 transistor.

CHAPTER 10

AMPLIFIER DESIGN

Previous chapters have primarily addressed device modeling. Now we direct our attention to the use of these devices in practical applications. To use a transistor as an amplifier, oscillator, filter, sensor, or any other application requiring an active device, it is necessary to add other active and passive elements for biasing. These biasing elements, such as resistors, can add thermal and excess noise to the circuit operation. Current sources and diodes can add shot, thermal, and excess noise. Capacitors as series or shunt elements can increase noise while decoupling capacitors bypass noise.

The active device models previously described apply to integrated amplifiers as well as discrete amplifiers. Often the noise contribution of active biasing elements also must be modeled in terms of the transistor noise model. Passive biasing and load resistances often contribute less noise than active devices but usually require additional processes and extra chip area.

All but the lowest-noise integrated amplifiers use differential stages with primarily active biasing. In this chapter the discussion will begin with single-ended amplifiers and discrete stages followed by a derivation of the noise of the differential amplifier stage. Single-ended amplifiers will inherently have 3 dB less noise than the same transistor used in the best differential configuration. For the ultimate noise performance a single-ended transistor input must be used. To characterize the performance of an IC process, it is often necessary to include discrete test transistor devices on the chip. Discrete devices are also added to a chip for modeling purposes to obtain the noise parameters of the process. The following discussion on single-stage biasing will aid in the testing and evaluation of these devices.

This chapter includes an element-by-element discussion of the additional noise contributions of the biasing elements to the equivalent input noise. First, a mathematical expression is derived; then a design example is given. PSpice simulations are used to illustrate the method for analyzing a specific circuit application. Noise contributions of all circuit elements can be calculated directly with PSpice as shown in Chap. 4.

10-1 TRANSISTOR CONFIGURATIONS

There are three useful connections in which bipolar and field effect transistors can be operated: common emitter (CE) or common source (CS), common base (CB) or common gate (CG), and common collector (CC) or common drain (CD).

Each configuration offers approximately the same power gain–bandwidth product and the same equivalent input noise. Therefore, configuration selection is based on the requirements for the gain, frequency response, and impedance levels of the amplifying system. The CE and CS configurations are used most often because they provide the highest power gain; the other configurations have advantages for certain applications. When a current amplifier is desired with high input impedance and low output impedance, a CD or CC stage can be used. Conversely, when a voltage amplifier is required with low input impedance and high output impedance, a CG or CB stage is a solution.

Because noise is not dependent on configuration, the circuit designer has the option of low-noise operation with the simultaneous freedom to select terminal impedance levels and other nonnoise characteristics. Although parameters such as E_{ni}, E_n, and I_n are usually measured in the CE–CS configuration for convenience, their values apply equally well to the CB–CG and CC–CD orientations, provided that frequencies are limited to those for which the collector–base internal feedback capacitance can be neglected. The equivalent input noise is the same for each configuration; but the output noise is not the same.

In the following derivations, an expression for the transfer gain K_t and equivalent input noise E_{ni} are derived. By referring all noise sources to the input, E_{ni} allows us to determine the noise contribution of the various passive elements required for biasing and coupling, and the effect of those elements on the transistor noise parameters E_n and I_n. Although the limiting value for E_{ni} does not vary because of the configuration selected, E_{ni} is increased by biasing and coupling element noise in certain connections.

Additional information regarding the gain and terminal impedance properties of transistor circuits is available in texts devoted to transistor electronics [1].

In order to simplify the symbols necessary to represent electrical quantities in the complex circuits given in this chapter, some deviations from previously used nomenclature are made.

Symbols for the thermal noise voltages of resistances differ from the system noted in the Chap 1. Here we use E_A, E_B, and so forth to represent the rms values of noise voltages in resistors R_A, R_B, and so forth. Normally, a capital letter with a capital subscript stands for a dc value. However, as mentioned earlier, E represents a noise voltage to distinguish it from a sinusoidal signal represented by V. When the reader encounters E, noise voltage is being discussed.

Where more than one transistor is used, a numerical subscript indicates the stage number. Thus E_{n2}^2 is the mean square value of the noise voltage generator representing stage 2. Symbol R_{i2} represents the input resistance of stage 2. All noise generators are considered to be uncorrelated.

In the performance summaries given with each circuit design example, E_{nT} and I_{nT} are used to represent the total input noise voltage and current. All noise parameters are given on a per hertz$^{1/2}$ basis. Capacitors are selected to provide a minimal noise contribution at 10 Hz.

10-2 COMMON-EMITTER STAGE

A typical CE amplifier is shown in Fig. 10-1. This circuit is biased for low-noise operation from 10 Hz to 10 kHz. The performance parameters are shown in the table at the side of the figure.

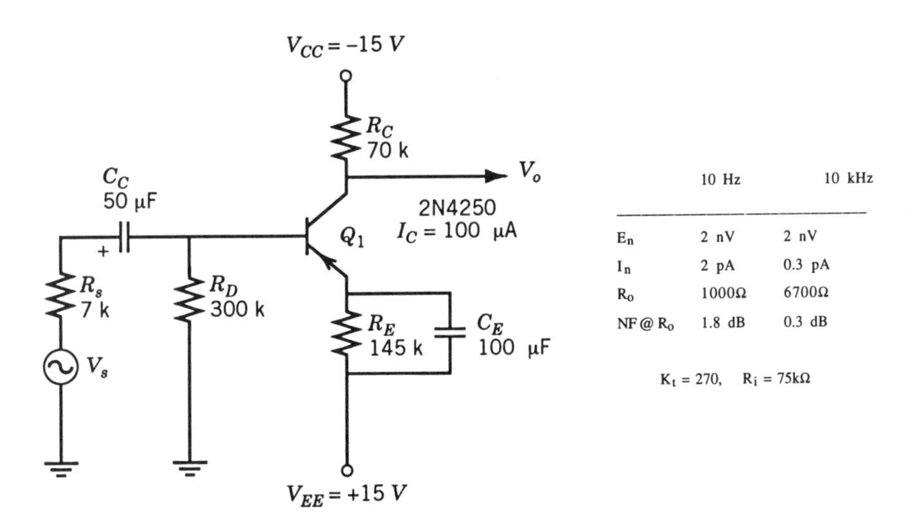

	10 Hz	10 kHz
E_n	2 nV	2 nV
I_n	2 pA	0.3 pA
R_o	1000Ω	6700Ω
NF @ R_o	1.8 dB	0.3 dB

$K_t = 270$, $R_i = 75$kΩ

Figure 10-1 Biased common-emitter amplifier.

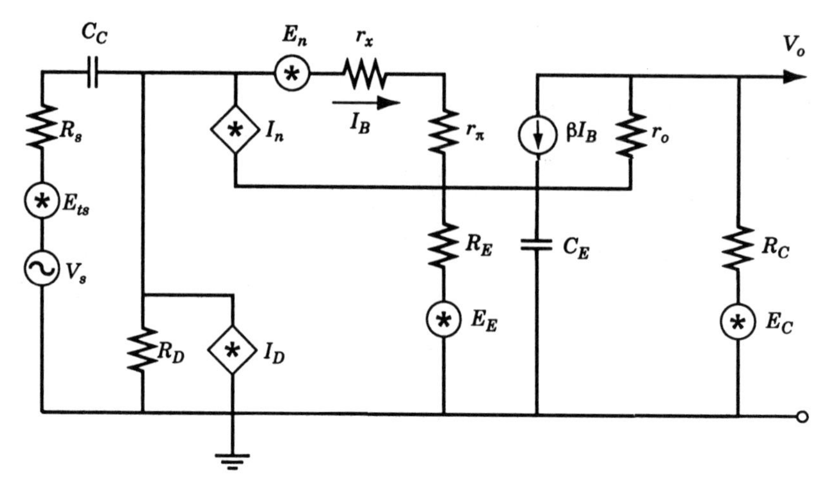

Figure 10-2 Noise and small-signal equivalent circuit for CE stage.

The small-signal equivalent circuit with noise sources is shown in Fig. 10-2. This circuit uses the hybrid-π BJT transistor model developed in Chap. 5 and adds the noise sources for the biasing elements.

The transfer voltage gain K_t is the ratio of the output signal voltage at the collector to V_s:

$$K_t \cong \left(\frac{-\beta R_L}{Z_i}\right)\left(\frac{Z_i \parallel R_D}{Z_s + Z_i \parallel R_D}\right) \tag{10-1}$$

The symbols used in deriving Eq. 10-1 are

$$Z_i = r_x + r_\pi + (\beta + 1)Z_E$$

$$R_L \cong R_C \parallel r_o \parallel R_{i2}$$

$$Z_E = R_E \parallel -jX_E$$

$$Z_s = R_s - jX_C$$

For noise analysis purposes our interest lies with the magnitude of K_t.

When there is negligible signal loss in biasing, coupling, and feedback elements, Eq. 10-1 simplifies to

$$K_t \cong -\frac{\beta R_L}{Z_s + r_x + r_\pi + \beta Z_E} \tag{10-2}$$

Under the assumptions that $Z_s \ll r_\pi$ and $Z_E \ll r_e$, the equation can be written in the classic form

$$K_t \cong -\frac{R_L}{r_e} = -g_m R_L \tag{10-3}$$

For simplicity, it is convenient to define an additional gain K_t' which removes the loading of following stages:

$$K_t' = K_t \quad \text{for} \quad R_L = R_C \tag{10-4}$$

The equivalent input noise E_{ni}^2 is determined from $E_{no}^2/(K_t')^2$. The result is

$$E_{ni}^2 \cong E_{ns}^2 + E_n^2\left(\frac{R_s + R_D}{R_D}\right)^2 + I_n^2(R_s - jX_C)^2 + I_D^2 R_s^2 + \frac{E_E^2}{1 + (\omega R_E C_E)^2} + \left(\frac{E_C}{K_t'}\right)^2 \tag{10-5}$$

Several observations can be made from Eq. 10-5. It is clear that the noise from each resistor increases E_{ni}. The transistor current noise generator I_n is increased by the reactance of the coupling capacitor C_C. The noise voltage of an unbiased emitter resistor directly adds to the input noise, so any resistor used for negative feedback must be kept small. An unbiased emitter resistor can be used to stabilize the stage gain; it must be kept much smaller than the source R_s to avoid adding noise. The voltage E_n is increased by the shunting effect of R_D, so R_D should be much larger than R_s for minimum noise. If R_D is made equal to the source resistance for impedance matching, the NF will be increased by 3 dB. The biasing resistor R_D generates thermal noise current in parallel with I_n, so it is important to keep R_D as large as possible without causing excess dc voltage offset ($V_{os} = I_B R_D$). A path for dc base current is always necessary. If it is permissible to pass the input bias current through the signal source R_s, then remove both C_C and R_D which will remove these two contributions to noise.

Be sure to bypass the power supply leads to reduce the noise of the power supply and any pickup on the supply lines. Because the CE connection provides the highest power gain, an amplifier with a CE input stage is not likely to have significant noise contributions from stages beyond the first. The input resistance is highly dependent on I_C. It turns out that the input resistance is also $R_o \beta_o^{1/2}$, which is always larger than the optimum source resistance R_o. This can be derived using Eq. 5-4 and Eq. 5-33.

10-2-1 Capacitor Selection

The bypass and coupling capacitors shown in Fig. 10-1 attenuate low-frequency signals and noise. The selection of values for these elements is the subject of this section.

The coupling or blocking capacitor C_C causes a low-frequency break in the response of the network at the value of f where its reactance is equal to the sum of R_s and the parallel combination of the amplifier input resistance and any biasing elements.

The noise requirement for C_C is more stringent; X_{CC} must be much less than R_s at the lowest frequency of interest. The element C_C is in series with R_s, and therefore the noise term $I_n R_s$ becomes $(I_n|R_s + jX_{CC}|)$. For low noise, $I_n X_{CC}$ must be much less than $I_n R_s$. If C_C sets the gain corner, $I_n X_{CC}$ will be a significant noise contribution. The problem is worse because I_n may be increasing with $1/f$ noise at low frequencies.

To remedy the problems noted in the preceding paragraph, C_C is often 100 times larger than the value defined by the gain criteria. Thus the input coupling capacitor *should not* be used as part of the frequency-shaping network. In fact, *if the input network is used as a filter, it will probably be the dominant source of noise in the system.*

The element C_E is used to provide a low-impedance bypass to ac so that the resistor R_E is effective only in the dc network. To maintain ac gain, the impedance of C_E must be low compared to r_e. Additional discussion of this cause of gain fall-off is given in the literature [1].

To meet the noise specification, C_E must effectively bypass the noise of R_E. Since the C_E–R_E network is effectively in series with the signal source, it is desired that

$$\frac{E_{tE}^2 + E_{xE}^2}{1 + \omega^2 R_E^2 C_E^2} \ll E_s^2 \tag{10-6}$$

where E_{tE} and E_{xE} are the thermal and excess noise voltages of R_E.

It is recommended that the low-frequency gain corner be set in a later stage rather than the input stage. In this manner $1/f$ noise originating in the input stage can be attenuated by the response-shaping network. When following this philosophy, the capacitance C_E must be very large and must satisfy Eq. 10-6.

10-3 SINGLE-SUPPLY COMMON-EMITTER STAGE

An amplifier stage can be operated with a single power supply by providing a virtual ground at node A as shown in the circuit of Fig. 10-3. This example is included to point out the noise effects of single-supply circuits. The equivalent input noise E_{ni} now includes the noise effects of R_A and R_B which are attenuated by capacitor C_B:

$$E_{ni}^2 \cong E_{ns}^2 + E_n^2\left(\frac{R_s + R_D}{R_D}\right)^2 + I_n^2(R_s - jX_C)^2$$

$$+\left[\frac{E_A^2}{1 + (\omega R_A C_B)^2} + \frac{E_B^2}{1 + (\omega R_B C_B)^2} + E_D^2\right]\frac{R_s^2}{R_D^2} + \frac{E_E^2}{1 + (\omega R_E C_E)^2} + \left(\frac{E_C}{K_t'}\right)^2 \tag{10-7}$$

The resistors R_A and R_B make a divider to establish a dc biasing voltage V_A at node A:

$$V_A = \frac{R_A V_{CC}}{R_A + R_B} \tag{10-8}$$

This is balanced by the voltage drop across R_E to bias the transistor. The bias resistor R_D must meet the noise criteria established in Sec. 10-2.

The bypass capacitor C_B ac grounds node A between R_A and R_B, thereby attenuating the noise generated in those resistances. Excess noise is generated in the divider resistors because of the dc drop. The noise voltages generated in R_A and R_B that appear across C_B are attenuated by $1/\omega R_A C_B$

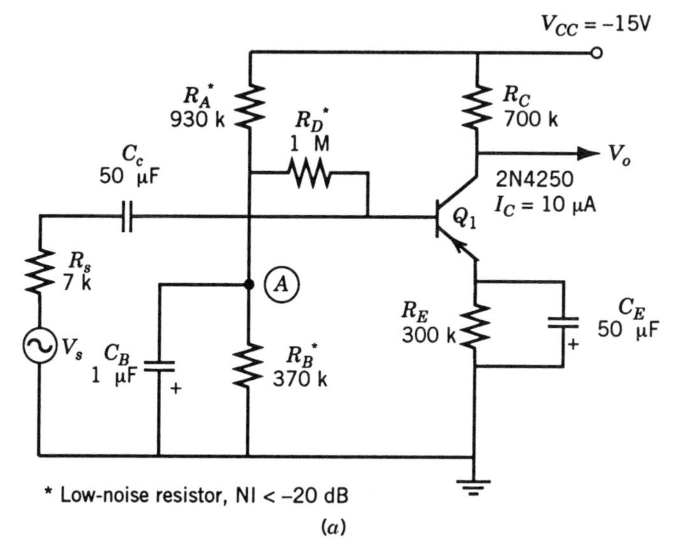

Figure 10-3 CE stage with a single power supply: (*a*) circuit diagram and (*b*) noise and small-signal equivalent circuit.

	10 Hz	10 kHz
E_n	4.5 nV	4.5 nV
I_n	0.3 pA	0.1 pA
R_o	10 kΩ	45 kΩ
NF at R_o	0.68 dB	0.35 dB

$K_t = 280$, $R_i = 780$ kΩ

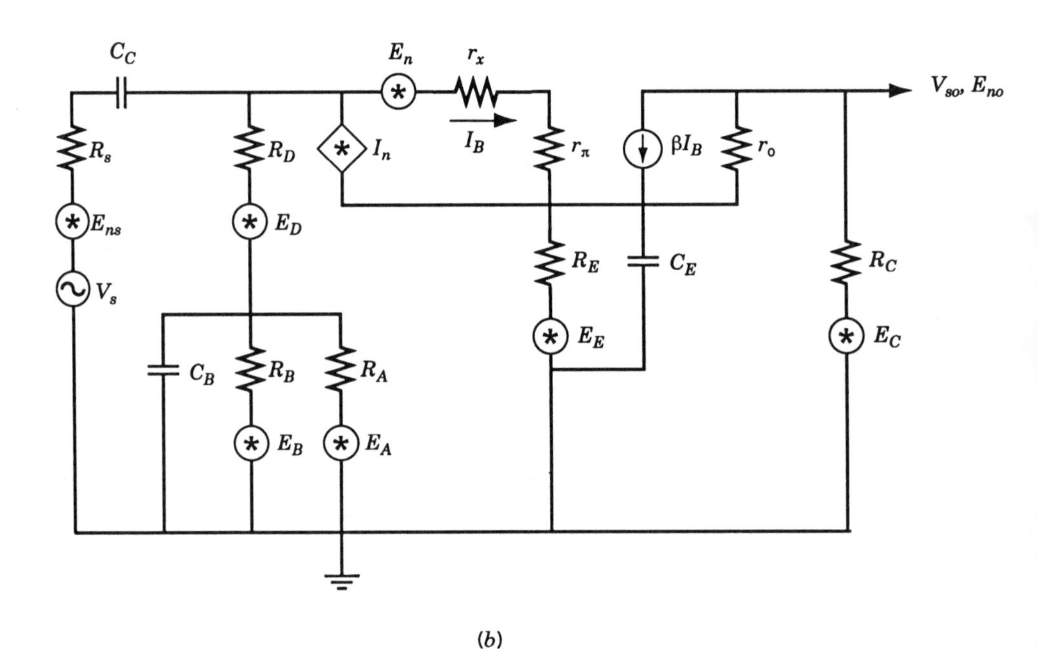

(b)

Figure 10-3 *(Continued)*.

and $1/\omega R_B C_B$. This noise across C_B is in series with resistance R_D. We desire that noise across C_B be negligible so that

$$E_{tD}^2 \gg \frac{E_{tA}^2 + E_{xA}^2}{1 + \omega^2 R_A^2 C_B^2} + \frac{E_{tB}^2 + E_{xB}^2}{1 + \omega^2 R_B^2 C_B^2} \tag{10-9}$$

where E_{tD} is the thermal noise of R_D.

A virtual ground can also be provided by several active circuits. The simplest of these is the Texas Instruments TLE 2425 [2]. This integrated circuit provides a low-impedance voltage at half V_{CC}. The output noise is high enough to require filtering. Circuits such as complementary transistors or an emitter follower can also be used to obtain a reference voltage at half V_{CC}.

10-4 NOISE IN CASCADED STAGES

Amplifier stages are cascaded to obtain the desired gain, impedance characteristics, frequency response, and power level. Although the noise of the first stage of a cascaded amplifier is usually dominant, subsequent stages can also contribute noise. In performing a system design, the noise of each stage, including biasing elements, must be considered.

A general block diagram of an amplifying system is shown in Fig. 10-4. The forward gain blocks are symbolized by K_{t1}, and so on. Noise in each stage is represented by E_n and I_n generators.

The equivalent input noise of the first stage alone is $E_{ns}^2 + E_{n1}^2 + I_{n1}^2 R_s^2$. The equivalent noise of the other stages can be determined in the same manner, by summing the E_n and $I_n R_s$ terms. For stages beyond the first, the effective source resistance is the output resistance of the preceding stage. Sum the total noise E_{no} at the output and divide by the total voltage gain of the system to determine the equivalent input noise E_{ni} of the system

$$E_{ni}^2 = E_{ns}^2 + E_{n1}^2 + I_{n1}^2 R_s^2 + \frac{E_{n2}^2 + I_{n2}^2 r_{o1}^2}{K_{t1}^2} + \frac{E_{n3}^2 + I_{n3}^2 r_{o2}^2}{K_{t1}^2 K_{t2}^2} + \cdots \quad (10\text{-}10)$$

where r_{o1} and r_{o2} are the dynamic output resistances of stages 1 and 2, respectively.

The very nature of a system of cascaded stages implies that the problem of noise analysis is complex. We have three tools at our disposal. Given a system, first we can perform a noise analysis using network analysis to derive performance relations. Second, we can utilize a personal computer and

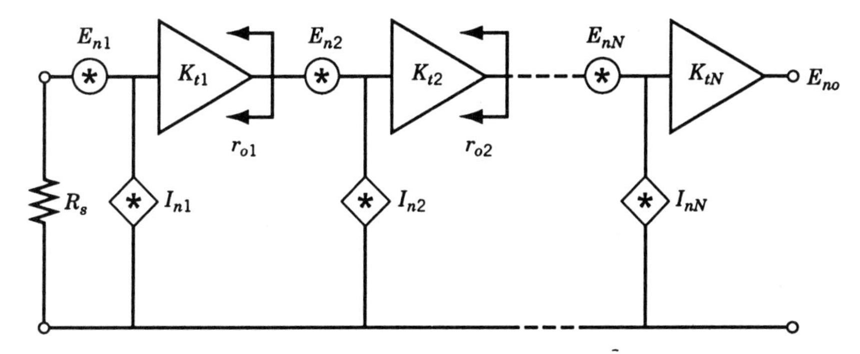

Figure 10-4 Cascade stage amplifier.

PSpice to determine numerical values for noise quantities. Third, we can resort to an experimental analysis of the system. Often the design engineer uses all three techniques in the pursuit of knowledge about a system.

It may be of value at this point to suggest a "trick" useful in the analysis of complicated systems. When making a computer analysis, the effect on the output noise of a noise source at any location internal to the system can be demonstrated by inserting a noise voltage (or current) generator at that location and by calculating the output noise from that generator alone. The generator can have a test value of 1 V or 1 A, for example, and can be a function of frequency as well. The same technique can be used in the laboratory. The effect of noise in any component can be accentuated by inserting a signal generator in series or in parallel with the component. The generator signal must override circuit noise in order to permit easy evaluation of its effect, but it must not be large enough to overdrive the active devices. This technique is useful, for example, when one seeks the noise contribution of a Zener diode used in a power supply or as a reference.

Integrated amplifiers usually are made up of a number of stages to provide voltage or current gain as well as impedance or voltage level transformations. Although these "stages" can be analyzed individually, some of the two-stage combinations are so common that they are considered as a single unit. In the following sections we analyze the CS–CE, CC–CE, and CE–CB, or cascode, circuits.

10-5 COMMON-SOURCE – COMMON-EMITTER PAIR

In the common-source (CS) configuration, the field effect transistor (FET) stage provides high voltage gain along with high input impedance. An example of this connection is the CS–CE pair circuit shown in Fig. 10-5. This

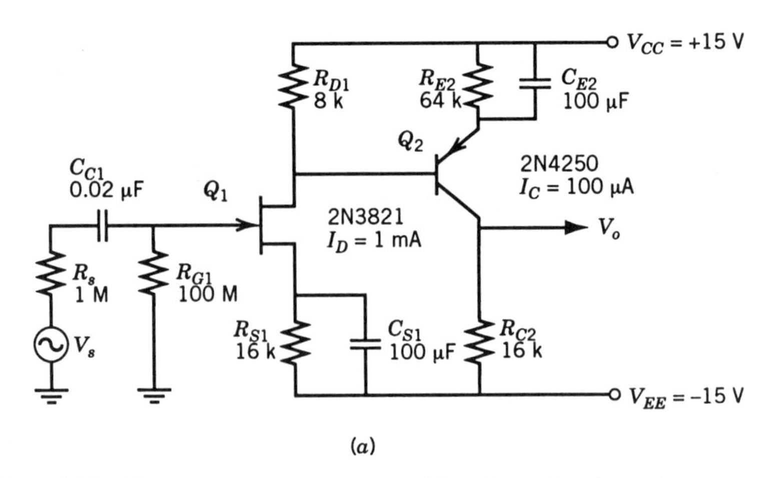

(*a*)

Figure 10-5 Common-source–common-emitter stages showing noise sources.

	10 Hz	10 kHz
E_n	8 nV	4 nV
I_n	7 fA	7 fA
R_o	1.1 MΩ	570 kΩ
NF at R_o	0.03 dB	0.015 dB
K_{t1} = 16 (FET only)		R_i = 100 MΩ
K_{tc} = 2300		

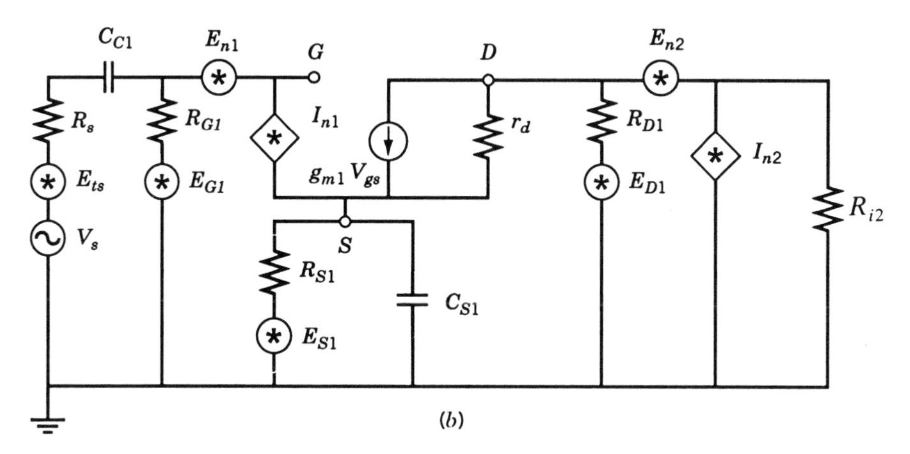

Figure 10-5 *(Continued)*.

example uses a junction FET but it applies to the MOSFET as well. The voltage gain provided by the CS stage is

$$K_{t1} \cong -\frac{g_{m1}R_{L1}}{1 + g_{m1}Z_{S1}} \tag{10-11}$$

where

$$R_{L1} = R_{D1} \parallel r_d \parallel R_{i2}$$

$$Z_{S1} = R_{S1} \parallel -jX_{S1}$$

For the pair, the overall midband gain from V_s to the collector of Q_2 is

$$K_{tc} \cong \frac{g_{m1}R_{L1}\beta R_{C2}}{(1 + g_{m1}Z_{S1})(r_{x2} + r_{\pi 2})} \tag{10-12}$$

For $r_{\pi 2} \gg R_{D1}$ and $R_{C2} \gg r_{o2}$, K_{tc} reduces to

$$K_{tc} \cong \frac{g_{m1} R_{D1} R_{C2}}{r_{e2}} \tag{10-13}$$

To minimize the E_{n1} nose contribution of the FET, a large drain current I_{D1} is used as explained in Chap. 6. A large I_{D1} requires a lower value of R_{D1} and gives a lower gain than obtained with the typical BJT stage.

The expression for the equivalent input noise is

$$\begin{aligned} E_{ni}^2 = E_{ns}^2 + E_{n1}^2 \left(\frac{R_{G1} + R_s}{R_{G1}} \right)^2 + I_{n1}^2 (R_s - jX_{C1})^2 + \frac{E_{G1}^2 R_s^2}{R_{G1}^2} \\ + \frac{E_{S1}^2}{1 + (\omega R_{S1} C_{S1})^2} + \frac{1}{K_{t1}^2} (E_{D1}^2 + E_{n2}^2 + I_{n2}^2 R_{D1}^2) + \frac{E_{C2}^2}{K_{tc}^2} \end{aligned} \tag{10-14}$$

From Eq. 10-14, it can be seen that the noise performance is similar to the CE stage. The bias resistor R_{G1} must be large enough to reduce its effect on E_n and to reduce its thermal noise contribution. The coupling capacitor C_{C1} must be large enough not to influence the I_{ni} term. Since I_n for a FET is much less than I_n for a BJT stage, C_{C1} is smaller. If the small bias current can flow through the source, remove C_{C1} and R_{G1}.

10-6 COMMON-COLLECTOR – COMMON-EMITTER PAIR

The CC–CE configuration offers low input noise with an E_n slightly larger than that of the CE stage along with significantly higher input resistance and lower input capacitance. For a general-purpose instrumentation amplifier application, this pair or the comparable FET pair CD–CS is a good compromise to achieve both low noise and high input impedance. A sample design is shown in Fig. 10-6. Since the gain of the first stage is approximately one, the second-stage noise also adds to the first stage so the term E_{nT} is used in the table for the total E_n.

Stage Q_1 provides a voltage gain of less than one. The transfer voltage gain of the pair is

$$K_{tc} = -\frac{\beta_1 \beta_2 R_{L1} R_{C2}}{(R_s + r_{x1} + r_{\pi 1} + \beta_1 R_{L1})(r_{x2} + r_{\pi 2} + \beta_2 Z_{E2})} \tag{10-15}$$

where

$$R_{L1} = R_{E1} \parallel (r_{x2} + r_{\pi 2} + \beta_2 Z_{E2})$$

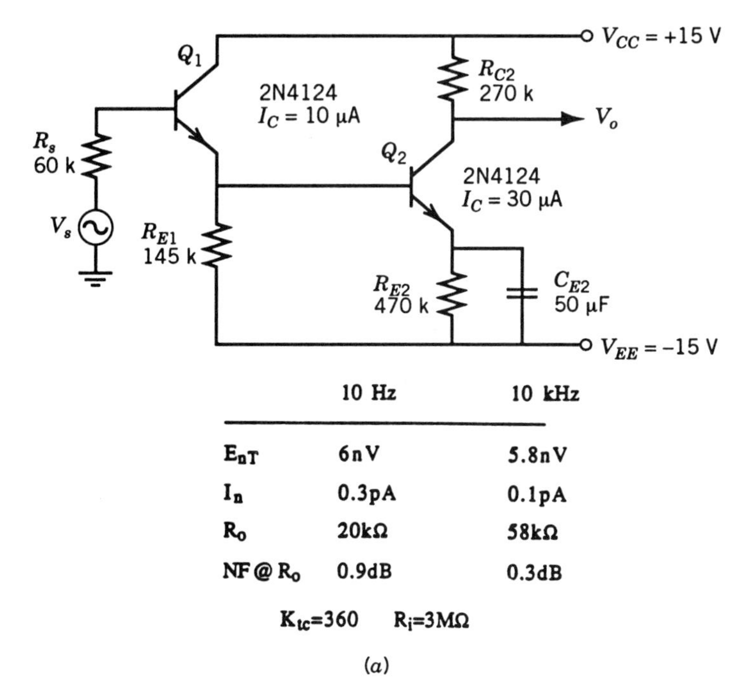

	10 Hz	10 kHz
E_{nT}	6 nV	5.8 nV
I_n	0.3 pA	0.1 pA
R_o	20 kΩ	58 kΩ
NF @ R_o	0.9 dB	0.3 dB

K_{tc}=360 R_i=3MΩ

(*a*)

Figure 10-6 Common-collector–common-emitter configuration.

Equation 10-15 can be simplified for the typical case of $\beta_1 R_{L1} \gg (R_s + r_{x1} + r_{\pi 1})$ and $r_{\pi 2} \gg r_{x2} + \beta_2 Z_{E2}$. Then we have

$$\boxed{K_{tc} \cong -\frac{R_{C2}}{r_{e2}}} \tag{10-16}$$

The equivalent input noise is

$$\boxed{E_{ni}^2 = E_{ns}^2 + E_{n1}^2 + I_{n1}^2 R_s^2 + \frac{E_{n2}^2}{K_{t1}^2} + \left(\frac{I_{n2} R_{E1}}{K'_{t1}}\right)^2 + \left(\frac{E_{E1}}{K'_{t1}}\right)^2 + \frac{E_{C2}^2}{K_{tc}^2}} \tag{10-17}$$

where K'_{t1} is the gain of a CE stage with R_{E1} as its load:

$$K'_{t1} = \frac{\beta_1 R_{E1}}{R_s + r_{x1} + r_{\pi 1}} \cong \frac{R_{E1}}{r_{e1} + R_s/\beta_1} \tag{10-18}$$

The input noise in Eq. 10-17 contains the noise of Q_1, but, since the gain of stage Q_1 is nearly unity, noise E_{n2} is as critical as E_{n1}. On the other hand, the coefficient of I_{n2}, $(r_{e1} + R_s/\beta_1)$, is much smaller than the I_{n1} term.

Another significant difference is the noise contribution of the emitter load resistor R_{E1}. The CC or CD amplifier is the one exception to the rule that the emitter resistor should be kept as small as possible to minimize noise. The noise voltage E_{E1} of the emitter resistor R_{E1} is attenuated by a gain factor K'_{t1}, which is equivalent to the gain of a CE stage in which R_{E1} is the load resistor. In addition, the noise current contribution $I_{n2}R_{E1}$ is attenuated by the same gain factor.

10-7 COMMON-EMITTER – COMMON-BASE PAIR

The CE–CB pair, known as the cascode circuit, is useful because of its low input capacitance and high output impedance. Because of the low input resistance of the second stage, the first-stage voltage gain is small which reduces high-frequency feedback through C_μ (the Miller effect). The input capacitance is much smaller than for a regular CE stage making it more useful at high frequencies.

A discrete version of the cascode circuit is shown in Fig. 10-7. Most of the voltage gain of the pair is obtained from Q_2. Although Q_1 provides little voltage gain, it does raise the power level of the signal. Again, E_{nT} is the total input E_n.

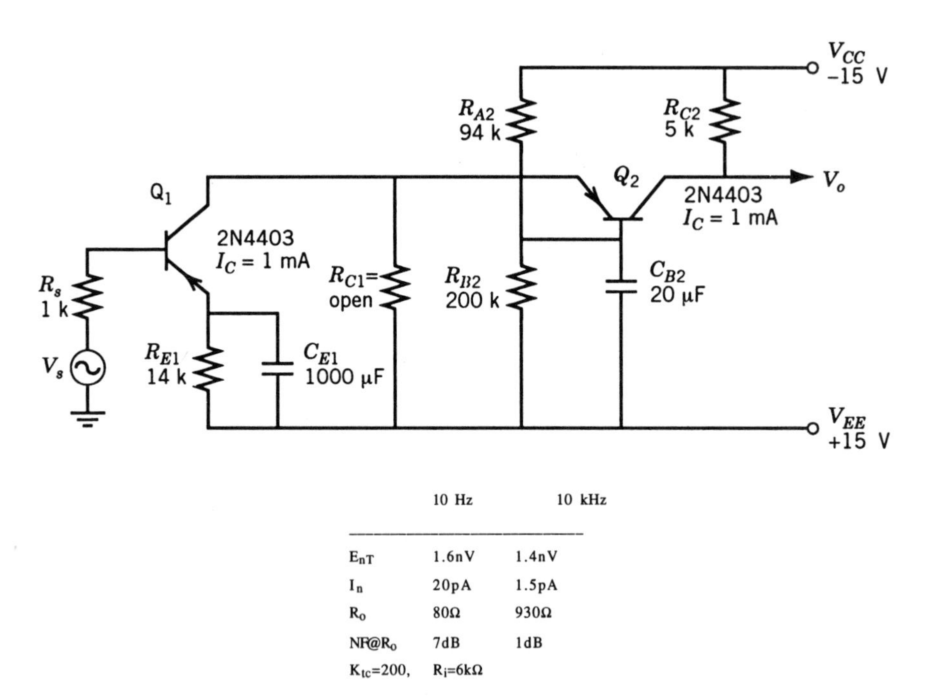

	10 Hz	10 kHz
E_{nT}	1.6nV	1.4nV
I_n	20pA	1.5pA
R_o	80Ω	930Ω
NF@R_o	7dB	1dB
K_{tc}=200,	R_i=6kΩ	

Figure 10-7 CE–CB pair cascode amplifier.

The cascode voltage gain K_{tc} of the pair is the product of the single-stage gains

$$K_{tc} = -\frac{\beta_1 \beta_2 R_{L1} R_{C2}}{(r_{x1} + r_{\pi 1} + R_s + \beta_1 Z_{E1})(r_{\pi 2} + r_{x2} + Z_{B2})} \tag{10-19}$$

where

$$R_{L1} = R_{C1} \parallel \frac{r_{x2} + r_{\pi 2} + Z_{B2}}{\beta_2} \cong r_{e2} \tag{10-20}$$

For the case of $R_s = 0$ and $R_{C1} \gg r_{\pi 2}/\beta_2$, the gain of this pair can be approximated by

$$\boxed{K_{tc} \cong -\frac{R_{C2}}{r_{e1}}} \tag{10-21}$$

The equivalent input noise E_{ni} is

$$\boxed{\begin{aligned} E_{ni}^2 = E_{ns}^2 + E_{n1}^2 + I_{n1}^2 R_s^2 + \frac{1}{K_{t1}^2}\left[E_{C1}^2 + E_{n2}^2 + I_{n2}^2 (R_{L1} + Z_{B2})^2\right] \\ + \frac{E_{A2}^2}{K_{t1}^2[1 + (\omega C_{B2} R_{A2})^2]} + \frac{E_{B2}^2}{K_{t1}^2[1 + (\omega C_{B2} R_{B2})^2]} + \frac{E_{C2}^2}{K_{tc}^2} \end{aligned}} \tag{10-22}$$

The gain K_{t1} is R_{L1}/r_{e1} where R_{L1} is the input impedance of the second stage. Thus the gain of the first stage is approximately one. The total noise voltage of the amplifier is the sum of the first- and second-stage noise voltages. The resistor R_{C1} carries additional collector current for Q_2 when it is desirable to operate Q_2 at a higher current level such as for increased gain–bandwidth.

10-8 INTEGRATED BJT CASCODE AMPLIFIER

The cascode amplifier is one of the most useful integrated inverting amplifiers [1]. It can be fabricated in BJT and BiCMOS technologies. Figure 10-8 shows the BJT configuration.

Comparing the integrated circuit of Fig. 10-8 with the discrete version of Fig. 10-7, Q_1 is the input common-emitter stage and Q_2 is the common-base gain stage in both. The load resistor R_{C2} is replaced by the active load resistance of Q_3. As noted on Fig. 10-8, the low resistance point is the input to the emitter of Q_2. The high resistance point is the output where the loads

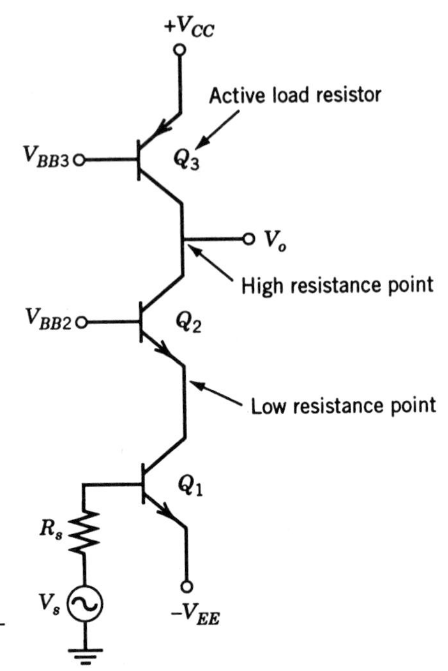

Figure 10-8 Integrated BJT cascode amplifier.

are connected. The overall gain K_{tc} is again

$$K_{tc} \cong -\frac{r_{o3}}{r_{e1}} \tag{10-23}$$

where r_{o3} is the dynamic output resistance of Q_3.

The equivalent input noise E_{ni} is

$$E_{ni}^2 = E_{ns}^2 + E_{n1}^2 + I_{n1}^2 R_s^2 + \frac{1}{K_{t1}^2}\left[E_{n2}^2 + I_{n2}^2(r_{e2} + Z_{B2})^2\right] + \frac{E_{o3}^2}{K_{tc}^2} \tag{10-24}$$

where K_{t1} is r_{e2}/r_{e1}. Since the collector currents are equal, $K_{t1} = 1$. Thus the noise voltage contribution of the second stage is equal to that of the first stage. The impedance Z_{B2} is the impedance of the base bias V_{BB2}, which should be very low, and r_{e2} is very small so the I_{n2} contribution is negligible. The noise voltage E_{o3} is the noise contribution of Q_3:

$$E_{o3} = E_{n3} K_{t3} \tag{10-25}$$

and

$$K_{t3} = \frac{r_{o2}}{r_{e3}} \cong K_{tc} \tag{10-26}$$

because Eq. 10-26 is the same as Eq. 10-23. This makes the equivaient input noise reduce to

$$\begin{aligned} E_{ni}^2 &= E_{ns}^2 + E_{n1}^2 + I_{n1}^2 R_s^2 + E_{n2}^2 + E_{n3}^2 \\ &\cong E_{ns}^2 + 3E_n^2 + I_{n1}^2 R_s^2 \end{aligned} \tag{10-27}$$

Thus it is especially important to have devices with a low E_n noise and low $1/f$ noise corner frequency.

10-9 DIFFERENTIAL AMPLIFIER

The most useful two-stage amplifier pair is the differential amplifier. It is particularly valuable because of its compatibility with integrated circuit technology and its ability to amplify differential signals. To illustrate this gain, take two arbitrary signals V_1 and V_2. Separate them into a *difference-mode* signal V_D and a *common-mode* signal V_C. Since each signal has half of the difference signal and all of the common signal,

$$V_1 = \frac{V_D}{2} + V_C \tag{10-28}$$

$$V_2 = -\frac{V_D}{2} + V_C \tag{10-29}$$

We can now define V_D and V_C as

$$V_D = V_1 - V_2 \tag{10-30}$$

and

$$V_C = \frac{V_1 + V_2}{2} \tag{10-31}$$

The objective of the differential amplifier is to amplify only the difference signal and to reject the common signal.

Now that we have defined the function of a differential amplifier, consider now the noise model. A differential amplifier is composed of two amplifier gain blocks with two inputs, one for each of the signals V_1 and V_2. Extending the general noise model derived in Chap. 2 for a single amplifier, we get the equivalent circuit of Fig. 10-9. This contains a noise voltage generator E_n in series and a noise current generator I_n in parallel with each side of the amplifier.

In the equivalent system diagram of Fig. 10-9, each of the differential amplifier inputs has its own set of noise generators, represented by E_{n1}, E_{n2},

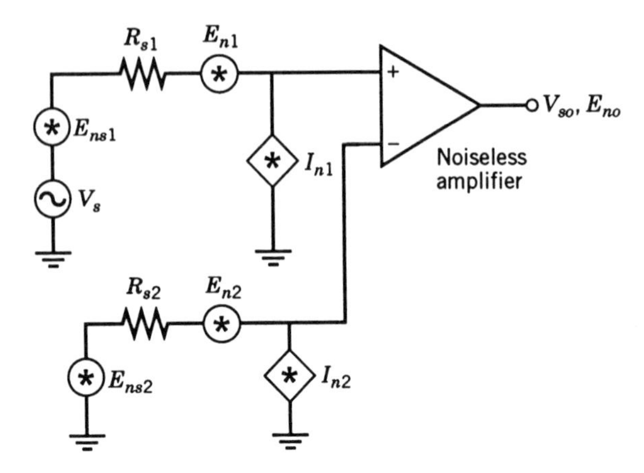

Figure 10-9 Differential amplifier noise model.

I_{n1}, and I_{n2}. Since the two sides of the amplifier are mirror images, the noise generators of both sides are equal. The resistor R_{s1} represents the source resistance in the noninverting (positive) input terminal. The inverting (negative) terminal of the differential amplifier is returned to ground through R_{s2}. The resistances R_{s1} and R_{s2} are not necessarily equal. The single-signal source V_s is in the noninverting input. This noise equivalent circuit is the most commonly used. It applies to negative feedback amplifiers as derived in Chap. 3. The resistor R_{s1} is the source resistance and R_{s2} is the biasing resistance or feedback resistance of a non-inverting feedback amplifier.

The equivalent input noise for the differential amplifier is the sum of the noises of the two halves of the pair:

$$E_{ni}^2 = E_{ns1}^2 + E_{ns2}^2 + E_{n1}^2 + E_{n2}^2 + I_{n1}^2 R_{s1}^2 + I_{n2}^2 R_{s2}^2 \tag{10-32}$$

Should $R_{s2} = 0$, the final term is eliminated, but E_{n2} is still present in E_{ni}.

Now consider the case when the source is ungrounded (a floating source). The noise model of Fig. 10-10 applies.

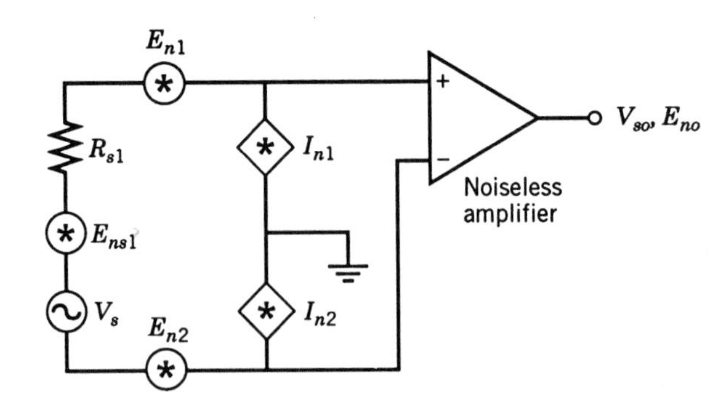

Figure 10-10 Differential amplifier with a floating signal source.

Now we define an equivalent input noise for a differential amplifier with an ungrounded floating source. First, the noise voltages E_{n1} and E_{n2} are summed to give a single amplifier noise voltage E_{nT}. This value, E_{nT}, is typically the one given on an op amp spec sheet as E_n:

$$E_{nT}^2 = E_{n1}^2 + E_{n2}^2 = 2E_{n1}^2 \tag{10-33}$$

The noise current contributions of I_{n1} and I_{n2} are halved since each effectively sees half of the source resistance. The total noise current $I_{nT}R_s$ becomes

$$I_{nT}^2 R_s^2 = I_{n1}^2\left(\frac{R_s}{2}\right)^2 + I_{n2}^2\left(\frac{R_s}{2}\right)^2 = \frac{I_{n1}^2}{2} \tag{10-34}$$

Assuming the noise is the same for each half, the equivalent input noise E_{ni} for the ungrounded source is

$$E_{ni}^2 = E_{ns}^2 + 2E_{n1}^2 + \frac{I_{n1}^2 R_s^2}{2} \tag{10-35}$$

For the ungrounded source, the noise current I_{nT} is 0.7 of the noise current for the grounded input case, and the noise voltage E_{nT} is 1.4 times the single-stage noise voltage E_{n1}.

10-9-1 Circuit Noise Components

Let us now concentrate on the noise sources within the differential amplifier. Typical circuits are shown in Figs. 10-11 and 10-12. For design purposes and for understanding, we need to develop the equivalent input noise model for the circuits showing the noise contribution of each element.

Since noise is referred to the input, it is first necessary to define several gain expressions for the circuit of Fig. 10-11. This approach is based on the work of Meindl [3].

The differential gain K_{dm} is the ratio of the differential output signal $(V_{o2} - V_{o1})$ to the differential input signal $(V_{s2} - V_{s1})$. The output is taken between collectors. From Fig. 10-11 we obtain the differential noise gain

$$K_{dm} = \frac{V_{o2} - V_{o1}}{V_{s2} - V_{s1}} = \frac{-2R_C}{\dfrac{1}{g_m} + R_E + \dfrac{R_s}{\beta}} \tag{10-36}$$

where $g_m = 1/r_e$ for each transistor separately. The transistors are assumed to have identical parameters, and the circuit elements for each channel are equal: $R_{C1} = R_{C2} = R_C$, $R_{E1} = R_{E2} = R_E$, and $R_{s1} = R_{s2} = R_s$. For the

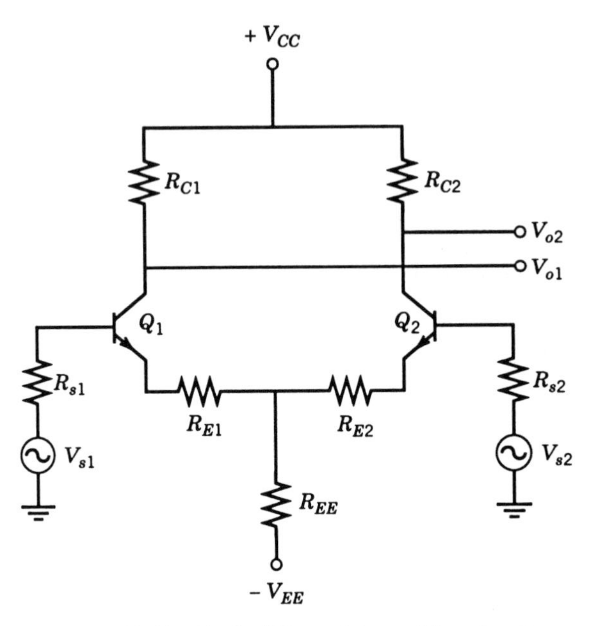

Figure 10-11 Basic differential amplifier circuit.

typical case when $R_E = 0$ and $R_s \ll r_\pi$, Eq. 10-36 reduces to

$$\boxed{K_{dm} \cong \frac{-2R_C}{r_e}} \tag{10-37}$$

The common-mode voltage gain K_{cm} is defined as the ratio of the common-mode output signal to a common-mode input. If the two signal sources V_{s1} and V_{s2} are equal and in phase, then

$$K_{cm} = \frac{V_{o2} + V_{o1}}{V_{s2} + V_{s1}} = \frac{-R_C}{\dfrac{1}{g_m} + R_E + \dfrac{R_s}{\beta} + 2R_{EE}} \tag{10-38}$$

When R_{EE} is very large,

$$\boxed{K_{cm} \cong \frac{-R_C}{2R_{EE}}} \tag{10-39}$$

More significant is the gain K_{dc} that produces a differential output signal for a common-mode input signal. This gain is highly dependent on circuit balance:

$$K_{dc} = \frac{V_{o2} - V_{o1}}{2V_s} = \frac{R_{C1}\left(\dfrac{R_{s2}}{\beta_2} + \dfrac{1}{g_{m2}} + R_{E2}\right) - R_{C2}\left(\dfrac{R_{s1}}{\beta_1} + \dfrac{1}{g_{m1}} + R_{E1}\right)}{2R_{EE}\left(\dfrac{R_{s2}}{\beta_2} + \dfrac{1}{g_{m2}} + R_{E2} + \dfrac{R_{s1}}{\beta_1} + \dfrac{1}{g_{m1}} + R_{E1}\right)} \tag{10-40}$$

This equation cannot be simplified by approximations because when the parameters of each channel are all equal, $K_{dc} = 0$. Note, to reduce K_{dc} requires large R_{EE} and good matching of R_{C1}, R_{C2}, and all the transistor parameters.

If a transistor is used as a constant-current sink, its output resistance r_{o3} is used in place of R_{EE}, then R_{EE} is replaced by r_{o3} in the equations.

The preceding gain equations apply to input signals that are correlated; that is, either dc signals or ac signals from the same source. The uncorrelated noise voltage E_n in series with each input is not attenuated by the gain K_{dc}. It is amplified by the differential gain K_{dm}. Each input noise source produces an output independent of the others.

Now we can write the expression for the equivalent input noise of the circuit of Fig. 10-11 in terms of the preceding gain expressions

$$\boxed{\begin{aligned} E_{ni}^2 &= E_{s1}^2 + E_{s2}^2 + E_{n1}^2 + E_{n2}^2 + I_{n1}^2R_{s1}^2 + I_{n2}^2R_{s2}^2 \\ &\quad + E_{E1}^2 + E_{E2}^2 + \frac{E_{C1}^2 + E_{C2}^2}{K_{dm}^2} + \frac{E_{EE}^2 + E_{VEE}^2 + E_{VCC}^2}{K_{dc}^2} \end{aligned}} \tag{10-41}$$

where E_{VEE} and E_{VCC} are the noise variations on the power supply lines at the R_{EE} location, E_{s1} is the thermal noise of R_{s1}, E_{E1} is the thermal noise of R_{EE}, and E_{C1} is the thermal noise of R_{C1} or the noise of an active load resistance. Ideally, the emitter resistances R_{EE1} and R_{EE2} will be zero since they contribute directly to the input noise. The gain K_{dc} attenuates V_{CC} noise, but it may be possible to reduce the V_{CC} noise further by coupling a portion of the noise to the inverting input or even to the base of the transistor acting as a constant-current sink (Q_3).

For low-noise operation a differential amplifier should be followed by a second differential stage. If the first stage is single-ended (output taken from one side of R_C), the differential gain expressions are replaced by their common-mode expressions, and there may be little or no rejection of the common-mode noise voltages E_{E1}, E_{E2}, E_{VEE}, and E_{VCC}.

The noise of the second differential amplifier stage is less critical because of the first-stage gain. The second-stage noise can be divided by the first-stage differential gain K_{dm}.

10-9-2 Integrated BJT Differential Amplifier

The differential amplifier can be readily integrated. By integrating all devices on one chip, the offset and thermal drift are reduced and common-mode rejection is improved. The common-mode noise rejection is improved significantly by the process of integration since the rejection is primarily determined by the balance of the differential stages. While overall geometry may vary, all adjacent devices on a wafer will be nearly identical. On the other hand, the noise performance of an IC will usually be poorer than that of discrete devices. In part, this is a result of design compromises in the IC fabrication process, such as long isolation diffusions, and because of the use of active loads and current sources.

Figure 10-12 is the integrated version of the differential amplifier shown in Fig. 10-11. Q_1 and Q_2 are the input gain stages, Q_3 and Q_4 are the active

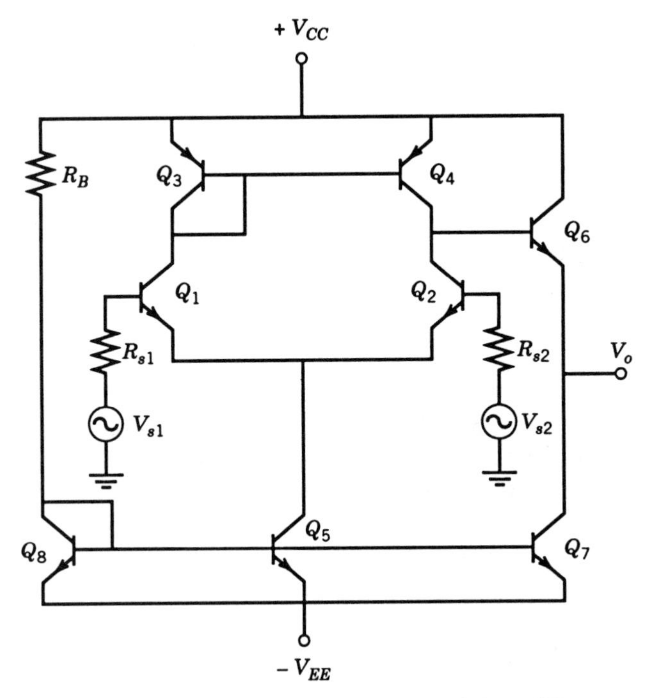

Figure 10-12 Integrated BJT differential amplifier.

load resistors, Q_5 and Q_7 are the current mirrors for biasing, R_B and Q_8 are the voltage references for the current mirrors, and Q_6 is the output buffer stage. Differential-mode and common-mode gains are as derived in Eqs. 10-36 through 10-40. We caution the reader that Fig. 10-12 is not a finished design. It does not include any current limit protection for Q_6 and Q_7. Furthermore, it will have a nonzero input offset voltage. It is shown here in this simplified schematic to present and discuss its noise characteristics.

Noise of the integrated BJT differential amplifier follows from the expression in Eq. 10-41 as

$$\boxed{\begin{aligned} E_{ni}^2 &= E_{s1}^2 + E_{s2}^2 + E_{n1}^2 + E_{n2}^2 + I_{n1}^2 R_{s1}^2 + I_{n2}^2 R_{s2}^2 \\ &\quad + E_{n3}^2 + E_{n4}^2 + \frac{E_{n5}^2 + E_{VEE}^2 + E_{VCC}^2}{K_{dc}^2} \end{aligned}} \tag{10-42}$$

The equivalent input noise is composed of the source thermal noise E_s and the E_n and I_n of the input stages as before. Now, the noise of the active loads also adds to the input. Since the active load resistors Q_3 and Q_4 have the same gain K_{dm} from their base to the output as the input signal, their E_n's contribute directly to the input noise. Since the impedances at the bases are low, I_{n3} and I_{n4} do not contribute noise. Any noise in the bias circuit for the loads would also contribute noise directly. The noise of the current sink Q_5 and noise from the power supplies are attenuated by the common-mode gain K_{dc}. The noise contribution of Q_5 is input E_{n5} plus biasing network noise times the gains, K_5:

$$K_5 = r_{e2}/2r_{e5} = 0.5 \tag{10-43}$$

Since the common-mode rejection is very high and the active circuits all have the same geometry and noise mechanisms, E_{ni} of Eq. 10-42 reduces to (when $R_{s1} = R_{s2}$)

$$\boxed{E_{ni}^2 = 2E_s^2 + 4E_n^2 + 2I_n^2 R_s^2} \tag{10-44}$$

10-10 PARALLEL AMPLIFIER STAGES

When the sensor resistance is very small, less than 100 Ω, an input coupling transformer can be utilized to match the source resistance to the R_o of the amplifier. Another method to accomplish matching is to reduce the amplifier's optimum noise resistance by paralleling several amplifying devices. This technique can provide matching for source impedances as small as a few ohms.

Paralleling amplifier stages is equivalent to paralleling their E_n and I_n generators and summing their outputs as illustrated in Fig. 10-13.

The equivalent noise voltage E_n' and the equivalent noise current I_n' of a parallel system consisting of N identical stages are given by

$$E_n' = \frac{E_n}{\sqrt{N}} \tag{10-45}$$

and

$$I_n' = I_n\sqrt{N} \tag{10-46}$$

A new optimum source resistance R_o' can now be defined. Its relation to R_o is

$$R_o' = \frac{E_n'}{I_n'} = \frac{R_o}{N} \tag{10-47}$$

Thus R_o can be lowered in proportion to the number of parallel stages. A practical limit will be determined by chip area, cost, and input capacitance.

The minimum NF, F_{opt} is proportional to the product of E_n and I_n. Since this product is unchanged by paralleling,

$$F_{\text{opt}}' = F_{\text{opt}} \tag{10-48}$$

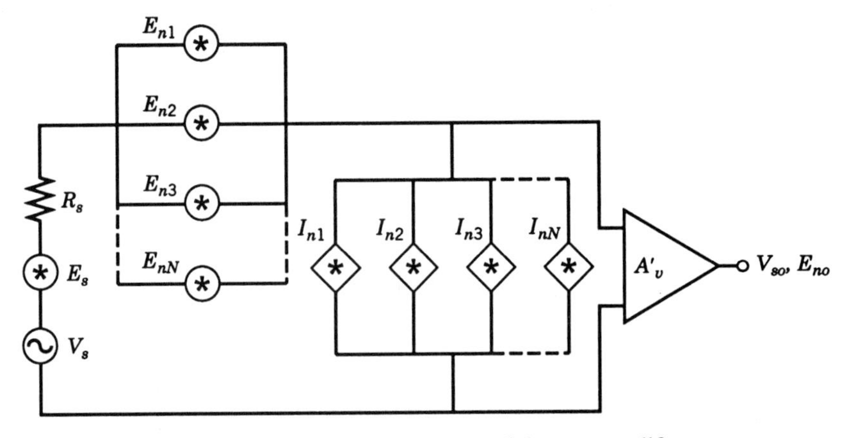

Figure 10-13 Noise model of parallel-stage amplifier.

Outputs from the amplifiers are summed so the voltage gain A'_v of the parallel amplifier is increased in proportion to the number of stages:

$$A'_v = NA_v \tag{10-49}$$

The limiting noise voltage E_n of a BJT transistor amplifier is determined by the base resistance r_x (Chap. 5). The base resistance can be reduced in a transistor layout by placing the base contact all around the emitter. Another method is to parallel stages as described previously. The limiting noise voltage E_n in field effect transistors is determined by channel resistance and g_m as derived in Chap. 6. To reduce the channel resistance and increase g_m, a high wide-to-length ratio is required. This requires a very wide structure, often serpentined, which is *equivalent* to paralleling many separate devices.

When paralleling stages, the input capacitance C'_π and the Miller effect increase in proportion to N because the inputs are paralleled. The output resistance r'_o is decreased by the number of parallel stages.

10-11 IC AMPLIFIERS

All integrated amplifiers must operate within the same general f_T and process requirements but there have evolved three generic types of integrated amplifiers based on application. These are operational amplifiers, instrumentation amplifiers, and audio amplifiers. So what is the difference, since the terms are used interchangeably?

An op amp is a combination of active devices characterized as a high-gain block with a well-defined gain-phase response containing a single dominant pole so it will be stable with large feedback (often 100%). Instrumentation amplifiers also have a large gain but do not have a single pole in the frequency response so will require external frequency compensation and less feedback. In return, we obtain a greater gain–bandwidth product for improved frequency response. For the integrated audio amplifier, the bandwidth is less critical so performance is optimized for low distortion and low power consumption. While op amps have a differential input, instrumentation and audio amplifiers are more likely to have single-ended inputs with reduced noise. This also requires external biasing.

Many linear ICs make use of the differential pair as shown in Fig. 10-12 and discussed in Sec. 10-9-2. In that circuit, transistors Q_1 and Q_2 provide the desired amplification. Transistor Q_5 forms part of a dc constant-current source to provide Q_1 and Q_2 with a constant level of emitter current to provide a stable operating point for those transistors. The noise sources of the two transistors in the monolithic differential amplifier are statically independent. Consequently, the pair exhibits at least twice the level of noise of a single transistor.

Noise mechanisms in monolithic transistors are the same as the FET and BJT models of Chaps. 5 and 6. The input noise current I_n is dependent on the shot noise of the dc gate or base currents. The ohmic portion of the channel and base resistances are responsible for the thermal noise of E_n. The shot noise of the collector current is responsible for the limiting E_n noise in BJT transistors. Low-frequency noise is also present. Expressions used to represent these mechanisms are the same for monolithic transistors and discrete transistors.

Popcorn or burst noise can also be presented in some units. The popcorn noise behavior of the IC transistor is discussed in Sec. 5-9. To represent the popcorn noise source a noise current generator as given in Eqs. 5-37 and 5-38 can be connected to the junction between active and inactive base resistances. The model is shown in Fig. 5-10.

In a differential input amplifier, popcorn noise would appear as a differential-mode signal, and could be located at the input of either transistor in the pair. The operation of any system would be severely affected by any popcorn noise generated in the initial stage. Popcorn noise in the second stage of an amplifier can seriously degrade performance also. Second-stage popcorn noise is a problem with lateral transistors because they are heavily doped which increases trapping and popcorn noise.

The E_n–I_n noise model and the concept of total equivalent input noise describe the noise behavior of IC amplifiers. These representations are used in the presentation of noise in selected linear ICs. Measured noise data on commercial integrated amplifiers is tabulated in App. A for reference. Methods for IC noise measurement are developed in Chap. 15.

SUMMARY

a. In a conventionally biased transistor stage, base-biasing resistors contribute thermal noise and excess noise to E_{ni}.

b. To refer noise from a resistor R_p shunting the signal source to E_{ni}, multiply the noise of that resistor by R_s/R_p.

c. Noise in the local feedback resistor R_E can be large; use of this kind of feedback is not recommended.

d. The capacitor bypassing an emitter resistance should be chosen to minimize the noise contribution of that resistance.

e. Frequency response shaping should usually be accomplished in stages beyond the input stage.

f. Selection of the configuration for a BJT input stage can be based on nonnoise characteristics.

g. E_{ni}, E_n, and I_n are basically identical for all BJT configurations. E_{ni} is increased by noise from biasing components and second-stage contributions.

h. Overall negative feedback does not affect E_{ni}, except for the added thermal and excess noise in the feedback elements. Hence there is no change in R_o.

i. Selection of the appropriate BJT pair depends on the requirements for the gain, input impedance, and frequency behavior. A low-noise figure is achievable for any pair if the proper precautions are included during design.

j. The FET–BJT pair can provide low-noise parameters, and is especially useful with sources of high internal resistance.

k. Noise in the differential amplifier is the result of many causes, including differential output for common-mode input. A second differential pair is recommended to minimize this noise input.

l. The parallel-stage amplifier is useful when a low R_o is needed.

PROBLEMS

10-1. **(a)** Verify $K_t = -254$ and $R_i = 87.2$ kΩ for the circuit of Fig. 10-1 at a frequency of 10 kHz. Use the 2N4250 data given in Table 5-1. Note that R_i is the input resistance seen looking into the base of the transistor. Assume $kT/q = 0.025$ V. Use the small signal model of Fig. 10-2.

(b) Determine which capacitor, C_C or C_E, sets the low -3-dB corner frequency. Then determine this frequency.

(c) Calculate the upper -3-dB frequency limit of this circuit.

(d) Finally, simulate the circuit and compare the simulation results with the calculated answers.

10-2. Modify the circuit in Fig. 10-1 by adding $R_F = 500\ \Omega$ between R_E and V_{EE}. Repeat Prob. 10-1. Find the new K_t, R_i, and lower and upper -3-dB corner frequencies. Compare with simulation results.

10-3. Perform the tasks given in Prob. 1 for the circuit shown in Fig. 10-3*a*.

10-4. Perform a noise analysis of the circuit of Fig. 10-3*a* in a noise bandwidth $\Delta f = 1$ Hz centered around frequencies of 10 Hz and 10 kHz. Consider the effects of excess noise as explained in Sec. 12-1 by letting the NI of the low-noise resistors be 0 dB. Evaluate each term in Eq. 10-7 to determine which are significant at the two specified frequencies. Then sum all terms to give the overall E_{ni}^2. Use the transistor data given in Table 5-1. What could be done to reduce E_{ni}^2 at each analysis frequency?

10-5. Calculate and compare E_{ni} for the circuits shown in Fig. P10-5 *a* and *b* in a noise bandwidth $\Delta f = 1$ Hz centered around frequencies of 10 Hz and 10 kHz. Assume $r_x = 200\ \Omega$ and $\beta = 217$ for both circuits. Let the NI = 0 dB for all resistors. By how much does the improved bias scheme reduce the equivalent input noise voltage?

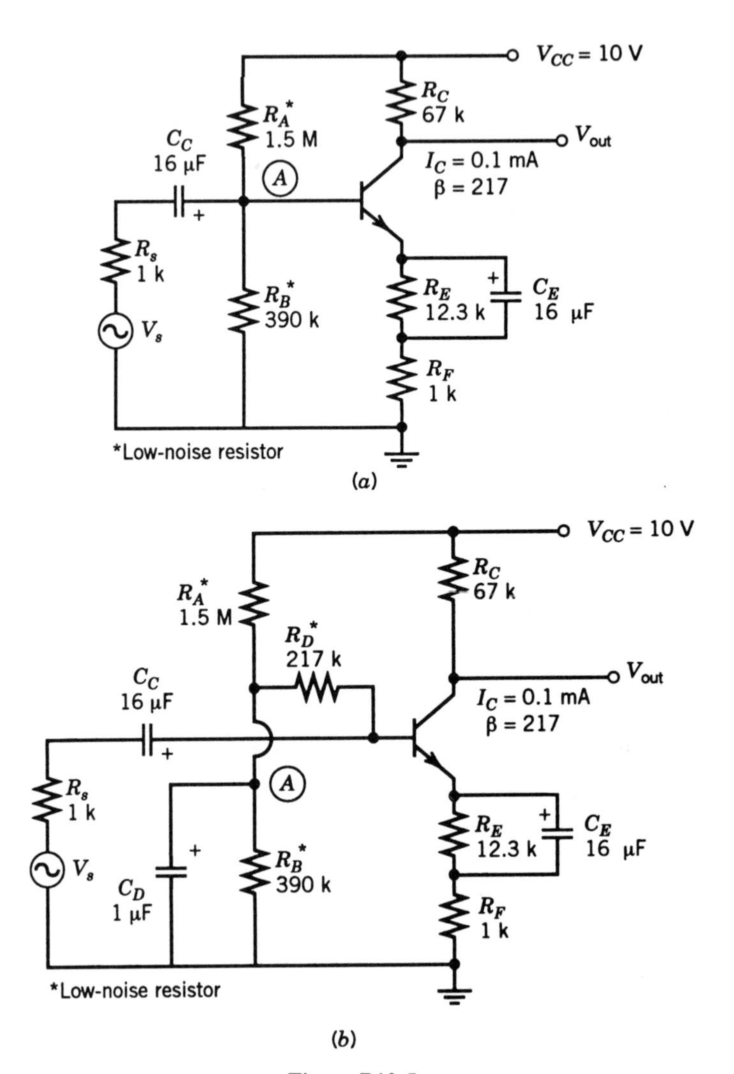

Figure P10-5

10-6. Repeat Prob. 10-4 by using Eq. 10-14 to determine the significant noise contributors to E_{ni}^2 for the circuit of Fig. 10-5*b*. Let NI = −5 dB for all resistors. Use the E_n and I_n data for the 2N3821 provided in Appendix B. Assume the data is for $I_D = 1$ mA.

10-7. Repeat Prob. 10-4 for the CC–CE pair of Fig. 10-6. Let NI = +10 dB.

10-8. Add a resistor R_E between the emitter of Q and $-V_{EE}$ of Fig. 10-8. Then design a bias circuit for this cascode amplifier so that all three transistors have dc currents of 1 mA. Bypass R_E with a large capacitor. Let $\beta = 200$, $R_s = 1$ kΩ, and $r_x = 100$ Ω. Calculate E_{ni}^2 for this circuit. Neglect excess noise. Let $V_{CC} = 15$ V and $V_{EE} = -15$ V.

REFERENCES

1. Geiger, R. L., P. E. Allen, and N. R. Strader, *VLSI Design Techniques for Analog and Digital Circuits*, Wiley, New York, 1990.
2. Texas Instruments Specifications for TLE2425.
3. Meindl, J. D., *Micropower Circuits*, Wiley, New York, 1969, p. 145.

CHAPTER 11

NOISE ANALYSIS OF D/A AND A/D CONVERTERS

Signal-processing applications of analog-to-digital (A/D) and digital-to-analog (D/A) converters continue to require finer and finer resolution. As resolution starts to approach the noise limits of a converter, the usefulness of the less significant bits becomes questionable since the probability of bit errors significantly increases. Effective bit resolution is determined by both interference-type noise as well as inherent fundamental noise within the converter circuitry.

Most interference-type noise can be reduced or eliminated through proper layout, shielding, and grounding techniques. At some point, however, the noise of a converter will be dominated by fundamental device noise. The reduction of fundamental noise requires converter operation at low temperatures which is often an impractical solution, or converter design which identifies dominant noise sources and seeks to reduce their contributions by changing transistor operating points or transistor types, lowering resistances, and optimizing the converter topology from a noise perspective.

Numerous articles and publications address noise models and noise analysis in integrated circuit devices, that is, resistors, diodes, bipolar junction

This chapter is extracted from the Ph.D. thesis of Katherine P. Taylor, who is now with the Department of Electrical and Computer Engineering Technology, Southern College of Technology, Marietta, Georgia.

transistors (BJTs), MOSFETs, and operational amplifiers (op amps) [1–9]. Relatively few references [10, 11] consider noise in more complex circuits such as A/D and D/A converters, and none of these seeks to model converters with anything more than a single comparator and one noise source.

Consequently, there is a need for noise analysis and noise modeling methodologies in A/D and D/A converters using more complete circuit models. This chapter describes a methodology of noise analysis to identify dominant noise sources in D/A and A/D converters. The noise modeling methodology can be adapted to other converter topologies with little modification provided that the noise contributions from individual converter elements are known. In addition, the modeling methodology is not technology specific and is therefore applicable to noise simulation of bipolar, MOS, and GaAs converters. The methodology is demonstrated using a simple voltage-scaled D/A topology and a flash A/D converter.

While the voltage-scaled D/A is not often implemented by integrated circuit manufacturers, it is chosen here because this topology is often used to explain D/A converter concepts and is well defined and understood. The flash A/D converter is chosen because of its simple topology with no D/A converter to compound the noise analysis.

This chapter first presents a noise analysis of binary-weighted and R–$2R$ resistor networks commonly found in D/A converters. Four different op amp summation schemes for producing the analog output of the D/A from the resistor networks are then considered. The noise from a voltage reference, a resistive divider, and comparators of a flash A/D are examined in terms of equivalent noise sources placed at the inputs of the comparators.

SPICE input files developed in conjunction with this methodology constitute a noise model of a particular converter topology whose noise parameters may be changed to match actual converter noise behavior more closely. SPICE simulations are used to verify the analytical results and to determine typical noise levels in the circuits considered here. The SPICE noise models allow easy identification of the dominant noise sources within a given topology for a given set of noise parameters. The effects of changing the parameters of the dominant noise sources (e.g., using components designed for low noise) are examined to address the design implications of the noise analysis. Finally, the noise in a complete analog-to-digital-to-analog (A/D/A) converter system is examined.

11-1 RESISTOR NETWORKS FOR D/A CONVERTERS

Figures 11-1 and 11-2 show a simplified N-bit D/A converter using an R–$2R$ network and a binary-weighted network, respectively. The first part of the

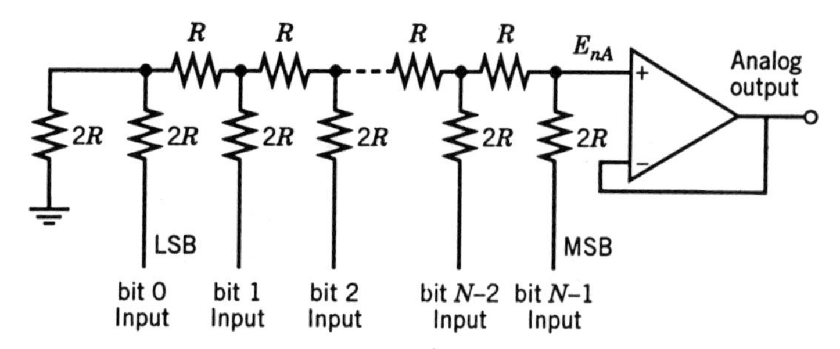

Figure 11-1 Simplified N-bit D/A converter with R–$2R$ network.

analysis considers only the resistors in the networks and excludes op amp noise sources and noise from R_F in the binary-weighted case.

Noise analysis of both resistor networks entails grounding the digital input bits and placing thermal noise voltage sources in series with each resistor. The value of the spectral density noise voltage of each source is $4kT$ times the corresponding resistance. The equivalent output noise voltage squared per hertz, E_{nA}^2, for the network is the sum of the individual resistor noise contributions evaluated through the appropriate voltage divider for each noise source.

The spectral density E_{nA}^2 of the R–$2R$ network is independent of the number of bits, N, and has a value of $4kTR$. The E_{nA}^2 of the binary-weighted

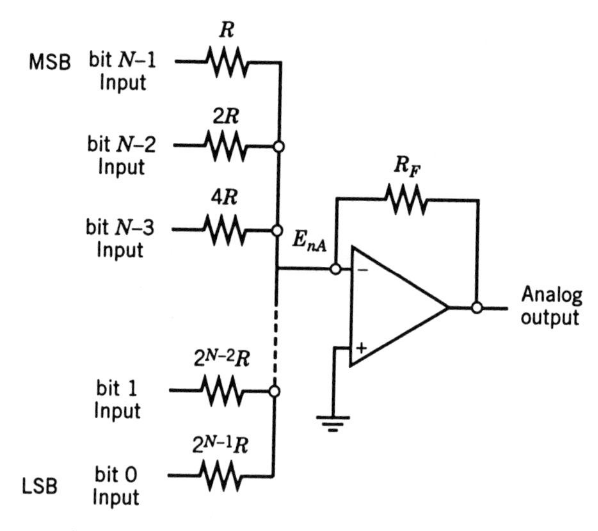

Figure 11-2 Simplified N-bit D/A converter with binary-weighted network.

network is given by

$$E_{nA}^2 = \frac{4kTR}{1 + \left[\dfrac{2^{N-1} - 1}{2^{N-1}}\right]} \tag{11-1}$$

As N increases, E_{nA}^2 rapidly approaches a value of $2kTR$. For $N = 6$, this approximation results in an E_{nA}^2 error of only 1.4%.

11-2 NOISE ANALYSIS OF D/A CONVERTER CIRCUITS

Four different op amp topologies which could be used in the simplified models of Figs. 11-1 and 11-2 are shown in Fig. 11-3. For the noise analysis of these topologies, the op amp noise was modeled with a noise voltage source E_n in series with the noninverting input terminal, and a noise current source I_n located between the two input terminals as shown in Fig. 11-4. Both E_n and I_n may be modeled as having flat noise spectra, $1/f$ noise spectra, or a combination of these two spectra such that the spectrum is flat for frequencies greater than a noise corner frequency f_{nc} and has a $1/f$ characteristic for frequencies below f_{nc}.

The contributions to E_{no}^2 from each noise voltage source are calculated by squaring the source noise voltage and multiplying by the square of the voltage gain from the source's location to the output. The I_n contribution to E_{no}^2 is calculated by squaring the value of I_n, multiplying by the square of the equivalent source resistance seen by I_n, and then multiplying by the square of the noninverting forward gain of the circuit.

For example, consider the R–$2R$ network with a noninverting amplifier shown in Fig. 11-3*c*. The resistor network has an equivalent resistance, R_{eq}, equal to R seen looking back into the network. The network also introduces a bit factor (BF) which is the voltage gain of the network if all of the inputs are connected together. This bit factor is given by

$$\text{BF} = \frac{2^N - 1}{2^N} \tag{11-2}$$

and is approximately equal to one.

Ignoring the input and output resistances of the op amp model shown in Fig. 11-5, the other circuit elements make the following contributions in units

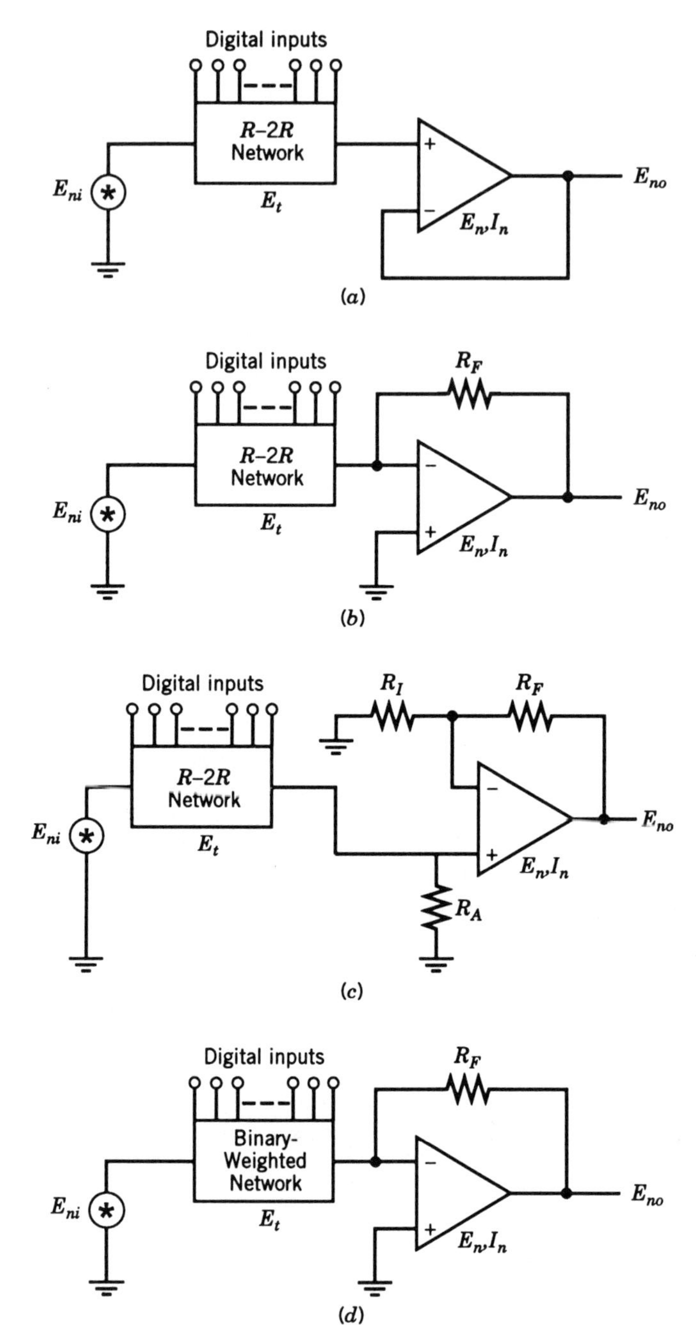

Figure 11-3 Four summation schemes: (*a*) *R*–2*R* network with voltage follower, (*b*) *R*–2*R* network with inverting amplifier, (*c*) *R*–2*R* network with noninverting amplifier, and (*d*) binary-weighted network with inverting summer.

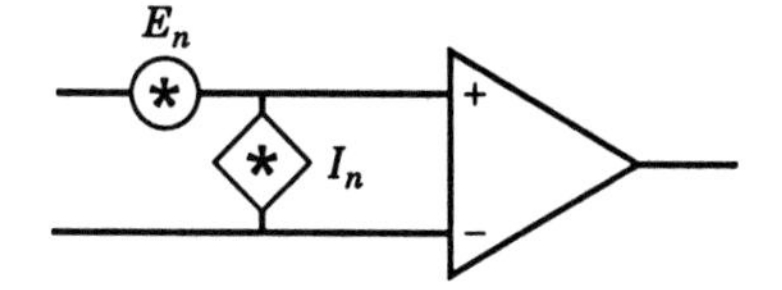

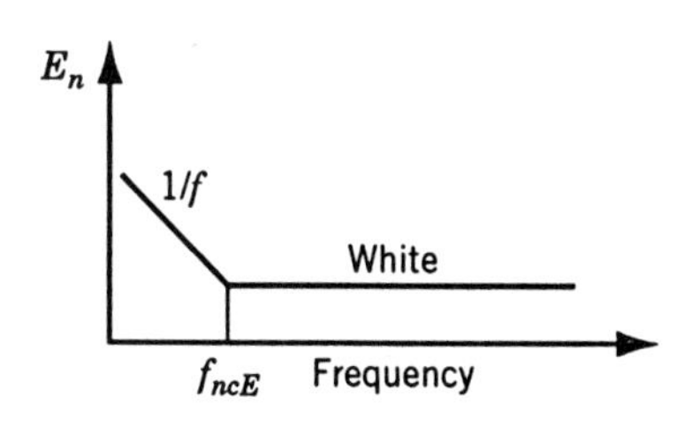

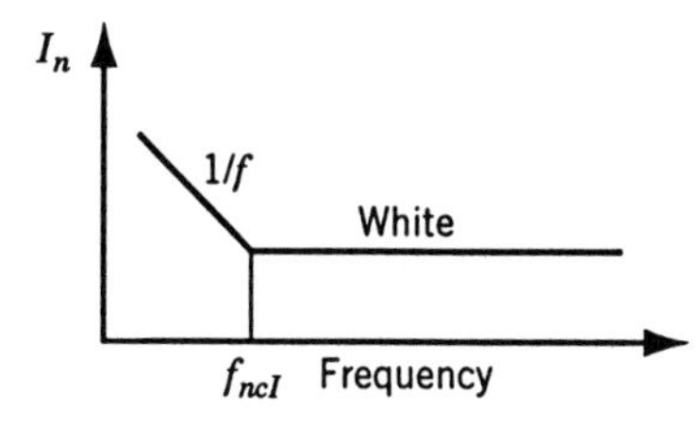

Figure 11-4 Operational amplifier noise sources.

of volts^2 per hertz to E_{no}^2:

R–$2R$ network:

$$4kTR_{\text{eq}}(\text{BF})^2\left[\frac{R_A}{R_A+R_{\text{eq}}}\right]^2\left[\frac{R_F+R_I}{R_I}\right]^2 \tag{11-3}$$

E_n:
$$E_n^2\left[\frac{R_F+R_I}{R_I}\right]^2 \tag{11-4}$$

R_A:
$$4kTR_A \tag{11-5}$$

R_I:
$$4kTR_I\left(\frac{R_F}{R_I}\right)^2 \tag{11-6}$$

R_F:
$$4kTR_F \tag{11-7}$$

I_n:
$$I_n^2\left[\frac{R_AR_{\text{eq}}}{R_A+R_{\text{eq}}}+\frac{R_FR_I}{R_F+R_I}\right]^2\left[\frac{R_F+R_I}{R_I}\right]^2 \tag{11-8}$$

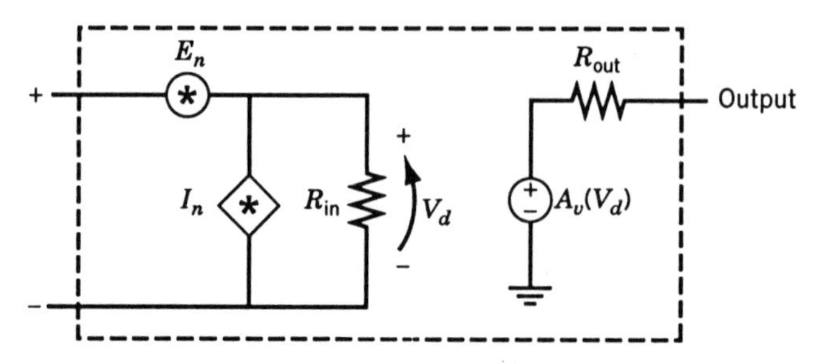

Figure 11-5 Operational amplifier model.

Table 11-1 lists the multiplication factor for each noise source's contribution to E_{no}^2 in the four topologies. The bit factors and R_{eq} for each topology are listed as well. Table 11-2 lists the multiplication factor for each noise source's contribution to E_{ni}^2 in the four topologies. E_{ni} is an equivalent noise voltage source located at the input to the resistor network whose squared value is equal to E_{no}^2 divided by the square of the voltage gain from the location of E_{ni} to the output. E_{no}^2/E_{ni}^2 for each topology is also given.

The analysis is simplified by considering that for equal digital inputs to the four topologies, the four analog outputs should be equal. Therefore, the following assumptions are made:

$$\text{BF} = 1 \tag{11-9}$$

R–2R networks:

$$R_F = R_I = R_A = R \tag{11-10}$$

Binary-weighted network:

$$R_F = R_{eq} = \frac{R}{2} \tag{11-11}$$

These assumptions make the transfer function from E_{ni} to E_{no} equal to one in all four topologies. Thus $E_{no}^2 = E_{ni}^2$. Equations 11-12 through 11-15 define E_{no}^2 for each topology. E_t^2 is equal to $4kTR$ in volts^2 per hertz.

R–2R with a voltage follower:

$$E_{no}^2 = E_n^2 + E_t^2 + I_n^2R^2 \tag{11-12}$$

R–2R with an inverting amplifier:

$$E_{no}^2 = 4E_n^2 + 2E_t^2 + I_n^2R^2 \tag{11-13}$$

R–2R with a noninverting amplifier:

$$E_{no}^2 = 4E_n^2 + 4E_t^2 + 4I_n^2R^2 \tag{11-14}$$

Binary-weighted with an inverting summer:

$$E_{no}^2 = 4E_n^2 + \tfrac{3}{2}E_t^2 + \tfrac{1}{4}I_n^2R^2 \tag{11-15}$$

TABLE 11-1 E_{no}^2 Terms

Topology	Bit factor	R_{eq}	E_n^2	I_n^2	Network E_t^2	E_t^2 of R_F	E_t^2 of R_I	E_t^2 of R_A
R–$2R$ with voltage follower	$\frac{2^N-1}{2^N}$	R	1	$(R_{eq})^2$	$(\text{BF})^2$	—	—	—
R–$2R$ with inverting amplifier	$\frac{2^N-1}{2^N}$	R	$\left(\frac{R_F+R_{eq}}{R_{eq}}\right)^2$	R_F^2	$(\text{BF})^2\frac{R_F^2}{R_{eq}^2}$	1	—	—
R–$2R$ with noninverting amplifier	$\frac{2^N-1}{2^N}$	R	$\left(\frac{R_F+R_I}{R_I}\right)^2$	$\left(\frac{R_AR_{eq}}{R_A+R_{eq}}+\frac{R_FR_I}{R_F+R_I}\right)^2 \times\left(\frac{R_F+R_I}{R_I}\right)^2$	$(\text{BF})^2\left(\frac{R_A}{R_A+R_{eq}}\right)^2 \times\left(\frac{R_F+R_I}{R_I}\right)^2$	1	$\left(\frac{R_F}{R_I}\right)^2$	1
Binary-weighted with inverting summer	1	$\frac{R}{1+\frac{2^{N-1}-1}{2^{N-1}}}$	$\left(\frac{R_F+R_{eq}}{R_{eq}}\right)^2$	R_F^2	$\frac{R_F^2}{R_{eq}^2}$	1	—	—

TABLE 11-2 E_{ni}^2 Terms

Topology	E_{no}^2/E_{ni}^2	E_n^2	I_n^2	Network E_t^2	E_t^2 of R_F	E_t^2 of R_I	E_t^2 of R_A
R–$2R$ with voltage follower	$(\mathrm{BF})^2$	$\dfrac{1}{(\mathrm{BF})^2}$	$\dfrac{R_{\mathrm{eq}}^2}{(\mathrm{BF})^2}$	1	—	—	—
R–$2R$ with inverting amplifier	$(\mathrm{BF})^2\dfrac{R_F^2}{R_{\mathrm{eq}}^2}$	$\left(\dfrac{R_F+R_{\mathrm{eq}}}{(\mathrm{BF})R_F}\right)^2$	$\dfrac{R_{\mathrm{eq}}^2}{(\mathrm{BF})^2}$	1	$\left(\dfrac{R_{\mathrm{eq}}}{(\mathrm{BF})R_F}\right)^2$	—	—
R–$2R$ with noninverting amplifier	$\left(\dfrac{(\mathrm{BF})R_A}{R_A+R_{\mathrm{eq}}}\right)^2\left(\dfrac{R_F+R_I}{R_I}\right)^2$	$\left(\dfrac{R_A+R_{\mathrm{eq}}}{(\mathrm{BF})R_A}\right)^2$	$\left(\dfrac{R_AR_{\mathrm{eq}}}{R_A+R_{\mathrm{eq}}}+\dfrac{R_FR_I}{R_F+R_I}\right)^2 \times\left(\dfrac{R_A+R_{\mathrm{eq}}}{(\mathrm{BF})R_A}\right)^2$	1	$\left(\dfrac{R_A+R_{\mathrm{eq}}}{(\mathrm{BF})R_A}\right)^2 \times\left(\dfrac{R_I}{R_F+R_I}\right)^2$	$\left(\dfrac{R_A+R_{\mathrm{eq}}}{(\mathrm{BF})R_A}\right)^2 \times\left(\dfrac{R_F}{R_F+R_I}\right)^2$	$\left(\dfrac{R_A+R_{\mathrm{eq}}}{(\mathrm{BF})R_A}\right)^2 \times\left(\dfrac{R_I}{R_F+R_I}\right)^2$
Binary-weighted with inverting summer	$\dfrac{R_F^2}{R_{\mathrm{eq}}^2}$	$\left(\dfrac{R_F+R_{\mathrm{eq}}}{R_F}\right)^2$	R_{eq}^2	1	$\left(\dfrac{R_{\mathrm{eq}}}{R_F}\right)^2$	—	—

These equations show that the R–$2R$ network with a voltage follower produces the lowest noise level. The largest noise level comes from the R–$2R$ network with a noninverting amplifier. The voltage follower topology produces the least noise not only because it has the fewest components but because its forward noninverting gain is equal to one. Thus there is no multiplication of the E_n and I_n terms contributing to E_{no}^2 as there are with the other topologies. These equations also show that if the amplifier's noise is predominantly due to I_n, then a binary-weighted network could possibly be better than the voltage follower topology because the I_n term is divided by four.

11-3 D/A SPICE SIMULATIONS

The circuits of Fig. 11-3 were simulated in PSpice with $N = 8$ and $R = 10\ \text{k}\Omega$. For Fig. 11-3*b* and *d*, R_F was equal to R. For Fig. 11-3*c*, $R_A = R$ and $R_F = R_I = 1\ \text{k}\Omega$. The model of Fig. 11-5 was used for the op amps. E_n and I_n were modeled as white-noise sources using the techniques outlined in Chap. 4. E_n was set equal to 20 nV/Hz$^{1/2}$, and I_n was set equal to 0.5 pA/Hz$^{1/2}$. These values are typical for a 741 op amp [1, 12]. Table 11-3 compares the PSpice E_{no}^2 (V^2/Hz) values and the theoretical E_{no}^2 (V^2/Hz) values derived from the results in Table 11-1 for each of the four topologies. Table 11-4 compares the PSpice E_{ni} (V/Hz$^{1/2}$) values with those derived from the results in Table 11-2.

Clearly, the correspondence between the theoretical results and the PSpice results is very good (less than 0.51% difference). Again, the R–$2R$ network with a voltage follower was found to produce the lowest noise. The R–$2R$ network with a noninverting amplifier has an E_{no}^2 which is nearly 3.4 times greater than the E_{no}^2 of the voltage follower configuration for the component values chosen. It should also be noted that the binary-weighted network with an inverting summer produces the second-lowest noise level for the component values chosen because of the reduction of the I_n term by a factor of 4.

TABLE 11-3 E_{no}^2 Comparison

Topology	SPICE (V^2/Hz)	Theoretical (V^2/Hz)
R–$2R$ with voltage follower	$5.92E-16$	$5.89E-16$
R–$2R$ with inverting amplifier	$1.96E-15$	$1.96E-15$
R–$2R$ with noninverting amplifier	$2.00E-15$	$1.99E-15$
Binary-weighted with inverting summer	$1.77E-15$	$1.76E-15$

TABLE 11-4 E_{ni} Comparison

Topology	SPICE (V/Hz$^{1/2}$)	Theoretical (V/Hz$^{1/2}$)
R–2R with voltage follower	2.44E – 8	2.44E – 8
R–2R with inverting amplifier	4.44E – 8	4.44E – 8
R–2R with noninverting amplifier	4.49E – 8	4.49E – 8
Binary-weighted with inverting summer	4.22E – 8	4.22E – 8

TABLE 11-5 Individual Contributions to E_{no}^2 (V^2/Hz)

Topology	E_n^2	I_n^2	Network E_t^2	E_t^2 of R_F	E_t^2 of R_I	E_t^2 of R_A
R–2R with voltage follower	4.0E – 16	2.5E – 17	1.64E – 16	—	—	—
R–2R with inverting amplifier	1.6E – 15	2.5E – 17	1.64E – 16	1.66E – 16	—	—
R–2R with noninverting amplifier	1.6E – 15	3.03E – 17	1.64E – 16	1.66E – 17	1.66E – 17	1.66E – 1
Binary-weighted with inverting summer	1.59E – 15	6.25E – 18	8.25E – 17	8.28E – 17	—	—

TABLE 11-6 Individual Contributions to E_{no}^2 Using Low-Noise Amplifier (V^2/Hz)

Topology	E_n^2	I_n^2	Network E_t^2	E_t^2 of R_F	E_t^2 of R_I	E_t^2 of R
R–2R with voltage follower	9.0E – 18	1.6E – 17	1.64E – 16	—	—	—
R–2R with inverting amplifier	3.6E – 17	1.6E – 17	1.64E – 16	1.66E – 16	—	—
R–2R with noninverting amplifier	3.6E – 17	1.94E – 17	1.64E – 16	1.66E – 17	1.66E – 17	1.66E –
Binary-weighted with inverting summer	3.59E – 17	4.0E – 18	8.25E – 17	8.28E – 17	—	—

Table 11-5 lists the contributions to E_{no}^2 made by each of the terms for $N = 8$, $R = 10\ \text{k}\Omega$, $E_n = 20\ \text{nV/Hz}^{1/2}$, and $I_n = 0.5\ \text{pA/Hz}^{1/2}$. The E_n term is the dominant noise source in all four topologies. If a low-noise amplifier such as the OP-37 [12, 13] with a much lower E_n is used, the noise is significantly reduced. The values in Table 11-6, calculated with an E_n of $3\ \text{nV/Hz}^{1/2}$ and an I_n of $0.4\ \text{pA/Hz}^{1/2}$, reflect this fact.

The low-noise amplifier significantly reduces the noise figure of the circuits because the optimum source resistance R_o is closer to the equivalent resistance R_s seen by the I_n noise source. For R_s to be equal to R_o, R_s must equal E_n/I_n which is 7.5 kΩ for the low-noise amplifier. For the voltage follower topology and $R = 10\ \text{k}\Omega$, the noisier op amp has a noise figure of 5.68 dB while the low-noise op amp has a noise figure of 0.63 dB. If the optimum source resistance is used (i.e., $R = 7.5\ \text{k}\Omega$ in the voltage follower topology) with the low-noise op amp, the noise figure is further reduced to 0.607 dB.

Up to this point two key noise sources have been excluded, the voltage reference noise E_{ref} and the resistor excess noise E_{ex}. E_{ref} is assumed to have a noise spectrum shape similar to that of the op amp E_n and a value of $100\ \text{nV/Hz}^{1/2}$ in the white-noise region. The resistor's noise index is assumed to be -10 dB. The D/A converter topologies that are considered with these sources are shown in Fig. 11-6.

With all digital inputs assumed to be logic 1's, E_{ref} is connected to each input resistor in the network. Recall that the R–$2R$ resistor network with the op amp introduces a bit factor (BF) which is the voltage gain of the network if all of the inputs are connected together (see Eq. 11-2). If the feedback resistor in the binary-weighted topology is chosen such that R_F is half of R, then the voltage gain to the output, if all inputs are connected together, is the same as that of the R–$2R$ topology. Otherwise, the voltage gain from the inputs to the output is given by

$$A_v = -\frac{R_F}{R_{\text{eq}}} \tag{11-16}$$

where R_{eq} is defined in Table 11-1. Therefore, the voltage reference noise contribution to E_{no}^2 is given by the product of E_{ref}^2 as a function of frequency and either $(\text{BF})^2$ or $(A_v)^2$ as appropriate. When some of the input bits are logic 0's, the E_{ref}^2 contribution to E_{no}^2 is weighted according to the following equation:

$$E_{\text{ref}}^2 \text{ contribution} = \left(\sum_{n=0}^{N-1} b_n \frac{1}{2^{N-n}} \right)^2 \left[E_{\text{ref}}^2 + \frac{E_{\text{ref}}^2(f_{nc\text{ref}})}{f} \right] \tag{11-17}$$

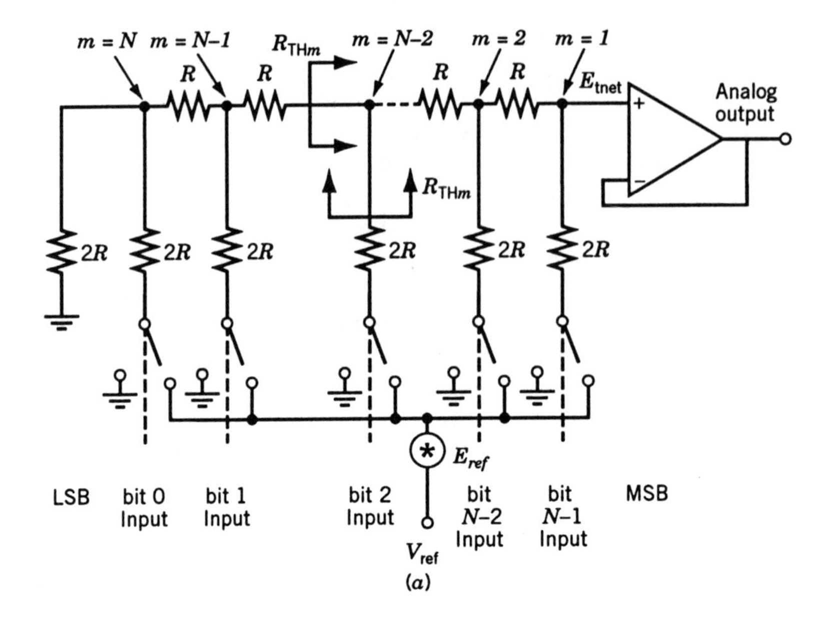

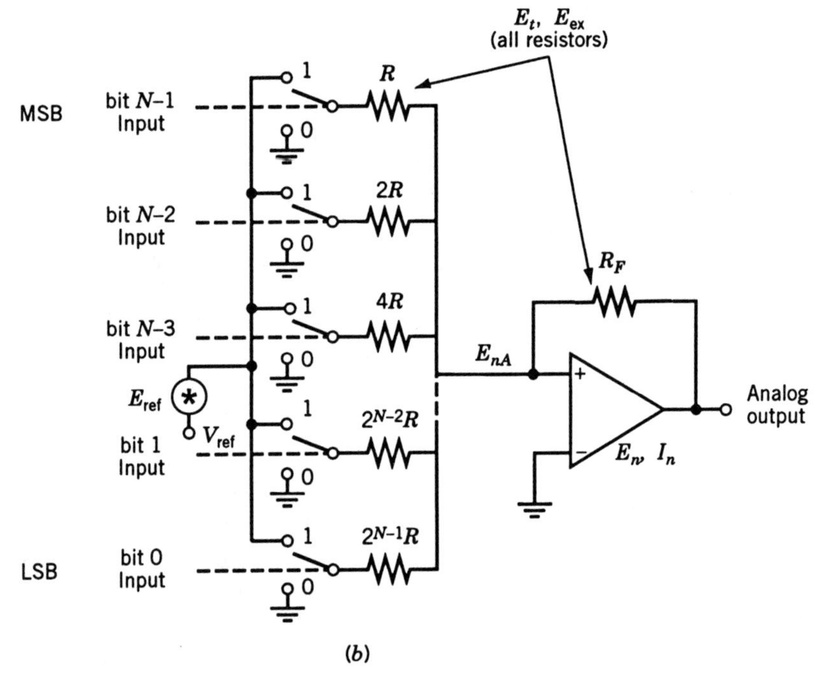

Figure 11-6 N-bit voltage-scaled D/A converter topologies: (a) R–$2R$ network and (b) binary-weighted network.

where N is the number of bits, n is the bit of interest, and b_n is the logic value of bit n (either 1 or 0).

The spectral density excess noise of a resistor, E_{ex}^2, is given by

$$E_{ex}^2 = \frac{10^{NI/10} \times 10^{-12}}{\ln 10}\left(\frac{V_{DC}^2}{f}\right) \tag{11-18}$$

where NI is the resistor noise index in decibels, f is the frequency in hertz, and V_{DC} is the dc voltage across the resistor. For the binary-weighted topology, the voltage that appears across an input resistor is zero if b_n is zero, and it is equal to V_{ref} if b_n is one. Equation 11-18 is used to calculate the spectral density excess noise of an input resistor and then this spectral density noise is multiplied by the square of the forward inverting gain of the op amp to get that input resistor's contribution to E_{no}^2. The excess noise from R_F makes a direct contribution to E_{no}^2 and is calculated using Eq. 11-18 where V_{DC} is the analog output voltage of the D/A converter, V_{out}.

For the R–$2R$ topology, a positional index, m, is used in Fig. 11-6a to designate specific nodes as shown. The voltage that appears across any resistor in the network is a function of V_{ref}, N, and m. The voltage, V_m, at any node m is given by

$$V_m = \sum_{j=1}^{m}\left(\frac{R_{THj}}{2R + R_{THj}} V_{ref} b_j\right)\left(\frac{1}{2^{m-j}}\right)$$

$$+ \sum_{j=m+1}^{N}\left[\left(\frac{R_{THj}}{2R + R_{THj}} V_{ref} b_j\right)\left(\prod_{p=m}^{j-1}\left(\frac{R_{THp}}{R_{THp} + R}\right)\right)\right] \tag{11-19}$$

where R_{THj} represents the Thevenin resistance looking into the jth node from the jth $2R$ input resistor and b_j is the logic value (0 or 1) of the jth bit. A resistance also equal to R_{THj} is seen looking into the jth node from the point just to the left of the jth node. These resistances are shown in Fig. 11-6a. For a converter of five or more bits, $R_{THm} \approx R$ for $m \geq 5$ with less than 0.3% error. The values for R_{TH1}, R_{TH2}, R_{TH3}, and R_{TH4} are $2R$, $1.2R$, $1.048R$, and $1.012R$, respectively.

After calculating each node voltage, the excess noise contributions can be calculated for each resistor using Eq. 11-18. The spectral density excess noise of each resistor is then multiplied by the appropriate square of the transfer function from the resistor of interest to the output. For the $2R$ resistors, the transfer function is given by

$$A_v = \frac{1}{2^m} \tag{11-20}$$

TABLE 11-7 Individual Contributions to E_{no}^2 (10^{-16} V^2/Hz): R–$2R$ Resistor Network

	Noise Source					
Frequency (Hz)	E_n^2	I_n^{2a}	Total E_t^{2a}	Total E_{ex}^{2a}	E_{ref}^2	Total E_{no}^2
1	44.01	—	—	1.47	1,092	1137
10	8.01	—	—	—	198.6	206.9
100	4.41	—	—	—	109.3	113.9
1000	4.05	—	—	—	100.3	104.6
10,000	4.01	—	—	—	99.45	103.6
100,000	4.01	—	—	—	99.37	103.5
1,000,000	4.01	—	—	—	99.36	103.5

[a]Contributions from these noise sources were always less than 0.16% of E_{no}^2 and consequently represent insignificant additions to the total E_{no}^2.

and for the R resistors, the transfer function is given by

$$A_v = \frac{1}{2^{m-1}} \tag{11-21}$$

Tables 11-7 and 11-8 show the individual noise source contributions to the output spectral density noise, E_{no}^2, for the R–$2R$ and binary-weighted converters, respectively, with the additional noise sources. In the converter with the R–$2R$ resistor network, there are two dominant contributors to E_{no}^2: voltage reference noise and op amp E_n noise. All other noise sources individually contribute less than 0.16% to the total output spectral density noise at any given frequency. For the converter with a binary-weighted resistor network, E_{no}^2 is significantly dominated by resistor excess noise at frequencies below 1 kHz and by voltage reference noise above 1 kHz.

At 1 Hz, the binary-weighted topology exhibits excess noise levels on the order of 5.76×10^{-12} V^2/Hz while the R–$2R$ topology has an excess noise level of 1.47×10^{-16} V^2/Hz. This significant difference in excess noise between the two topologies is caused by the difference in voltage across the resistors since excess noise is proportional to V_R^2. In the binary-weighted topology, all 10 V of the reference voltage appears across the input resistors, and nearly this same voltage (255/256 of V_{ref}) across the feedback resistor. In the R–$2R$ topology, only small fractions of the reference voltage appear across the resistors with the largest voltage drop being 6.66 V as previously noted. Furthermore, the excess noise contributed by this one resistor is significantly attenuated at the output of the converter. The binary-weighted converter produced slightly greater output noise at frequencies above 10 kHz than the R–$2R$ converter, but significantly greater output noise at frequencies below 10 kHz.

It must be remembered, however, that these simulations represent a worst-case analysis since the E_{ref} contribution to E_{no}^2 is dependent on the

TABLE 11-8 Individual Contributions to E_{no}^2 (10^{-16} V^2/Hz): Binary-Weighted Resistor Network

Frequency (Hz)	Noise Source					Total E_{no}^2
	E_n^2	I_n^{2a}	Total E_t^{2a}	Total E_{ex}^2	E_{ref}^2	
1	175.3	—	—	57,570	1,092	58,830
10	31.90	—	—	5,757	198.6	5,987
100	17.55	—	—	575.7	109.3	702.7
1,000	16.12	—	—	57.57	100.3	174.2
10,000	15.98	—	—	5.76	99.45	121.4
100,000	15.96	—	—	0.58	99.36	116.1
1,000,000	15.96	—	—	0.06	99.35	115.5

[a] Contributions from these noise sources were always less than 0.16% of E_{no}^2 and consequently represent insignificant additions to the total E_{no}^2.

TABLE 11-9 Individual Contributions to E_{no}^2 (10^{-16} V^2/Hz): Modified R–$2R$ Resistor Network

	Noise Source						Reduced NI
Frequency (Hz)	E_n^{2a}	I_n^{2a}	Total E_t^{2a}	Total E_{ex}^2	E_{ref}^2	Total E_{no}^2	Total E_{no}^2
1	—	—	—	14,381	221.0	14,610	1,667
10	—	—	—	1,438	40.21	1,479	184.7
100	—	—	—	143.8	22.13	166.2	36.77
1,000	—	—	—	14.38	20.32	34.96	21.96
10,000	—	—	—	1.44	20.14	21.84	20.54
100,000	—	—	—	0.14	20.12	20.52	20.39
1,000,000	—	—	—	0.01	20.12	20.39	20.38

[a]Contributions from these noise sources were always less than 0.82% of E_{no}^2 and consequently represent insignificant additions to the total E_{no}^2.

digital input. E_{ref} would have no contribution to E_{no}^2 if the digital inputs were all logic 0's. This leads to a design modification in which an extra bit is added to the converter, the reference voltage V_{ref} is doubled, and only the lower $N - 1$ bits have logic inputs. The most significant bit is always kept at a logic 0. Doubling the reference voltage keeps the output voltage range the same as in the original D/A converter.

Only the R–$2R$ converter model was modified and simulated with the lower-noise components because addition of another bit, or equivalently doubling the RF or halving R in the binary-weighted topology, would still necessitate doubling V_{ref} to keep the output voltage range the same. Doubling V_{ref} would increase the excess noise by a factor of 4, and excess noise was already a problem in this topology.

Table 11-9 shows the individual noise source contributions to E_{no}^2 when lower-noise components are used in the modified R–$2R$ converter. With E_n equal to 3 nV/Hz$^{1/2}$ and E_{ref} equal to 90 nV/Hz$^{1/2}$, the addition of another bit significantly reduces E_{no}^2 at frequencies above 100 Hz. At frequencies lower than 100 Hz, the output spectral density noise is increased dramatically because of greater excess noise from doubling V_{ref} and the addition of the two extra resistors. At frequencies above 10 Hz, the extra bit reduces the E_{ref} contribution to E_{no}^2 by a factor of 4 which significantly reduces E_{no}^2 since E_{ref} is nearly the sole contributor to E_{no}^2 above 100 Hz. If lower-noise index resistors (NI = -20 dB) are used in addition to lower E_{ref} and E_n, E_{no}^2 is greatly reduced. This is shown in the last column of Table 11-9.

These results show that the low-noise design of voltage-scaled D/A converters involves using a low-noise op amp and the lowest-noise voltage reference possible, in the R–$2R$ resistor network converter topology. If the full-scale output voltage range is 5 V or less, the addition of another bit whose input is grounded as the MSB of the converter is a viable method of significantly reducing E_{no}^2.

11-4 NOISE IN A FLASH A/D CONVERTER

The flash A/D converter of Fig. 11-7 has a labeled positional index, m, to designate the mth node of the circuit. A voltage reference V_{ref} is divided into equally spaced threshold voltages at the inverting inputs of the comparators. The analog input voltage is simultaneously compared to each of the threshold voltages. The voltage reference noise and the comparator noise sources exhibit voltage and current noise with white-noise spectra beyond a lower-noise corner frequency and $1/f$ spectra at frequencies below the noise corner frequency. The resistors in the divider generate thermal noise as well as excess noise with a $1/f$ spectrum. A bit error will occur if the sum of the instantaneous noise and the analog input voltage exceed the quantization interval in which the analog input occurs.

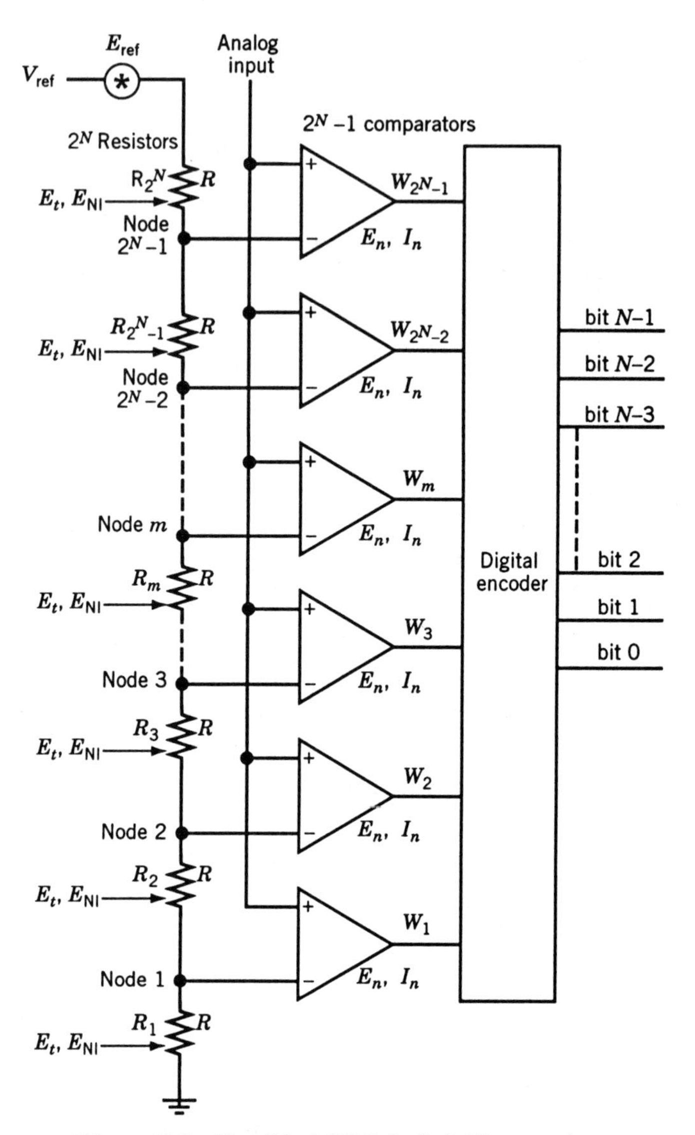

Figure 11-7 Simplified N-bit flash A/D converter.

It is possible to include a noise contribution from the digital encoder, E_{DNL}, by adding another noise source whose value corresponds to the encoder noise at the output of the comparator of interest multiplied by the reverse transmission gain of the comparator. Since the reverse transmission gain of a comparator is quite small, E_{DNL} is usually not significant in determining whether or not an error will occur in a comparator's decision. Thus E_{DNL} is excluded from this analysis. E_{DNL} becomes significant at the output of the converter only if it exceeds the noise margin of the decoder circuitry.

The noise generated by a comparator operating in its transition region can be modeled in terms of an E_n noise voltage source and an I_n noise current source in the same way that the noise of an op amp is described. Noise at the input to any one comparator has contributions from that comparator's E_n source, all I_n sources, each resistor in the divider, and the voltage reference. The other comparators' E_n sources do not contribute to the equivalent noise at the input of the comparator of interest because of the high input impedance of the comparators associated with the other E_n sources. Using the positional index m to designate the mth node at the input to the mth comparator, Eq. 11-22 gives the total spectral density noise voltage present at the input to the mth comparator E_{nim}^2:

$$E_{nim}^2 = \left[E_{nm}^2 + \frac{E_{nm}^2(f_{ncE})}{f}\right] + E_{tm}^2 + \sum_{j=1}^{2^N-1} E_m^2(I_{nj}) + \left(\frac{m}{2^N}\right)^2\left[E_{\text{ref}}^2 + \frac{E_{\text{ref}}^2(f_{nc\text{ref}})}{f}\right] + E_{\text{ex}m}^2 \tag{11-22}$$

for $m = 1, 2, 3, \ldots, 2^N - 1$.

E_{tm}^2 and $E_{\text{ex}m}^2$ are the total noise contributions at the mth node caused by thermal noise and resistor current noise, respectively. They are described by Eqs. 11-23 and 11-24 where E_t^2 is the thermal noise of a single resistor of resistance R, and E_{ex}^2 is the excess noise of a single resistor in the voltage divider:

$$E_{tm}^2 = mE_t^2\left(\frac{2^N - m}{2^N}\right)^2 + (2^N - m)E_t^2\left(\frac{m}{2^N}\right)^2 \tag{11-23}$$

$$E_{\text{ex}m}^2 = mE_{\text{ex}}^2\left(\frac{2^N - m}{2^N}\right)^2 + (2^N - m)E_{\text{ex}}^2\left(\frac{m}{2^N}\right)^2 \tag{11-24}$$

$E_m^2(I_{nj})$ is the noise voltage squared at the mth node caused by the jth I_n noise current source. The I_n contributions to the noise at node m are given by

$$E_m^2(I_{nj}) = \left[E_j^2(I_{nj})\right]\left(\frac{m}{j}\right)^2 \quad \text{if} \quad m \le j \tag{11-25}$$

or

$$E_m^2(I_{nj}) = \left[E_j^2(I_{nj})\right]\left[\frac{2^N - m}{2^N - j}\right]^2 \quad \text{if} \quad m \ge j \tag{11-26}$$

$E_j^2(I_{nj})$ is the noise at the jth node caused by the jth I_n noise current source and is equal to

$$E_j^2(I_{nj}) = \left[\frac{j(2^N - j)R}{2^N}\right]^2 \left[I_{nj}^2 + \frac{I_{nj}^2(f_{ncI})}{f}\right] \tag{11-27}$$

Determining which comparator has the greatest amount of equivalent noise at its input requires analysis of a complex function of m, N, frequency, and the spectra of the noise sources. As far as individual sources are concerned, the voltage reference contributes the most noise at the input to the $2^N - 1$th (i.e., the top) comparator. All other sources of noise contribute the most noise at the 2^{N-1}th (i.e., the middle) comparator.

11-5 A/D SPICE SIMULATION

This A/D analysis is concerned only with finding the equivalent input noise at the input to each comparator and not with modeling of the A/D function. The op amp model of Fig. 11-5 was used in the SPICE simulation connected as a voltage follower in place of each comparator. To simulate the equivalent noise at the input to a comparator, it is only necessary that the E_n and I_n noise spectra of the comparator be the same as the E_n and I_n noise spectra of the op amp used in the SPICE simulations. Then, the noise at the output of each op amp in a voltage follower configuration is equal to the noise at the input of each op amp and therefore *represents* the noise that would be present at the input to each comparator.

For the SPICE simulation, R was set equal to 100 Ω. E_{ref} was set equal to 100 nV/Hz$^{1/2}$ with a noise corner frequency, $f_{nc\text{ref}}$, of 10 Hz. These values are typical of manufacturers' voltage reference data [12–14]. Measured E_n data on four units of two commercially available comparators, LM-339's and L-161's, yielded values from approximately 10 to 30 nV/Hz$^{1/2}$ which are in the same general range of op amp E_n data. Thus a value of 20 nV/Hz$^{1/2}$ was chosen for E_n in the simulation. E_n noise corner frequencies ranged from approximately 1 kHz for the LM-339's to frequencies below 100 Hz for the L-161's. Consequently, a value of 100 Hz was chosen for the E_n noise corner frequency f_{ncE}. Since it was impossible to stabilize the comparator to measure I_n, I_n was set equal to 0.5 pA/Hz$^{1/2}$, and the I_n noise corner frequency, f_{ncI}, was set equal to 100 Hz.

The op amp was modeled as a subcircuit in the SPICE listing, and the subcircuit was called seven times in simulating the case of $N = 3$. The noise sources were modeled using diodes and dependent sources as outlined in [15]. The subcircuit approach is a much simpler way to simulate the numerous noise sources while keeping E_n and I_n sources uncorrelated. The same

TABLE 11-10 Total E_{ni}^2 (10^{-16} V^2/Hz) at Each Node Versus Frequency

Frequency (Hz)	Node 1	Node 2	Node 3	Node 4	Node 5	Node 6	Node 7
1	1,015	1,491	1,831	2,036	2,106	2,041	1,840
10	106.5	158.3	199.4	229.8	249.5	258.4	256.6
100	15.68	25.09	36.25	49.14	63.78	80.16	98.28
1,000	6.594	11.77	19.93	31.08	45.22	62.34	82.45
10,000	5.686	10.44	18.30	29.27	43.66	60.56	80.86
100,000	5.595	10.30	18.14	29.09	43.17	60.38	80.70
1,000,000	5.586	10.29	18.12	29.07	43.15	60.36	80.69

result would have occurred if 14 separate diode noise references were used for the E_n and I_n noise sources.

Table 11-10 shows the square of the total amount of equivalent input noise, E_{ni}^2, which is present at the inputs of all comparators as a function of the positional index m and frequency for a 3-bit flash A/D converter. The data in Table 11-10 show that the node with the most noise is a function of frequency. However, as N increases, the function which determines the node with the most noise will depend less on frequency since E_{ref} which is greatest at the top node will have a greater contribution to E_{ni}^2. Table 11-11 shows the individual noise source contributions to E_{ni}^2 at the middle comparator ($m = 4$) at node 4 as a function of frequency for a 3-bit flash A/D converter. Notice that the voltage reference noise E_{ref} is dominant at frequencies of 100 Hz and above, even at the middle comparator where all of the other noise sources have their maximum contribution. This dominance by E_{ref} explains why the greatest equivalent input noise is found at the top comparator for frequencies above 100 Hz in Table 11-10. The smallest equivalent noise is found at the first ($m = 1$) comparator since the contributions from all of the noise sources are at a minimum for $m = 1$.

TABLE 11-11 Contributions to the Total E_{ni}^2 (10^{-16} V^2/Hz) at the Middle Node for a 3-Bit Flash A/D Converter

Frequency (Hz)	Noise Source					
	E_{ref}	Thermal[a]	Excess	E_n	I_n^a	Total E_{ni}^2
1	275.0	—	1,357	404.0	—	2,036
10	50.03	—	135.7	44.01	—	229.8
100	27.53	—	13.57	8.005	—	49.14
1,000	25.28	—	1.357	4.405	—	31.08
10,000	25.06	—	—	4.045	—	29.27
100,000	25.04	—	—	4.009	—	29.09
1,000,000	25.03	—	—	4.006	—	29.07

[a]Contributions from these sources were insignificant.

TABLE 11-12 Contributions to the Total E_{ni}^2 (10^{-16} V^2/Hz) at the Top Node for a 5-Bit Flash A/D Converter

Frequency (Hz)	Noise Source					
	E_{ref}	Thermal[a]	Excess	E_n	I_n^a	Total E_{ni}^2
1	1,032	—	41.08	404.0	—	1,478
10	187.8	—	4.108	44.01	—	235.9
100	103.4	—	—	8.005	—	111.8
1,000	94.92	—	—	4.405	—	99.38
10,000	94.06	—	—	4.045	—	98.13
100,000	93.98	—	—	4.009	—	98.01
1,000,000	93.98	—	—	4.006	—	98.00

[a]Contributions from these sources were insignificant.

Tables 11-12 and 11-13 respectively show the individual noise contributions to E_{ni}^2 versus frequency for a 5-bit flash converter at the top ($2^N - 1$th $=$ 31st) comparator where E_{ref} is most significant, and at the middle (2^{N-1}th $=$ 16th) comparator where all other noise sources are most significant. Tables 11-14 and 11-15 show the individual noise contributions to E_{ni}^2 versus frequency for an 8-bit flash converter.

The total amount of equivalent input noise increases as N increases, but the percentage increase in E_{ni} per 1-bit increase in N decreases as N becomes large. The dominant noise source in all cases in the white-noise region of E_{ni} is the voltage reference. E_n is also a significant noise contributor with respect to the other noise sources, particularly at the top comparator. At the middle comparator, the E_n contribution becomes less significant as N increases. Increasing N to 8 or more bits causes the I_n contribution to become more significant than the E_n contribution. Resistor excess noise is

TABLE 11-13 Contributions to the Total E_{ni}^2 (10^{-16} V^2/Hz) at the Middle Node for a 5-Bit Flash A/D Converter

Frequency (Hz)	Noise Source					
	E_{ref}	Thermal	Excess	E_n	I_n^a	Total E_{ni}^2
1	275.0	—	339.2	404.0	—	1,020
10	50.03	—	33.92	44.01	—	128.3
100	27.53	—	3.392	8.005	—	39.10
1,000	25.28	—	0.339	4.405	—	30.18
10,000	25.06	0.133	—	4.045	—	29.29
100,000	25.04	0.133	—	4.009	—	29.20
1,000,000	25.03	0.133	—	4.006	—	29.19

[a]Contributions from these sources were insignificant.

TABLE 11-14 Contributions to the Total E_{ni}^2 (10^{-16} V^2/Hz) at the Top Node for an 8-Bit Flash A/D Converter

Frequency (Hz)	Noise Source					
	E_{ref}	Thermal[a]	Excess[a]	E_n	I_n^a	Total E_{ni}^2
1	1,092	—	—	404.0	—	1,497
10	198.5	—	—	44.01	—	242.7
100	109.3	—	—	8.005	—	117.3
1,000	100.3	—	—	4.405	—	104.8
10,000	99.46	—	—	4.045	—	103.5
100,000	99.36	—	—	4.009	—	103.4
1,000,000	99.36	—	—	4.006	—	103.4

[a]Contributions from these sources were insignificant.

TABLE 11-15 Contributions to the Total E_{ni}^2 (10^{-16} V^2/Hz) at the Middle Node for an 8-Bit Flash A/D Converter

Frequency (Hz)	Noise Source					
	E_{ref}	Thermal[a]	Excess[a]	E_n	I_n	Total E_{ni}^2
1	275.0	—	—	404.0	865.0	1,588
10	50.03	—	—	44.01	94.22	193.8
100	27.53	—	—	8.005	17.14	54.45
1,000	25.28	—	—	4.405	9.432	40.51
10,000	25.06	—	—	4.045	8.662	39.12
100,000	25.04	—	—	4.009	8.585	38.98
1,000,000	25.03	—	—	4.006	8.577	38.96

[a]Contributions from these sources were insignificant.

greatest when N is small (e.g., $N = 3$) because the dc voltage drop across the resistors in the divider is greater when N is small.

11-6 CONVERTING ANALOG NOISE TO BIT ERRORS

In 1974, Bernard Gordon presented a probabilistic analysis of the noise effects on A/D conversion accuracy [10, 11] which was based on a converter model of a single comparator and one equivalent noise source at the comparator input. The action of the comparator in the presence of noise ultimately determines whether or not there are any digital coding errors. Gordon models all of the converter noise sources by an equivalent noise source at the input to the comparator. While Gordon's analysis is quite useful in evaluating the effects of noise on conversion accuracy, it is first necessary to determine a realistic value for the equivalent noise source σ

from circuit topology or device selection to use in the analysis. A methodology is developed in this chapter [16].

In his analysis, Gordon reduces a converter down to a single comparator conceptually, and then examines the probability that the comparator will make an erroneous decision in the presence of noise given a certain-size quantum interval (one LSB of a converter) and level of rms noise (σ). All noise sources within a converter are modeled as a single equivalent noise source at the input to the comparator which is added to the input signal. The noise is assumed to have a Gaussian distribution. The reference input to the comparator is considered to be noise free and constant.

If the input voltage signal is constrained to always be centered within the quantum interval, then the probability of an error by the comparator is determined solely by the ratio of the quantum interval to the level of noise.

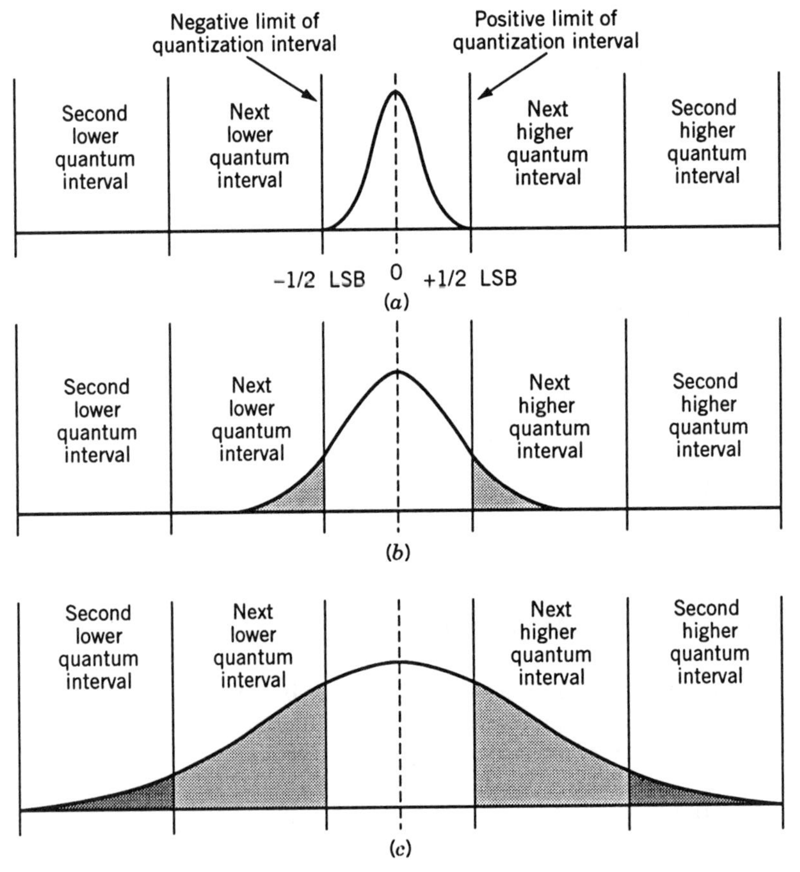

Figure 11-8 Comparator inputs centered in the quantum interval with added noise: (*a*) noise does not exceed quantum interval boundaries, (*b*) noise exceeds interval boundaries by less than 1 LSB, and (*c*) noise exceeds interval boundaries by more than 1 LSB.

Three different cases are shown in Fig. 11-8. In Fig. 11-8*a*, $\pm 3\sigma$ added to the input causes a small enough variation in the input signal such that the input signal always remains within the quantum interval boundaries. Thus there will be no comparator error (i.e., neither this comparator nor adjacent comparators in a flash A/D converter will be affected adversely by noise). In Fig. 11-8*b* and *c*, there is some nonzero probability that thc quantum interval boundaries will be exceeded when noise is added to the input signal. This probability is equal to the sum of the shaded areas under the Gaussian curve. If the ratio of σ to the size of the quantum interval is large as in Fig. 11-8*c*, more than one comparator in a string of comparators may have an error. The probability that two comparators will be in error is the sum of the darker shaded areas while the probability that only one comparator will be in error is the sum of the lighter shaded areas.

In a flash A/D converter, this probabilistic analysis is applicable to any of the comparators and their associated quantum interval. In Sec. 11-5 the comparator of interest was designated the mth comparator and was attached to the mth node of the resistive divider (see Fig. 11-7). Noise causes one of two types of errors to occur. First, if the input signal plus noise exceeds the upper quantum interval boundary, then the mth comparator decides correctly while more significant comparator(s) incorrectly produce a logic 1 output. The number of comparators above the mth comparator that produce an incorrect logic 1 output is dependent on the amount of noise present and the converter's resolution. If the input signal plus noise exceeds the lower quantum interval boundary, then the mth comparator generates an incorrect logic 0, and, depending on the level of noise, one or more of the comparators below the mth comparator may also generate an incorrect logic 0.

The case where the input is no longer constrained to be centered in the interval is now considered. Here, the input can be considered to be shifted from the center by a specific amount, E, where E is defined to be less than 1/2 LSB in magnitude. Figure 11-9*a* and *b* shows the cases of E equal to +1/4 LSB and −1/4 LSB, respectively.

Figure 11-9 shows that a shift of the input from the center of the quantum interval will reduce the probability of one of the types of errors while increasing the probability of the other type of error. In any case, the probability that an error of either type will occur is the sum of the shaded areas. For every possible shift of E of the input from the quantum interval center, there is a different probability of an error. Thus the overall probability of an error is the sum of the shaded areas that occur for every possible value of E. These concepts are mathematically represented in the double integral of the following equation:

$$P_E = \int_{-(\mathrm{LSB}/2)}^{+(\mathrm{LSB}/2)} P(X|E)P(E)\,dE$$

$$= \frac{1}{(\mathrm{LSB})}\int_{-(\mathrm{LSB}/2)}^{+(\mathrm{LSB}/2)}\left[1 - \frac{1}{\sqrt{2\pi}}\int_{-\mathrm{LSB}/2-E}^{\mathrm{LSB}/2-E}\exp\left(\frac{-x^2}{2}\right)dx\right]dE \quad (11\text{-}28)$$

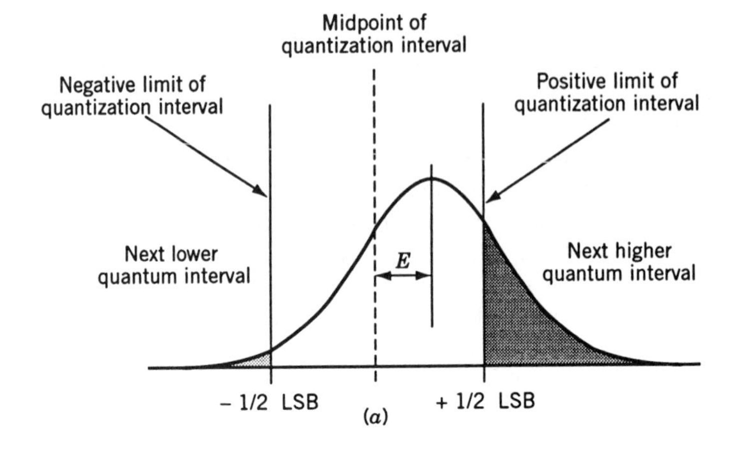

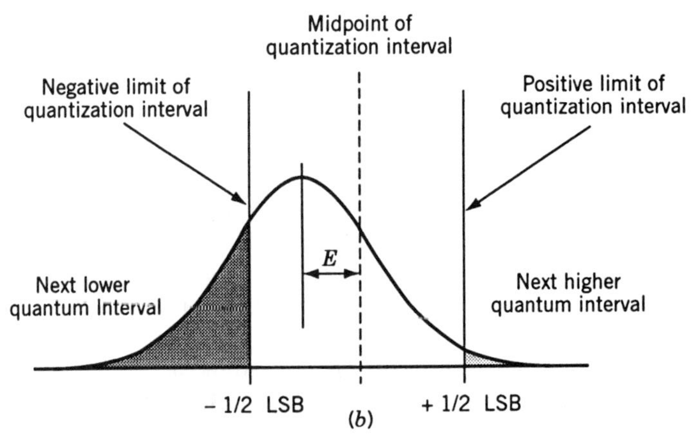

Figure 11-9 Non-centered comparator inputs with added noise: (*a*) input shifted $+1/4$ LSB from center of quantum interval and (*b*) input shifted $-1/4$ LSB from center of quantum interval.

P_E is the probability of a bit error, $P(E)\,dE$ is the probability of the analog input being shifted from the center of the quantization interval by an amount between E and $E + dE$, and $P(X|E)$ is the probability that the Gaussian noise $P(X)$ will exceed the quantization interval when it is superimposed on V with its mean at E.

The integrand of the outer integral represents the area under the Gaussian curve which exceeds the quantization interval boundaries for a fixed value of E (i.e., the shaded areas). For a particular value of E expressed as a fraction of a LSB, this area can be found using a look-up table of areas under a Gaussian curve. The outer integral is then evaluated numerically to give the total probability of a bit error for an analog input occurring anywhere within the quantization interval (i.e., $|E| \leq 1/2$ LSB). Equation 11-22 can be evaluated in a spreadsheet program using a rectangular approximation by dividing the $\pm 1/2$ LSB interval into 1000 equally spaced segments ($\Delta E = 0.001$).

11-7 BIT ERROR ANALYSIS OF THE FLASH CONVERTER

Since the largest noise is typically at the $2^N - 1$th (top) comparator, the greatest probability of a bit error occurs when the analog input voltage, V, is in the range of $[V_{ref}\ (2^N - 2)/2^N]$ to $[V_{ref}\ (2^N - 1)/2^N]$ where V_{ref} is the

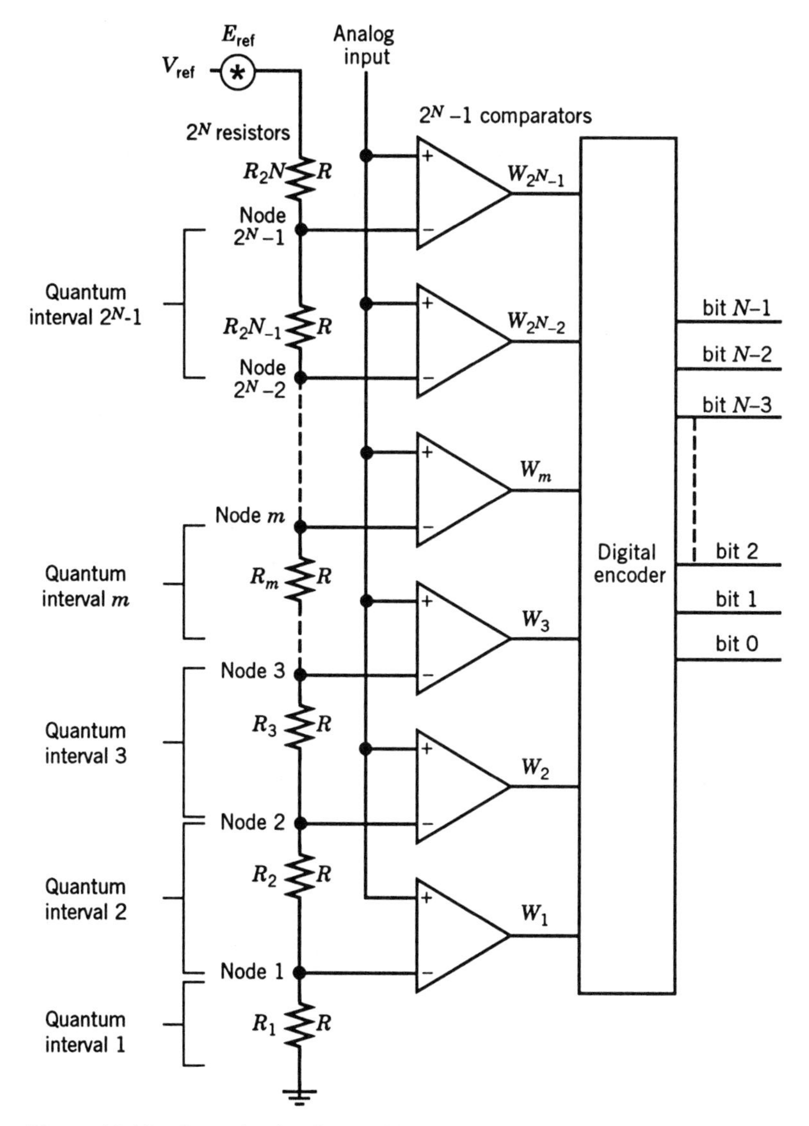

Figure 11-10 Quantization interval boundaries for a flash A/D converter.

TABLE 11-16 Comparator Error Probabilities (%)

Number of Bits	$V_{FS} = 5$ V	$V_{FS} = 10$ V
3	0.20	0.033
5	4.23	1.04
8	100	77.6

reference voltage. These are the lower and upper limits of the quantization interval at the top comparator, respectively. $[V_{\text{ref}}\,(2^N - 2)/2^N] + 1/2$ LSB is the center of the quantization interval. This is illustrated in Fig. 11-10.

It is assumed that the noise has a Gaussian amplitude distribution which is centered at the analog input voltage and that there is a uniform distribution of shift, E, of V from $[V_{\text{ref}}\,(2^N - 2)/2^N] + 1/2$ LSB.

Table 11-16 gives the maximum probability of a comparator error (i.e., error probability is less over other quantization intervals because the noise is less) for converters of 3, 5, and 8 bits for full-scale voltages of 5 V and 10 V using the noise parameter values in Sec. 11-5. For an 8-bit converter using a full-scale voltage of 5 V, the probability of a comparator error is 100% if the analog input can have a value anywhere within the quantization interval (i.e., within ± 0.5 LSB of the quantization interval center value). If the input voltage is constrained to be within ± 0.3 LSB of the center of the quantization interval, however, the probability of an error is on the order of 1% to 2%. Table 11-16 shows that as N decreases, the probability of a comparator error decreases significantly.

The effects of changing the noise parameters of various noise sources in the model are now examined to evaluate the converter noise performance improvement that could be obtained with low-noise design techniques. The noise parameters of the three most significant noise sources at the top node are considered for the three values of N previously simulated (i.e., 3, 5, and 8).

For a given value of N, if E_n was a dominant noise source, a new lower value of 3 nV/Hz$^{1/2}$, which is typical of low-noise op amp values, is used [12, 13]. A noise index of -20 dB (i.e., 10 dB lower) is used where the resistor current noise was a dominant source. In all simulations, the voltage reference was the greatest noise contributor in the white-noise region of E_{ni}^2. Consequently, E_{ref} is reduced by 10% to 90 nV/Hz$^{1/2}$.

Broadband noise present at the top node for the original noise parameters is compared with broadband noise generated with the lower-noise components in Table 11-17. These numbers are based on a comparator noise bandwidth of 220 MHz determined by computer simulation of comparator macromodels. Reducing the noise of the three most dominant sources results in a broadband noise decrease of approximately 12% for all values of N. The probabilities of a comparator error (for the same quantum interval previously discussed) shown in Table 11-18 exhibit much less change with lower noise

TABLE 11-17 Broadband Noise Comparison (μV)

Number of Bits	Full-Noise Model	Lower-Noise Model
3	1333.6	1171.2
5	1468.4	1295.0
8	1508.2	1331.5

components. For inputs over the entire quantum interval, the error probability is still 100% for the 8-bit converter with $V_{FS} = 5$ V.

The probability of a bit error is reduced by decreasing the ratio of the noise present at the comparator input to the resolution of the converter. This is done either by increasing V_{FS} at the expense of the converter's resolution, or by reducing the input noise level. Reduction of the noise entails using the lowest-noise voltage reference available, using low values of R (100 Ω or less), using comparators with low E_n values, particularly if N is small (e.g., $N = 3$), and using comparators with low I_n values if N is large (e.g., $N = 8$). Which approach is most feasible for reducing noise-generated coding errors and what constitutes an acceptable error probability will depend on the application for which the converter is used. As long as the ratio of (3 × the rms noise) to the converter resolution is less than one, any coding error that does occur will affect only the least-significant bit of the converter's digital output.

It is now appropriate to ask, "What does a probability of a bit error of, for example, 100% as given by Eq. 11-28 mean?" What it does not mean is that the probability of a bit error is 100% for any one conversion. For any one conversion, the probability of a bit error is dependent on the noise level, the converter resolution, and where the input occurs within the quantum interval, that is, on the amount of shift from the center of the quantum interval. This probability will always be less (usually significantly less) than the overall probability of a bit error. However, over a very large number of input samples to the converter, the probability of a bit error will approach that given by Eq. 11-28. In the example of this probability being 100%, most, but probably not all, of the conversions would be in error. However, without

TABLE 11-18 Comparator Error Probabilities (%) for Lowered Noise

Number of Bits	$V_{FS} = 5$ V	$V_{FS} = 10$ V
3	0.17	0.026
5	3.92	0.90
8	100	68.4

some a priori knowledge of the noise-free input signal, it is impossible to know which of the conversions are erroneous and which are not.

Thus, if the overall probability of a bit error is high, the validity of the lower bit(s) at the output of an A/D converter is questionable, and the effective resolution of the converter is reduced. The next section examines the reduction in converter resolution that occurs for the noise levels determined by the simulations described in Sec. 11-5.

11-8 A/D/A NOISE ANALYSIS

The effect of noise on an 8-bit A/D/A system is examined for a full-scale voltage of 10 V and an input voltage of 10 V minus 1.5 LSB, or 9.9414 V. The comparators of the flash A/D section are assumed to have a noise bandwidth of 2.75 MHz, and R is 100 Ω. The D/A converter uses a voltage follower and an R–$2R$ network with $R = 10\ \text{k}\Omega$. The op amp is assumed to have a noise bandwidth of 1.57 MHz. The comparators and the op amp have $E_n = 20\ \text{nV/Hz}^{1/2}$ and $I_n = 0.5\ \text{pA/Hz}^{1/2}$ and noise corner frequencies of 100 Hz.

From the SPICE simulation, the rms noise present at the input to the top comparator is $101.7\ \text{nV/Hz}^{1/2}$ or 168.65 μV over the noise bandwidth of the comparator. This magnitude of noise leads to a digital output of $(254)_{10}$ 99.7% of the time. Excluding noise in the D/A converter, this digital output corresponds to an analog voltage of 9.9219 V.

The rms noise voltage from the R–$2R$ network and the voltage follower is equal to 30.6 μV over the 1.57-MHz noise bandwidth. Thus 99.7% of the time, the D/A converter will add an amount of noise whose magnitude is less than or equal to 91.7 μV to the output of the A/D/A system. Thus the sum effect of noise on the overall system will cause the output to be approximately 9.9219 V, and the noise of the D/A converter is negligible.

If the analog input is allowed to be anywhere within the LSB interval (i.e., between 9.9219 V and 9.9609 V), the digital outputs will be between $(253)_{10}$ and $(255)_{10}$ 99.7% of the time. The analog output range will be between 9.8828 and 9.9609 V. The maximum error caused by noise is ±1 LSB.

The error caused by fundamental noise in a D/A or A/D converter can be predicted using the methodology described in this chapter. This error can then be included in a converter's error budget along with errors due to offsets, nonlinearities, and so on, and a more accurate determination of a converter's true resolution can be made which includes the effects of fundamental noise. Other noise sources (perhaps artificial noise) can easily be added to the noise model if their values at some point in the converter are known.

SUMMARY

a. Noise sources in bipolar, MOS, and GaAs converters can be modeled using SPICE. The effects of various noise sources on converter performance can be examined and dominant noise sources determined.

b. A noise analysis of an R–$2R$ network and a binary-weighted network illustrates the noise analysis methodology. Four different op amp summation schemes for voltage-referenced D/A converters and the noise at the inputs to the comparators of a flash A/D converter were also analyzed.

c. The SPICE files used in the noise analysis constitute noise models of the converter topologies. The noise parameters of the models can be adapted for more involved noise analysis of converters or to match the models' noise behavior to actual converter noise performance. The SPICE noise models will identify the dominant noise sources within a given topology.

d. The D/A noise model simulations show that the R–$2R$ network with a voltage follower produces the lowest noise. The binary-weighted network with an inverting summer produces the second-lowest noise level because of the reduction of the I_n term by a factor of 4.

e. Using a low-noise op amp significantly reduces the output noise of a D/A converter since the op amp E_n is by far the dominant noise source among the sources considered.

f. In the flash A/D converter, fundamental noise is dominated by voltage reference noise in the flash A/D converter while the importance of other noise sources is dependent on the number of bits, the frequency, the node of interest, and the converter topology.

g. The results of this analysis show that fundamental noise limits are an important consideration in the design of A/D and D/A converters.

h. The methodology and models can be used in conjunction with Gordon's probabilistic error analysis to predict more accurately the probability of a bit error in a converter. Thus the effects of noise can be taken into account in determining the true resolution of a converter.

PROBLEMS

11-1. Prove that E_{nA}^2 for the R–$2R$ network of Fig. 11-1 is $4kTR$.

11-2. Verify Eq. 11-1 for the equivalent noise voltage of the binary-weighted D/A network of Fig. 11-2.

11-3. Consider an 8-bit flash A/D converter of the type shown in Figure 11-7. Let $V_{ref} = 8.995$ V and $R = 100\ \Omega$.

(a) Calculate the reference voltage at the inverting input to each comparator.

(b) Now suppose the lowest resistor is changed to 50 Ω. Repeat part (a).

(c) Finally, suppose both the topmost and bottommost resistors are each 50 Ω and all others are still 100 Ω. Repeat part (a).

11-4. Study the 3-bit flash A/D converter example summarized in Table 11-11. At low frequencies the excess noise dominates.

(a) Determine how the excess noise present at 1 Hz would change under the condition that the bottommost resistor was changed to 50 Ω, with all the remaining resistors set to 1000 Ω. Let V_{ref} = 8.5 V.

(b) Now consider the noise due to the reference, E_{ref}. Determine the change in this noise contribution at 10 kHz, where this source is dominant.

REFERENCES

1. Motchenbacher, C. D., and F. C. Fitchen, *Low-Noise Electronic Design*, Wiley, New York, 1973.
2. Buckingham, M. J., *Noise in Electronic Devices and Systems*, Halstead Press, New York, 1983.
3. Rohrer, R., L. Nagel, R. Meyer, and L. Weber, "Computationally Efficient Electronic-Circuit Noise Calculations," *IEEE J. Solid-State Circuits*, **SC-6** (August 1971), 204–213.
4. Jaeger, R. C., and J. Brodersen, "Low-Frequency Noise Sources in Bipolar Junction Transistors," *IEEE Trans. Electron Devices*, **ED-17** (February 1970), 128–134.
5. Brodersen, A. J., E. R. Chenette, and R. C. Jaeger, "Noise in Integrated-Circuit Transistors," *IEEE J. Solid-State Circuits*, **SC-5** (April 1970), 63–66.
6. Meyer, R. G., L. Nagel, and S. K. Lui, "Computer Simulation of $1/f$ Noise Performance of Electronic Circuits," *IEEE J. Solid-State Circuits*, **SC-8** (June 1973), 237–240.
7. Bilotti, A., and E. Mariani, "Noise Characteristics of Current Mirror Sinks/Sources," *IEEE J. Solid-State Circuits*, **SC-10** (December 1975), 516–523.
8. Nicollini, G., D. Pancini, and S. Pernici, "Simulation-Oriented Noise Model for MOS Devices," *IEEE J. Solid-State Circuits*, **SC-22** (December 1987), 1209–1212.
9. Trofimenkoff, F. N., and O. A. Onwuachi, "Noise Performance of Operational Amplifier Circuits," *IEEE Trans. Education*, **E-32** (February 1989), 12–16.
10. Gordon, B. M., "Noise-Effects on Analog to Digital Conversion Accuracy." *Computer Design* (March 1974), 65–76.

11. Gordon, B. M., "Noise-Effects on Analog to Digital Conversion Accuracy. II," *Computer Design* (April 1974), 137–145.
12. *Integrated Circuits Databook*, Vol. 1, Analog Devices Inc., 1988.
13. *Analog IC Databook*, Precision Monolithics Inc., 1988.
14. *Linear Databook*, Vol. 2, National Semiconductor Corp., 1988.
15. Scott, G. J., and T. M. Chen, "Addition of Excess Noise in SPICE Circuit Simulations," *IEEE SOUTHEASTCON Conference Proceedings*, Vol. 1, Tampa, FL, April 5–8, 1987, 186–190.
16. Connelly, J. A., and K. P. Taylor, "An Analysis Methodology to Identify Dominant Noise Sources in D/A and A/D Converters," *IEEE Trans. Circuits and Systems* (October 1991), 1133–1144.

PART IV

LOW-NOISE DESIGN APPLICATIONS

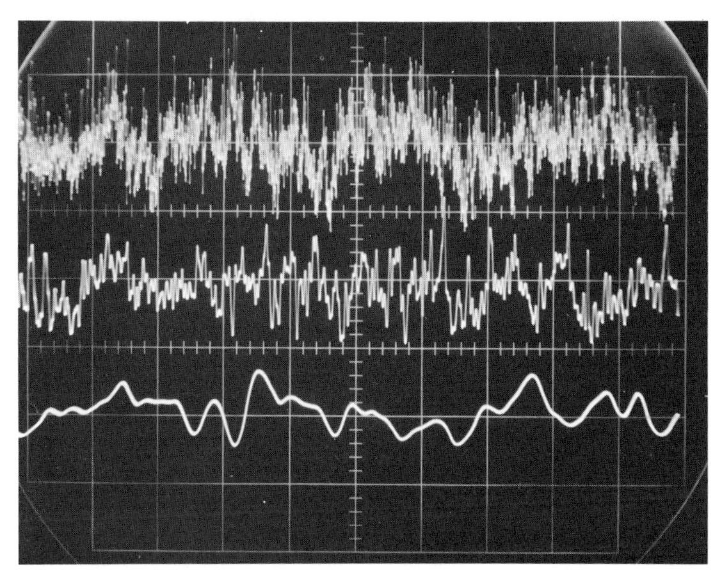

These traces show the effect of bandwidth limiting of $1/f$ noise. Top waveform, bandwidth is 2 kHz; middle waveform, bandwidth is 200 Hz; bottom waveform, bandwidth is 20 Hz. Horizontal sensitivity is 50 ms/cm. Note that the peak amplitude is not proportionately reduced by bandwidth limiting.

CHAPTER 12

NOISE IN PASSIVE COMPONENTS

The input transistor and the source resistance should be the dominant noise sources in an amplifier. Unfortunately, as we have seen, transistors are not the only circuit elements that generate noise. Passive components located in the low-signal level portions of the electronics also can be major contributors. This chapter discusses noise in passive components: resistors, capacitors, diodes, batteries, and transformers. The noise mechanisms of these components are described, and methods for minimizing their effects are considered.

12-1 RESISTOR NOISE

There are several commonly used types of fixed resistors, both discrete and integrated. Some of these are: carbon composition, deposited carbon, cermet, thick film, metal film, thin film, metal foil, and wirewound. Each has particular characteristics suitable for specific applications in electronic circuitry.

The total noise of a resistor is made up of thermal noise and excess noise as discussed in Chap. 1. All resistors have a basic noise mechanism, thermal noise, caused by the random motion of charge carriers. This noise voltage is dependent on the temperature T of the resistor, the value of the resistance R, and the noise bandwidth Δf of the measurement. In every conductor above absolute zero, there are charge carriers excited by thermal energy, jumping about within the conductor. Each jump is equivalent to a small burst of current through the resistance producing a voltage drop E_t. The mean squared value of this thermal noise voltage of a resistance R is

$$E_t^2 = 4kTR\,\Delta f \qquad (12\text{-}1)$$

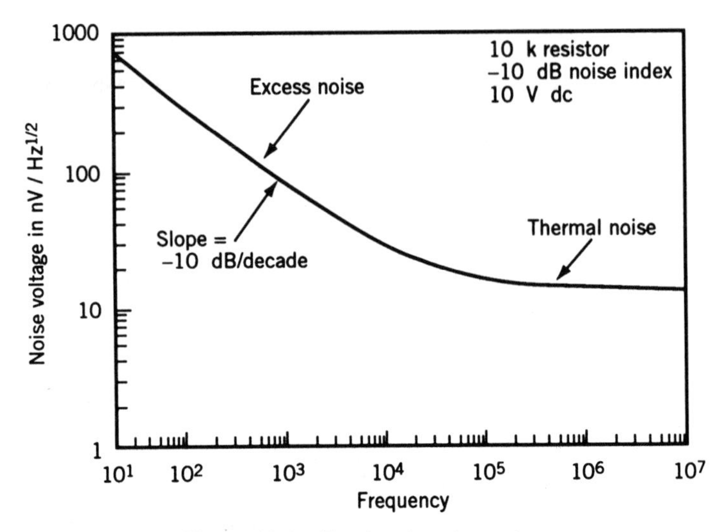

Figure 12-1 Total noise of a resistor.

Often an excess noise is generated when a direct current (dc) flows through a resistor. Excess noise gets its name because it is found to exist *in addition to* the fundamental thermal noise of the resistor. It is also referred to as *contact noise* or *current noise* because it occurs at the contact point between conductors when a dc current is flowing. Excess noise usually occurs when a current flows in any discontinuous conductor. The magnitude of the current noise is dependent on some inherent properties of the resistor such as resistive material, uniformity, processing, fabrication, and size and shape of the resistor. Although much research has been conducted on the source of this excess noise, there is no apparent functional relationship between the noise and physical properties. It is necessary to measure and characterize the excess noise of each type of resistor.

Equation 12-1 shows that thermal noise power is proportional to bandwidth (white noise). On the other hand, excess noise has been measured to have a $1/f$ power spectrum. Since noise power varies inversely with frequency, noise voltage increases as the square root of decreasing frequency.

The total noise generated in a resistor is the rms sum of the thermal and excess noises as illustrated in Fig. 12-1. The thermal noise component dominates at high frequencies, and the excess $1/f$ noise component dominates at low frequencies.

An experimentally determined equation for $E_{ex}(f)$, the excess noise voltage in a resistor R at frequency f, is

$$E_{ex} = \frac{mI_{DC}R}{\sqrt{f}} = \frac{mV_{DC}}{\sqrt{f}} \tag{12-2}$$

where m is a constant, dependent on the manufacturing process, and I_{dc} is the direct current flowing in the resistor R. The total excess noise over any bandwidth Δf is

$$E_{ex}^2(\Delta f) = \int_{f_1}^{f_2} E_{ex}^2(f)\, df \tag{12-3}$$

Substituting Eq. 12-2 into 12-3 and solving,

$$E_{ex}^2(\Delta f) = \int_{f_1}^{f_2} \frac{m^2 V_{DC}^2}{f}\, df = m^2 V_{DC}^2 \ln \frac{f_2}{f_1} \tag{12-4}$$

Equation 12-4 shows that excess noise power is proportional to the ratio of upper and lower cutoff frequencies. There is equal noise power in each frequency octave or frequency decade.

12-1-1 Noise Index

A standardized test method for measuring the "excess" current noise of a fixed resistor in terms of a noise index (NI) was developed by the National Bureau of Standards (NBS). Described in the report "A Recommended Standard Resistor-Noise Test System" by G. T. Conrad Jr., N. Newman, and A. P. Stansbury [1, 2], this test is also used in Mil-Std-202, Method 308, dated 29 November 1961.

The *noise index is defined as the rms value (in* μV) *of the noise in a resistor for each volt of dc drop across the resistor in one decade of frequency*. Using the results of Eq. 12-4, even though the excess noise is caused by direct-current flow, the definition of NI results in an expression that is independent of I_{DC} and R. We obtain

$$\mathrm{NI} = \frac{E_{ex}}{V_{DC}} \tag{12-5}$$

NI is usually expressed in decibels:

$$\mathrm{NI} = 20 \log \frac{E_{ex}}{V_{DC}} \qquad \mathrm{dB} \tag{12-6}$$

where E_{ex} is the microvolts per frequency decade. The equation gives NI = 0 dB when $E_{ex}/V_{DC} = 1\ \mu\mathrm{V/V}$. A method of measurement is outlined in the literature [1, 2].

Example 12-1 *Resistor noise calculation*. We are given a 10-kΩ composition resistor with a NI of 0 dB or 1 μV/V. It is used in a circuit with a frequency response from 10 Hz to 10 kHz, and the dc voltage drop across the resistor is 10 V. Consequently, the resistor has 1 μV of noise for each volt of dc in each decade of frequency. Since excess noise increases linearly with supply voltage, there is a 10 μV/decade for a 10-V dc drop. In the 3 decades from 10 Hz to 10 kHz, there is equal noise power in each decade of frequency. The total noise in 3 decades is the rms sum of the noise components in each decade or $\sqrt{3}$ times the noise in 1 decade. The total excess noise voltage across the 10-kΩ resistor is

$$1(\mu\text{V/V/decade}) \times 10(\text{V}) \times 1.732(\text{decade}) = 17.32\ \mu\text{V}$$

The *thermal noise* in a 10-kHz bandwidth (from Eq. 12-1) is only 1.25 μV. You can see that the excess current noise is considerably greater than the limiting thermal noise. The total noise of this 10-kΩ resistor over the 10-Hz-to-10-kHz bandwidth is the square root of the mean square sum of the excess noise and thermal noise, or

$$\sqrt{\left[1.25(10)^{-6}\right]^2 + \left[17.3(10)^{-6}\right]^2} = 17.4\ \mu\text{V}$$

12-1-2 Bandwidth Correction

Now we can derive the expressions for excess spot noise and broadband noise in terms of the measured noise index NI. Since the NI is noise/per decade, select a 1-decade frequency range, $f_2/f_1 = 10$. Then Eq. 12-4 becomes

$$E_{\text{ex}}^2(\Delta f) = m^2 V_{\text{DC}}^2 \ln 10 = 2.303 m^2 V_{\text{DC}}^2 \tag{12-7}$$

From Eq. 12-5,

$$E_{\text{ex}}^2 = \text{NI}^2 V_{\text{DC}}^2 \tag{12-8}$$

Substituting Eq. 12-7 into 12-8,

$$2.303 m^2 V_{\text{DC}}^2 = \text{NI}^2 V_{\text{DC}}^2$$

and solving for m,

$$m = \frac{\text{NI}}{\sqrt{2.303}} = 0.659\text{NI} \tag{12-9}$$

Now we get the expression for excess noise at any frequency by substituting for m in Eq. 12-2:

$$E_{\text{ex}}(f) = \frac{0.659\text{NI}V_{\text{DC}}}{\sqrt{f}} \qquad \frac{\mu\text{V}}{\sqrt{\text{Hz}}} \tag{12-10}$$

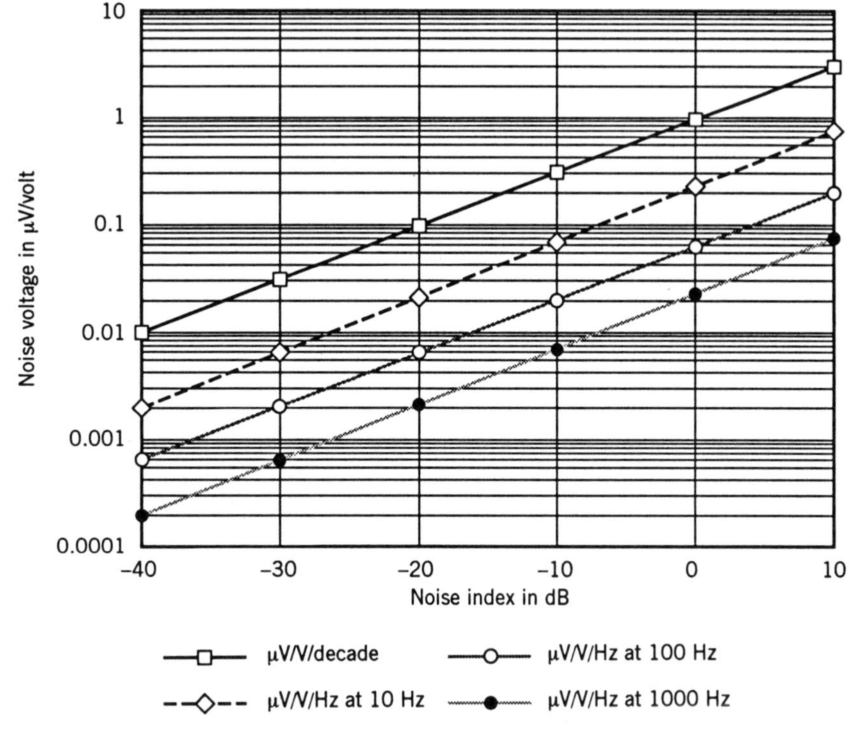

Figure 12-2 Conversion from NI to $\mu V/V$.

The total excess noise over the bandwidth from f_1 to f_2 in terms of the noise index NI is obtained by substituting for m in Eq. 12-4:

$$E_{ex}(\Delta f) = 0.659 \mathrm{NI} V_{DC} \sqrt{\ln(f_2/f_1)} \qquad \mu V \qquad (12\text{-}11)$$

The conversion from NI in decibels to the noise spectral density at 10 Hz, 100 Hz, and 1 kHz is shown in Fig. 12-2.

12-1-3 Noise Index of Commercial Resistors

The noise index is important when selecting a resistor for use in high-gain circuits, low-level audio frequencies, and other low-frequency applications. Excess noise from resistors may mask the desired signals if the resistors are not chosen for their low-noise characteristics.

Remember that excess noise is dependent on the voltage drop across the resistor. *If there is no voltage drop, you do not need a low-noise resistor.* Excess noise in a resistor is only significant at low frequencies and when there is a dc voltage drop across the resistor.

It is difficult to obtain exact noise information from resistor manufacturers. This is mostly economic; a noise test is an extra measurement not usually made by the manufacturer since most applications do not have a critical noise requirement. Typically, there is a 20-dB spread in the noise index within any type and value. If the excess noise is critical, place a maximum noise specification on the resistor and pay the additional cost for the testing.

Noise variations depend on the type of resistor, manufacturing processes, and process control. It has been shown that the greater the noise variation, the lower the reliability [3, 4]. Any resistor within a process lot that shows a significantly increased noise will probably have some manufacturing defect. Resistors with high values of NI also tend to be less stable.

The amount of excess resistor noise depends on the type of resistor and its manufacturing process. As pointed out earlier, excess noise is caused by current flowing through a discontinuous conductor. In general, the more uniform and defect free the resistive material, the lower the noise index. The contact or termination to the resistive material is also very critical.

There are several types of resistors available on the market. The quietest are the new bulk metal foil and wirewound resistors. Also fairly quiet are the evaporated-film and sputtered-film resistors. Metal film and oxide-film (tin oxide) resistors are about equivalent to the evaporated-film resistors. Thick-film and cermet resistors produce more noise while the common carbon composition and carbon film resistors are the noisiest.

Bulk metal foil and wirewound resistors have the capability of generating minimal excess noise, only thermal noise. Manufacturing defects such as mechanical damage to the conductor, bad welds, or partially shorted turns will cause excess noise to be generated. Metal foil resistors can be considered as flat wire resistors since they can be fabricated from a foil of bulk metal such as nickel–chromium (about 0.1 mil. thick) bonded to ceramic. A serpentine pattern is then etched to obtain the desired resistance. Metal foil resistors are available with a guaranteed noise index specification of less than −32 dB. Wirewound resistors can also have minimal excess noise. Their principal noise source is at the end termination. Those with crimped terminations can be noisy; the ones with solder or welded terminations will probably be low noise. The disadvantages to wirewound resistors are: They tend to be expensive, limited to the lower values of resistance, and have poor frequency response because they are inductive.

Thin-film resistors are formed with an atom-by-atom deposition process, such as vacuum depositing, cathode sputtering, and vapor plating. Thin films are deposited on an insulating substrate and protected by an insulating coating. Terminals are formed at ohmic contacts on the film made through etched apertures in the coating. Film thicknesses are about 5000 Å. The more common thin-film materials are nickel–chromium alloys and tantalum. The decision of material type to be used for a given application depends on factors such as temperature coefficient, resistivity, and stability with time.

Metal film resistors are a good choice for low-noise applications. They are made by evaporating a thin film of metal on a ceramic core. The resistive path is frequently helixed to increase the resistance. Not all metal film resistors are low noise. For instance, those with poor end terminations generate significant excess current noise at the end contacts. Poor helixing also makes them noisy. Care is necessary in the handling of the glazed-type metal film resistors. With these units, if the leads are pulled too hard or bent very close to the body, the seal will crack and the resistor will become noisy. The molded metal film resistors are not as susceptible to this problem.

Thick-film devices are made by screen deposition of an ink, and the curing or firing of this ink in a furnace. Thick-film resistors are fired in an atmosphere usually between 500 and 1100°C. At these temperatures, the solids oxidize, melt, and sinter together to solidify the film, level the surface, and bond the film to the substrate. Organic binders and vehicle solvents are evaporated off. Thick film is a relative term since the films are only a few thousandths of an inch thick. Cermet thick films are metal–ceramic combinations such as chromium, palladium, silver, and silicon monoxide, or glass on a ceramic tube.

A carbon composition resistor is made of carbon granules mixed with a binder, molded around lead wires and heat fused. Direct current flows unevenly through the resistor because of variations in conductivity. There is something like microarcs or a microplasma discharge between the carbon gains resulting in spurts or bursts of current flow. These bursts of current cause the excess noise. The more uniform the resistor, the less the excess

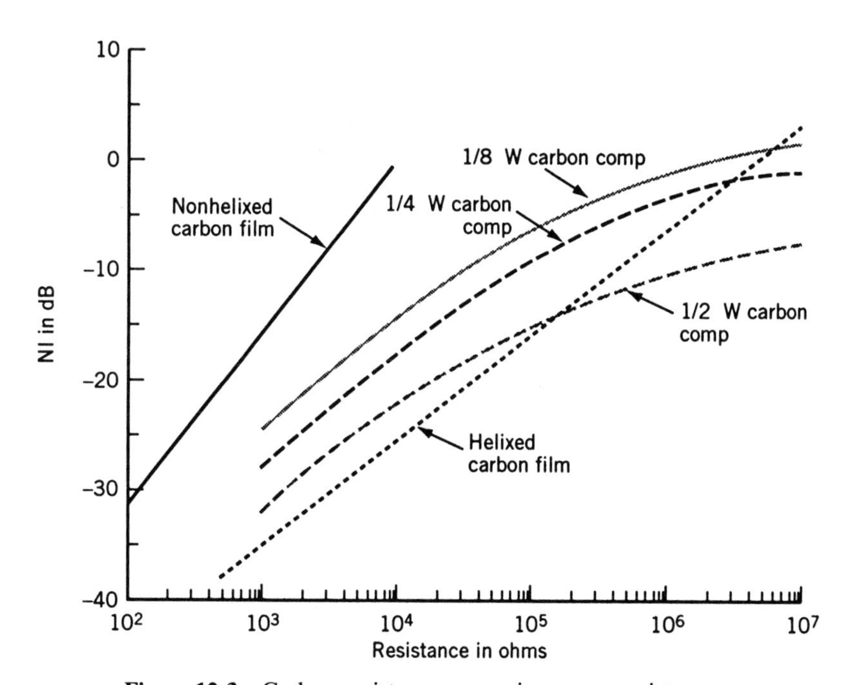

Figure 12-3 Carbon resistor excess noise versus resistance.

noise. Carbon resistors have the advantage of being the least-expensive type of resistor and are the most commonly used.

In general, high-wattage resistors have less noise than low-wattage resistors; also, high-resistance values have a higher NI than low-resistance values. For example, a high-resistance metal film resistor has a very thin film of metal on a substrate and so it is more affected by surface variations in the substrate. Similarly, a high-resistance composition carbon resistor has less carbon and more binder so it is less uniform. This variation of NI with resistance and resistor wattage is graphically shown in Fig. 12-3 by the dashed curves. Note the decrease of noise index with increasing wattage. Data for this graph were obtained from the Allen Bradley catalog for hot-molded carbon composition resistors [5].

Variation in noise index versus resistance of two different process types of carbon film resistors is shown in Fig. 12-3 as the solid lines. Note that a helixed resistor will have a lower NI than a nonhelixed resistor. The resistance per square of a nonhelixed resistor is much higher, because it is a thinner film; hence the noise is greater. Note also the increasing NI with increasing resistance value.

12-1-4 IC Resistors

Monolithic resistances fabricated during the base diffusion or the emitter diffusion operations are normally isolated from other IC components by junction isolation. These reverse-biased *pn* junctions exhibit some shot noise. Excess noise will be present primarily at the resistor contacts. Resistances made from FET-like structures can have noise sources as discussed in Chap. 6.

Noise measurements have been made on *n*-diffused and *p*-diffused resistors fabricated with a digital MOSIS process [6]. These show a noise index of -10 dB and -5 dB, respectively. Polysilicon-deposited resistors in the same process show a NI of -15 dB. Integrated resistors fabricated with a low-noise analog IC process will show a lower NI.

12-2 NOISE IN CAPACITORS

In general, capacitor noise is not a problem in circuit design. An ideal capacitor is noiseless. A pure reactive impedance does not produce thermal noise. A real capacitor, however, is not perfectly lossless. It has a certain amount of series resistance and shunt leakage resistance. These real components of the impedance contribute thermal noise, but they are usually negligible.

The noise model of a capacitor is a noise current generator shunted by the capacitor's impedance. The main noise contribution is at low frequencies, where the reactance of the capacitor does not effectively shunt the internal noise. Low-frequency excess noise is the dominant noise mechanism.

For low-frequency circuit applications, large values of capacitance are usually required. Electrolytic capacitors are often used because of their smaller physical size. Tantalum electrolytics have been used successfully in low-noise circuits. Aluminum electrolytics with their higher leakage and higher forming curents are less desirable.

Electrolytic capacitors are used in two principal ways: bypass capacitors and coupling capacitors. No instances have been observed when a bypass capacitor added noise to the circuit. Its own impedance effectively shunts the noise generators.

Coupling capacitors occasionally add noise. After an electrolytic capacitor has been reverse-biased, it generates bursts of noise for a period of time, from a few minutes to several hours. Usually, a transient causes the reverse bias. Occasionally, during turn-on, the coupling capacitors are reverse-biased as one stage turns "on" faster than the next. After this happens, the capacitor may be noisy for a while. There are three ways of avoiding this problem. One is to design the circuit so that there is no reverse bias at any time and no turn-on transient. A second solution is to place a low-leakage silicon diode in parallel with the capacitor. Any reverse-bias forward-biases the diode and prevents the capacitor from breaking down. A third method is to use two electrolytic capacitors of twice the size connected back to back (in series). This forms a nonpolar electrolytic capacitor. Regardless of the polarity of applied voltage, the proper capacitor takes the voltage drop.

12-3 NOISE OF REFERENCE AND REGULATOR DIODES

In a forward-biased diode the principal noise is the shot noise attributed to the dc. There is also excess noise in the forward-biased diode; as discussed in Chap. 5, the base–emitter diode of a transistor exhibits excess noise. *A reverse-biased diode has two breakdown mechanisms that exhibit different noise properties. These are the Zener and avalanche mechanisms.*

Internal field emission, the Zener effect, is the predominant mechanism in diodes with low (7 V or less) reverse-breakdown voltages. At a very thin junction, the electric field can become large enough for electrons to jump the energy gap. The Zener mechanism exhibits shot noise. There is little excess noise. Some Zener diodes have $1/f$ noise corners lower than 10 Hz. Being shot noise limited, the rms noise current can be calculated from the direct current by using Eq. 1-50 for shot noise current from Chap. 1:

$$I_{\text{sh}}^2 = 2qI_{\text{DC}}\,\Delta f \tag{1-50}$$

Zener noise voltage E_z is the product of shot noise current from Eq. 1-50 and the dynamic resistance r_d of the Zener diode at the bias point as shown in Eq. 12-12:

$$E_z = I_{sh} r_d \tag{12-12}$$

The Zener diode is a low-noise regulator or reference diode.

In a diode with a larger junction width, breakdown occurs at a lower voltage than can be explained by the Zener effect. This is avalanche breakdown. As the reverse bias on the diode is increased, carriers are accelerated to an energy level great enough to generate new hole–electron pairs on collision. These new pairs are accelerated by the electric field of the bias voltage and generate more hole–electron pairs. The process cascades and is self-sustaining as long as the field is maintained.

The noise of an avalanche breakdown is larger and more complicated than Zener noise. There is noise due to the avalanche mechanism and also a multistate noise mechanism. As with forward-bias shot noise, the avalanche mechanism results in the random arrival of carriers after crossing the junction. In this case, there are bundles of carriers that give an amplified shot noise. The avalanche noise has a flat frequency spectrum; hence it is white.

The more troublesome noise in a reverse-biased avalanche diode is the multistate noise. It gets its name from the fact that the noise voltage appears to switch randomly between two or more distinct levels. These levels may differ by many millivolts. As the reverse current of the diode is increased, the higher level is favored until finally it predominates. Although the average noise current is constant, the period is completely random ranging from microseconds to milliseconds. The switching time between levels is extremely fast. Multiple-level operation is apparently due to defects and localized inhomogeneities in the junction region. These can create a local negative-resistance area with either switching or microplasma oscillation.

Multistate noise has a wide frequency range, but it is predominantly $1/f$ noise. It is definitely an excess noise mechanism that does not need to exist. It is process dependent and varies from manufacturer to manufacturer. Lower-noise units can be selected from a production run. High multistate noise generally indicates lower device voltage reference stability. If a low-noise reference diode is needed, the most acceptable solution is to use the lower-voltage Zener diode. When the need exists for a reference in the avalanche region, greater than 5 V, select units for low noise.

The noise behavior of breakdown diodes is depicted in Fig. 12-4. For this diode series, units with breakdowns below 3 V exhibit very little noise; units breaking down above 5 V are quite noisy.

Breakdown diodes are not recommended as coupling or biasing elements in a low-noise amplifier because of their noise behavior. When a regular or reference diode is used in the power supply of a low-noise amplifier, it should be decoupled by an *RC* network or a capacity multiplier as described in Chap. 13. If a reference diode is used as a low-noise calibration source, select a low-voltage Zener type and shunt it with a large bypass capacitor.

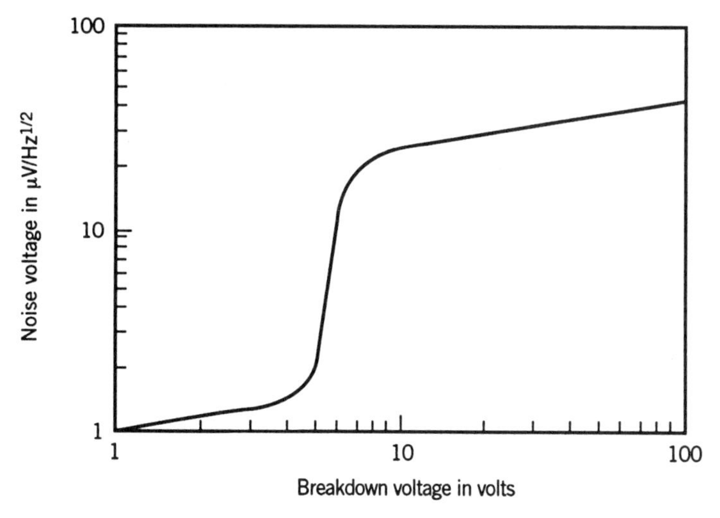

Figure 12-4 Noise voltage versus breakdown voltage for a family of diodes.

12-4 BATTERIES

Batteries are desirable power supplies for low-noise amplifiers for several reasons. The battery can be located with the amplifier in a shielded case to avoid pickup. An isolated battery power supply is less likely to have ground loops and 60-cycle pickup.

In general, batteries are not noise sources. Since a battery has current flowing and an internal impedance, there is some noise. A battery serves as a large capacitor and therefore shunts its own internal noise. Only when nearly exhausted does the noise rise. If the noise of a battery supply is a problem, bypass the supply or decouple the noise with a capacity multiplier. This also reduces the series impedance of the power supply and helps to prevent regeneration and motorboating in the amplifier.

12-5 NOISE EFFECTS OF COUPLING TRANSFORMERS

Transformers are used in several types of low-level applications: data acquisition, data transmission, geophysical measurements, dc chopper amplifiers, bridges, and so forth. They perform one or more of the following: isolation, impedance matching, noise matching, and common-mode rejection. Low-level transformers are used to discriminate against the interference that often accompanies signal voltages. These interfering signals come from stray magnetic fields, ground loops, common-mode signals, and machine-made noise.

Types of magnetic components used in low-level systems include input transformers, chopper input transformers, interstage transformers, output transformers, filter reactors, and low-pass filters. Our particular interest in this section is with input transformers and chopper transformers. An input transformer is used to match the amplifier noise characteristics to the sensor impedance. Thus it is possible to design an amplifier for operation at its optimum noise factor F_{opt} and then transform the sensor impedance to look like R_o. This transformer noise matching is discussed in Chaps. 8 and 9.

There are several second-order effects that can add noise to a low-level transformer circuit. Some of the considerations necessary in the proper application of transformers are common-mode rejection, magnetic shielding, primary inductance, frequency response, and microphonics. The following sections discuss each of these effects.

12-5.1 Electrostatic Interwinding Shielding

All instrumentation transformers incorporate electrostatic Faraday shields between the windings for isolation and common-mode rejection. This is illustrated in Fig. 12-5. This interwinding shield may consist of a copper foil interleaved between the layers, or it may be a complete box-type construction as shown in Fig. 12-5. Consider that a common-mode voltage between the sensor and ground is present at the primary terminals of the input transformer. Any capacitive coupling between the primary and secondary windings couples part of the common-mode voltage to the secondary. The interwinding Faraday shield terminates the electrostatic field from the primary. The more the shield encloses the primary winding, the better the common-mode rejection.

The effectiveness of the electrostatic shielding can be tested. A common-mode voltage V_{cm} is connected between the primary and ground. The secondary voltage V_s is measured. Any coupling impedance, X_{CM}, is considered to be an equivalent primary-to-secondary capacitance. Therefore, the

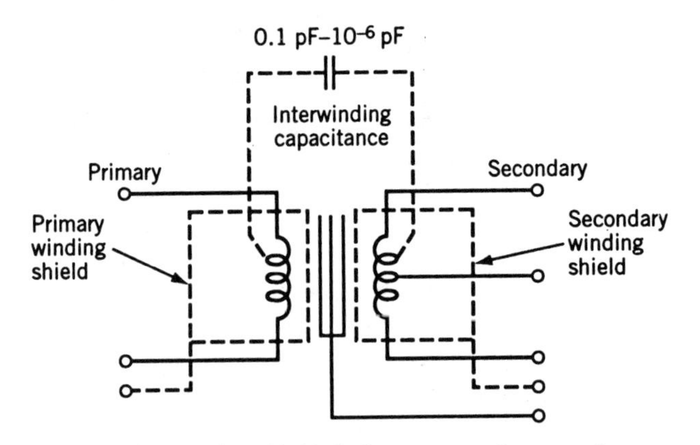

Figure 12-5 Winding shields in instrumentation transformers.

equation $V_s/V_{cm} \equiv R_L/X_{CM}$ can be solved for the interwinding capacitance. Typical capacitances can be as high as 10 pF or as low as 5×10^{-7} pF.

In addition to low interwinding capacity, common-mode rejection requires a balanced primary with symmetrical capacitances. If there is unbalanced shunt capacity between the center tap and the two ends of the primary, a common-mode voltage injected in the center tap causes unbalanced voltages in the primary winding. This, then, is a true differential-mode signal. The common-mode rejection of a good transformer is difficult to measure and even more difficult to specify.

12-5.2 Magnetic Shielding

For low-level use, a transformer must be shielded from externally caused magnetic fields. Magnetic pickup can also be reduced by coil construction techniques. External fields are rejected by the "hum-bucking" winding technique. "Hum-bucking" implies that the transformer has two secondary windings oppositely wound on the core. An externally generated magnetic field induces equal and opposite voltages in the windings. This is most effective when the field is distant and the transformer can be spatially oriented.

Magnetic shielding is accomplished by surrounding the transformer with multiple layers of mu-metal interleaved with heavy copper shorting layers. Multiple layers provide higher attenuation than a single layer of equal thickness. Magnetic shielding provides a low reluctance path around the transformer core for the interfering signal. Any fields that penetrate the first

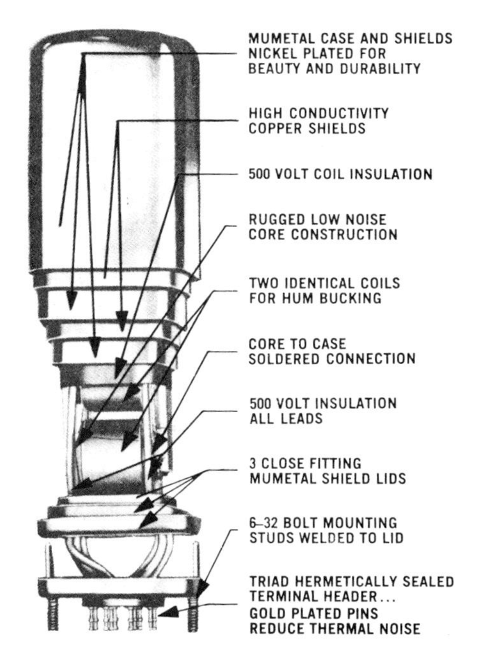

Figure 12-6 Exploded view of a shielded instrumentation transformer.

magnetic layer induce eddy currents in the copper layer beneath it. These currents generate a reverse magnetic field to oppose the interfering field. Multiple layers can be noted in the expanded view of a transformer given in Fig. 12-6.

Magnetic-shielding effectiveness is the ratio of the pickup in a transformer with and without shielding. It can be measured by placing the transformer between two large Helmholtz coils. The coils are driven with an ac that sets up a known magnetic-field intensity between them. This test is described in MIL-T-27.

12-5-3 Transformer Primary Inductance

Transformer primary winding inductance is important because it determines the low-frequency response of the transformer, and reducing inductance increases the equivalent input noise voltage of the amplifier. The low-frequency cutoff of the transformer occurs when the primary impedance equals the parallel resistance of the source and reflected load resistance. Below this frequency, the equivalent noise voltage generator of the amplifier is increased as shown in Chap. 7.

To measure the inductance of a low-level transformer accurately, use a low-frequency impedance bridge with very small drive signals. Measure the inductance at the lowest frequency used.

It is difficult to control the primary inductance of a transformer. The number of turns can be held constant, but the core permeability and stacking efficiency of the laminations vary from unit to unit. Permeability can vary over 4:1 range. The permeability of the core is more constant if an air gap is introduced, but this increases the size of the transformer and makes it more susceptible to magnetic pickup. Because of permeability tolerances, the minimum acceptable inductance must be determined. One of the differences between transformer manufacturers is how well they handle their stacking factors. If they stack tighter and increase the permeability of the core, the transformer has greater inductance or smaller size. This reduces the cost as well.

12-5-4 Frequency Response

The low-frequency response of a transformer is determined by the shunting effect of its primary inductance. The high-frequency cutoff is determined by the shunting effect of its winding capacitance. The widest-frequency response is obtained when the primary and secondary inductances are terminated in their nominal impedances. A transformer can be used, of course, at other than its nominal impedances. Usually, low-frequency response can be improved by lowering the source impedance below nominal. The converse also applies.

In general, the equivalent input noise increases when operating outside the passband. Beyond roll-off, amplifier noise remains constant while the signal passed by the transformer falls off; therefore, the equivalent input noise increases. In addition, core losses and transformer noise increase with frequency.

12-5-5 Microphonics and Shock Sensitivity

High-inductance transformers generate extraneous voltages when shocked or vibrated. This can be caused by changes in the permeability of the core in the presence of an internal or an external field, or by a change in coupling with the internal or external magnetic field. If a transformer is vibrated, the laminations may shift in position or the core may be stressed. In either case, the permeability of the core is modulated. If there is a magnetic field present, the changing inductance generates a voltage. With very high inductance windings, just squeezing the transformer with your fingers generates a voltage.

An external dc magnetic field is also a problem. A power supply choke is one source of such a field. The earth's field is another. If the transformer vibrates in the field or if the magnetic-field source vibrates, a voltage can be generated directly in the coil or indirectly by inducing a voltage in the shield and then into the coil. Motion of a magnetized shield also induces a voltage in the transformer windings.

Laminations and shielding cans should be demagnetized by the manufacturer before assembly in order to minimize the residual flux in those parts of the transformer assembly. Once it is assembled it is difficult for the user to degauss a transformer adequately. Inadvertent magnetization can result from the passing of a dc bias current through a winding or from measuring the winding continuity with an ohmmeter. A small dc in the windings, such as a transistor base current, can make the transformer more microphonic. If a transformer is inadvertently magnetized, it can be degaussed by passing low-frequency ac through the windings and slowly decreasing the amplitude of this wave to zero. To minimize shock sensitivity the following can be done:

1. Have the core vacuum-impregnated so that the laminations cannot move with respect to one another.
2. If space permits, mechanically isolate the coil and laminations from the shields and the shields from each other.
3. Specify that the core and shields are to be demagnetized before assembly.
4. Shock-mount the transformer on the chassis.
5. Shock-mount any other magnetic-field sources such as power transformers or chokes.
6. **Do not** check the dc continuity of any of the transformer windings unless they can be adequately demagnetized!

Chopper transformers also have a problem with thermoelectric voltages. In a circuit formed with wires of different metals, a temperature difference across the junction will generate a thermoelectric voltage. This thermally generated voltage (thermal EMF) can add or subtract from the dc signal being modulated and cause an error. Therefore, locate the transformers away from any heat sources.

SUMMARY

a. In addition to thermal noise, resistors exhibit excess noise when dc is present.
b. Excess noise can be minimized by proper selection of the resistor manufacturing process. Metal foil and film units are usually superior.
c. The noise index is given in units of microvolts per volt per decade, or

$$\mathrm{NI} = 20 \log\left(\frac{E_{\mathrm{ex}}}{V_{\mathrm{DC}}}\right) \quad \mathrm{dB}$$

over the frequency decade used to determine E_{ex}.
d. Capacitors usually do not present a noise problem because their own capacitance shunts the internal noise. Electrolytics with high leakage can be troublesome.
e. Low-voltage Zener diodes are lower-noise devices. Avalanche-breakdown devices can be noisy, and selected units may have to be used if the application is critical.
f. Breakdown diodes do not make good coupling or biasing elements in a low-noise system.
g. Batteries are not a source of noise except when nearly exhausted.
h. Coupling transformers should have electrostatic interwinding shielding and magnetic shielding.
i. To control microphonics in interstage transformers, manufacturing techniques and handling and inspection must be tightly controlled.

PROBLEMS

12-1. Find the total mean squared noise voltage in a 620-kΩ resistor in a decade of frequency geometrically centered at 1 kHz. The noise index of the resistor is 0 dB, and there is 5 V of dc voltage across the resistor.

12-2. A noise test on an integrated resistor yields the following data: $R = 100$ kΩ; total noise voltage (including thermal) is 600 nV for a noise bandwidth of 1 Hz; and $V_{DC} = 4$ V. Determine the noise index in decibels and the noise index in $\mu V/V_{DC}/Hz^{1/2}$ at a frequency of 100 Hz.

12-3. A 1-kΩ resistor has a noise index of 0 dB with 1 V of dc bias voltage across it. Determine the noise bandwidth, Δf, for which the thermal noise is equal to the excess noise.

12-4. How much forward biasing current must be supplied to a diode in order for it to generate the same amount of noise as a 1-kΩ resistor?

REFERENCES

1. Conrad, G. T., Jr., N. Newman, and A. P. Stansbury, "A Recommended Standard Resistor-Noise Test System," *IRE Trans. Component Parts*, **CP-7**, 3 (September 1960), 1–4.
2. Stansbury, A., "Measuring Resistor Current Noise," *Elec. Equipment Eng.* **9**, 6 (June 1961), 11–14.
3. Curtis, J. G., "Current Noise Tests Indicate Resistor Quality," *Int. Elec.*, **7**, 2 (May 1962).
4. Anderson, C. V., "Metal Film Resistors," *Electronic Design*, **17**, 16 (August 1978), 122–126.
5. Allen Bradley Co. Resistor Catalog, 1992.
6. Taylor, K. P., "Noise Models of A/D and D/A Converters," Ph.D. Thesis, (unpublished) Georgia Institute of Technology, Atlanta, Georgia, August 1991.

CHAPTER 13

POWER SUPPLIES AND VOLTAGE REFERENCES

In the design of low-noise systems, the noise contributions of all parts of the system, including power supplies, must be minimized. When designing a power source for low-noise electronics, the supply must be quiet and unvarying. Since many low-noise amplifiers are ac coupled, the supply may not need to be regulated, only isolated and well filtered.

A dc power source for low-noise electronics must remove the common-mode power line voltages that may be coupled into the supply, remove the ripple generated in the supply rectifiers, and remove any noise generated in the regulator. Common-mode voltages originating in the ac power lines can be isolated by decoupling and shielding. Cross-talk from other circuits can be minimized by layout and adequate ground connections. If the supply can be filtered, noise is not a serious problem, for the means of ripple removal also attenuates noise coming from regulators and components in the supply. When a regulated supply must be used, the regulator and voltage reference must be carefully chosen as described in the following sections.

13-1 TRANSFORMER COMMON-MODE COUPLING

On most power lines there is a lot of noise and other spurious signals that can modulate or add to the power frequency. There can also be a common-mode voltage CMV_1 present between the line and ground and a signal CMV_2 between power and earth ground, as shown in Fig. 13-1. These noise voltages are coupled into the circuit power supply by the transformer primary-to-

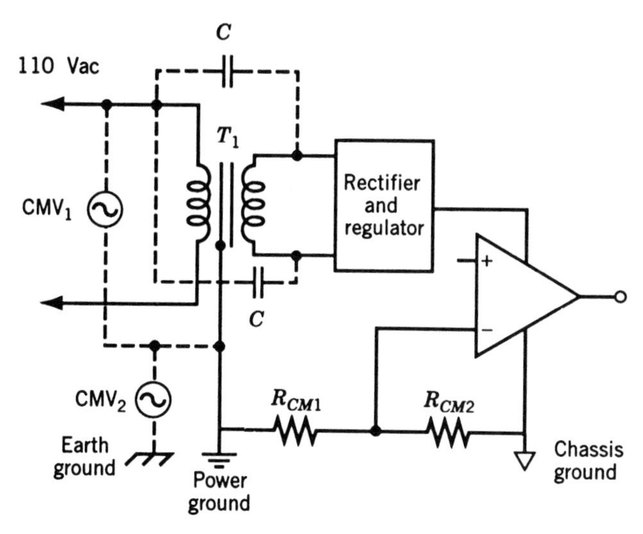

Figure 13-1 Power line common-mode voltage coupling.

secondary interwinding capacitances C. Typically, this noise can be coupled to the amplifier as an extra noise mechanism. In medical equipment, these voltages can be large enough to be life threatening.

There is another way for this common-mode signal to get into the amplifier. At some location the amplifier output is connected to ground or earth. This may be after some additional stages, but eventually the circuit ground is connected to the earth or power line ground. This provides two grounds in the system, one at the common-mode signal and the other through the amplifier, and we have a ground loop through R_{CM1} and R_{CM2}. This loop provides the potential for a small circulating current through the circuit and is an additional source of noise and line frequency pickup.

A very small amount of circulating current can cause a lot of pickup. As Ohm pointed out: "One microampere through 4 milliohms is 4 nanovolts," and that is equal to the thermal noise of a 1000-Ω resistor in a 1 Hz bandwidth.

The question that arises is, "How can we eliminate these sources of pickup?" A battery power supply will eliminate the coupling to the line but this is not always practical. To use an ac power supply, we must break the ground loops. This can be done with a shielded power transformer containing interwinding Faraday electrostatic shields built by Topaz and other transformer companies, as was discussed in Sec. 12-5.

The transformer isolates in two ways. First, it isolates by shielding. The primary shield of the transformer decouples from the circuit power supply, and the secondary shield decouples from the electronic system. The level of decoupling is dependent on the effectiveness of the shielding. When using this type of shielded power transformer do not bring the ac supply line inside the case. The primary transformer leads are shielded and can be brought out to a bulkhead terminal or to a terminal strip external to the shielded enclosure of the dc power supply. Second, it limits the frequency response to prevent high-frequency transients from passing into the system.

All low-noise equipment needs an isolating transformer with decoupling shielding. If the equipment is used in life support medical applications, then it is probably engineering malpractice to design an instrument without an isolating transformer [1].

13-2 POWER SUPPLY NOISE FILTERING

Noise and stray signals can be coupled into the amplifier system through the biasing power supply. Because of the balanced differential style of op amps, signals on the power lines are strongly attenuated. Typically, low-noise operational amplifiers have a power supply noise rejection of 100 dB at low frequencies. This means that signals on the power lines are attenuated by 10^5 at the input. To achieve an equivalent input noise of 1 nV/Hz the power supply noise must be less than 100 μV/Hz. Commercial voltage regulators typically have an output noise of a few microvolts over a bandwidth of 0.1 to 10 Hz with reduced noise at high frequencies. For most operational amplifier applications, this is sufficient because the amplifier is differential and has high power supply noise rejection.

There are other applications, such as single-ended amplifiers, which do not exhibit the high power supply noise rejection. In these instances, additional filtering of the supply lines may be required. Biasing supplies for sensors and detectors are particularly critical for noise since the supply noise is divided and coupled directly to the input signal.

Since a highly regulated supply voltage is usually not needed for a low-noise ac amplifier or a bias supply, a filter stage can be added on the

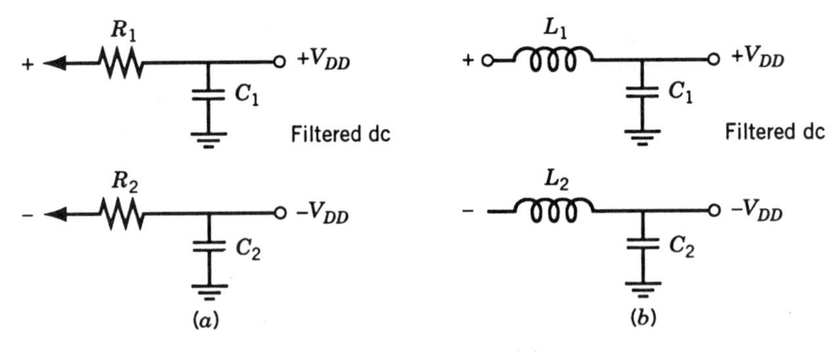

Figure 13-2 (*a*) RC filter and (*b*) LC filter.

board near the amplifier stage. The series impedance of the filter will usually increase the output impedance of the supply and decrease the regulation. By placing the filter physically close to the amplifier or sensor, ground loops are minimized and pickup on the supply line can be attenuated.

There are three principal methods of power supply filtering. The first is the straightforward *RC* or *LC* filter as illustrated in Fig. 13-2. A second is the capacity multiplier discussed in the next section and shown in Fig. 13-3. A third is the ripple clipper circuit shown in Fig. 13-4.

The ac filtering of Fig. 13-2*a* is proportional to the *RC* product. This may require large values of *R* or *C* or both.

13-3 CAPACITY MULTIPLIER FILTER

Using an active device as a capacity multiplier can reduce the size of the capacitor dramatically, as shown in Fig. 13-3.

The filtering of the V_{BB} source is determined by the parallel time constant of R_1, R_2, and C_1. Even if C_1 is small, the values of R_1 and R_2 can be large since the input impedance of the amplifier is large. The maximum values of R_1 and R_2 are determined by the offset current of the amplifier used. The total noise of the filtered output is the rms sum of the filtered noise of V_{BB}, amplifier noise voltage E_n, amplifier noise current $I_n Z_s$, and any noise of V_{DD} not rejected by the amplifier. Impedance Z_S is the parallel impedance of R_1, R_2, and X_{C1}. The amplifier noise current I_n will not be a factor if the impedance of C_1 is small. Resistors R_1 and R_2 will generate excess noise because of the dc drop. Therefore, use resistors with a low noise index.

Since this capacity multiplier circuit is supplying power to another circuit, select an amplifier with a low output resistance capable of driving a load. A series-pass transistor or a second buffer op amp can be added for additional load capacity.

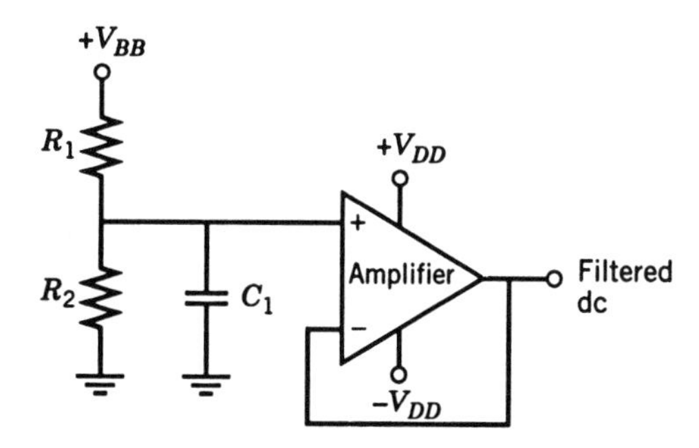

Figure 13-3 Active capacity multiplier.

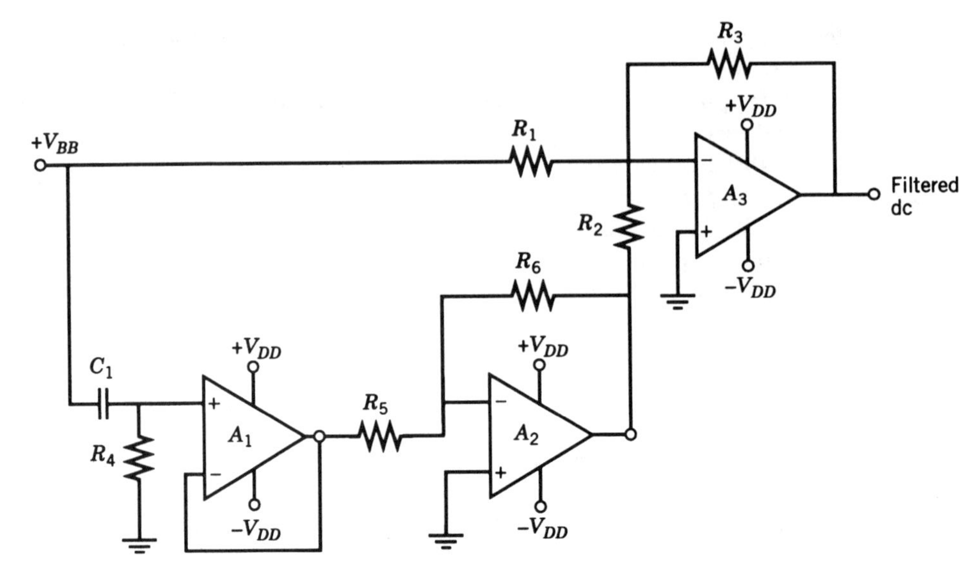

Figure 13-4 Noise clipper circuit.

13-4 NOISE CLIPPER

A third type of filtering circuit is the noise clipper shown in Fig. 13-4. This circuit strips the noise from the dc, inverts it, and *subtracts* the noise from the dc reference supply. The ac noise and ripple are coupled by C_1 into buffer amplifier A_1. The noise signal is then inverted in unity-gain amplifier A_2. For frequencies where $X_{C1} \ll R_4$, the output of A_2 is equal to and 180° out of phase with the input noise. Then the output of A_2 is summed through R_2 with the supply voltage and noise at the input of A_3. Since the ac signal from A_2 is equal to and out of phase with the incoming noise, it cancels ac fluctuations and ripple. This is limited at higher frequencies, where the frequency responses of A_1, A_2, and A_3 are not identical, and at very low frequencies, where the reactance of C_1 becomes too large with respect to R_4.

13-5 REGULATED POWER SUPPLIES*

Ideally, a perfectly regulated supply would have no noise. Given a noisy, unregulated input voltage, it would regulate to zero output variation. In practice it only regulates as well as its rejection of the noise in the unregu-

*The following sections on regulated supplies and voltage references were contributed by W. Tim Holman, Georgia Institute of Technology, Atlanta, GA 30332.

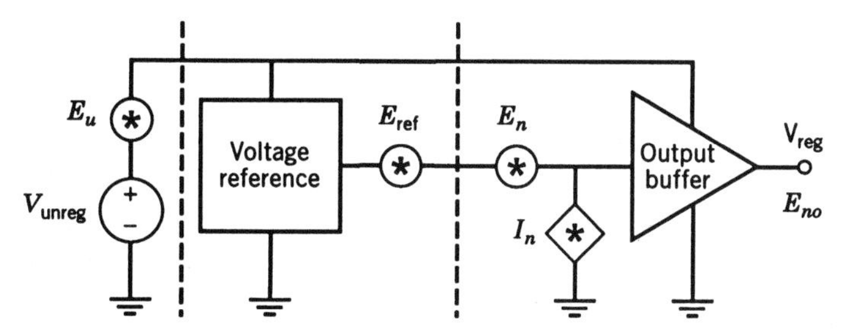

Figure 13-5 Block diagram of a regulated power supply.

lated input voltage plus any reference, amplification, and regulator noise. A block diagram of a typical regulated supply is shown in Fig. 13-5 [2].

For this circuit configuration, the output noise can be expressed as follows:

$$E_{no} = \sqrt{(K_r A_b E_u)^2 + (A_b E_{\text{ref}})^2 + (A_b E_n)^2 + (A_b r_r I_n)^2 + (K_b E_u)^2} \tag{13-1}$$

where

E_u = noise voltage generated by the unregulated supply
E_{ref} = output noise voltage generated by the voltage reference
E_n = equivalent input noise voltage of the output buffer
I_n = equivalent input noise current of the output buffer
K_r = voltage gain of unregulated supply noise through the voltage reference
K_b = voltage gain of unregulated supply noise through the output buffer
r_r = output resistance of the voltage reference
A_b = voltage gain of the output buffer

In practical terms, a good low-noise voltage regulator should have high power supply noise rejection (K_r and K_b approaching zero), and the internal noise generated by the reference and output buffer circuits should be made as low as possible. Other common specifications for voltage regulators include the following [3–5]:

Accuracy. Each regulator should generate a specific voltage with a minimum (or absence) of trimming and should maintain that voltage with a minimum of long-term drift.

Temperature regulation. The reference voltage should be independent of the operating temperature. Temperature regulation is usually expressed as a temperature coefficient (ppm/°C).

Power supply regulation. The change in dc output voltage with respect to a change in dc supply voltage should be as small as possible.

Load regulation. If the regulator is designed to sink and source current, the change in output voltage with respect to a change in load current should be as small as possible.

Power consumption and efficiency. The quiescent power consumption of the regulator should be minimized. If the circuit is designed to sink or source current, the ratio of maximum output power to quiescent power should be as large as possible.

Stability and transient response. The regulator should be stable and should respond to changes in supply or load conditions as quickly as possible.

Power supply noise rejection is generally frequency dependent and will decrease with increasing frequency. The designer of a discrete voltage regulator usually has the option of adding filtering at critical nodes to limit the output noise bandwidth of the circuit. Integrated circuit designers, on the other hand, cannot rely on the availability of large capacitors or inductors, since these components are generally expensive or impractical to fabricate in typical IC processes. Also, high levels of $1/f$ noise in a voltage regulator will result in excessive output noise in a range of frequencies (typically 10 Hz and below for BJT regulators and 1 kHz and below for CMOS regulators) that is difficult to filter even with discrete components.

In most voltage regulator applications, $1/f$ output noise will have a much greater effect on the output voltage error than thermal or shot noise. Reduction of $1/f$ noise should be a prime consideration in the design of a high-performance low-noise voltage regulator. However, some high-frequency low-noise systems may be very sensitive to the level of white noise in the regulated power supply, so this component of the output noise should not be ignored.

Voltage regulators generally fall into two categories: linear regulators and switching regulators. Switching regulators have the advantage of higher efficiency than linear regulators, but they are also extremely noisy due to the clock signal and switching transients, especially in high-current applications [6]. Unless extensive external filtering and shielding are available to the designer, good low-noise regulator design will require a linear approach.

13-6 INTEGRATED CIRCUIT VOLTAGE REFERENCES

In broad terms, a voltage *regulator* can be considered a voltage *reference* with a low-impedance output buffer. The voltage reference is usually the dominant source of noise in a regulator circuit. Since the design of low-noise amplifier (buffer) stages has been discussed in the previous chapters, we shall focus on a noise analysis of commonly used references in integrated circuit design.

As stated previously, an ideal voltage reference should have infinite rejection of supply noise. However, if the power supply is well regulated and well filtered, a voltage reference with poor power supply noise rejection can still be useful in many applications. For example, a designer may wish to generate a stable bias voltage from a regulated low-noise power supply. In this case, reduction of noise generated by the reference circuit itself is more important than the ability to reject external noise.

13-6-1 Resistor Voltage Divider

One of the simplest possible voltage references in an integrated circuit technology is two resistors connected as a voltage divider [3] (Fig. 13-6). This circuit has the advantage of a low temperature coefficient (provided the resistors have similar geometries and are made of the same material), and the output voltage can be set to any selected voltage for any supply voltage. Disadvantages include poor output and supply regulation. Noise from the V_{DD} supply is only attenuated at the output by the ratio of R_2 to the sum of R_1 and R_2. However, given a well-filtered and well-regulated supply voltage and a high-impedance load, the resistor divider can be an excellent low-noise reference limited by the thermal noise of the resistors. By adding a low-noise buffer at the divider's output, a high-performance "virtual ground" can be created.

The intrinsic output noise of a resistor divider can be expressed as

$$E_{no} = \sqrt{\frac{(R_2 E_{R1})^2 + (R_1 E_{R2})^2}{(R_1 + R_2)^2}} \qquad (13\text{-}2)$$

where E_{R1} and E_{R2} are the intrinsic noise sources of resistors R_1 and R_2, respectively. These noise sources are the rms sums of the thermal noise and

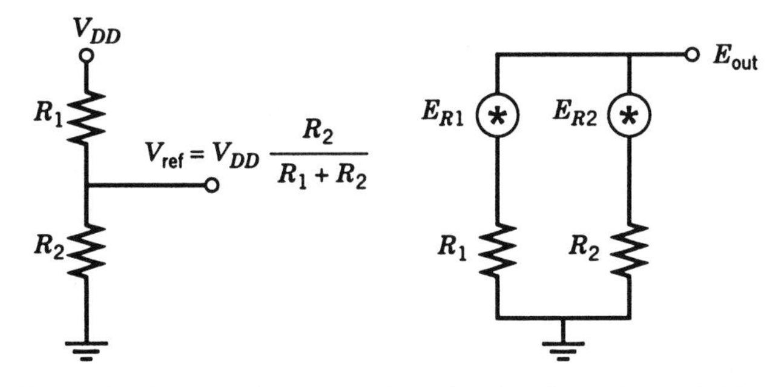

Figure 13-6 dc and ac equivalent circuits of a resistor divider.

the excess noise of the resistors. The noise index for polysilicon resistors in integrated circuits is typically about −30 dB, which means that, for large supply voltages, excess noise will dominate. For lower supply voltages, thermal noise will be dominant. The limiting noise of the divider is the thermal noise of the two resistors in parallel,

$$E_{no} = \sqrt{4kT\,\Delta f(R_1 \parallel R_2)} \tag{13-3}$$

If a discrete divider is built using resistors with a high noise index, for example, carbon composition types, the excess noise will be the dominant output noise mechanism.

Example 13-1 What are the excess and the thermal noise of a 10,000 Ω resistor with a noise index of −30 dB and a dc voltage drop of 10 V across the resistor (a) in a bandwidth from 0.01 to 10 Hz? (b) in a bandwidth from 10 to 10,000 Hz?

Solution (a) From Eq. 13-3, the thermal noise of a 10-kΩ resistor is 12.6 nV/Hz. In a 10-Hz bandwidth it is multiplied by $\sqrt{10}$ to give 40 nV of thermal noise. Referring to Sec. 12-1, a −30-dB NI translates to 30 nV/V/decade of frequency. The frequency range from 0.01 to 10 Hz is 3 decades of frequency. The excess noise of the 10-kΩ resistor with a 10-V drop is then $E_{ex} = 30 \text{ nV} \times 10 \times \sqrt{3} = 520$ nV. So the excess noise is 520 nV, compared to 40 nV of thermal noise.

(b) The thermal noise of the 10-kΩ resistor in a 9990-Hz bandwidth is 1.26 μV. The excess noise is 30 nV × 10 × $\sqrt{3}$ = 520 nV.

13-6-2 CMOS Voltage Divider

A simple voltage reference can be constructed in a CMOS technology by connecting a PMOS and an NMOS transistor in series as shown in Fig. 13-7 [3]. Ignoring the effects of channel length modulation, the reference voltage can be expressed as

$$V_{\text{ref}} = \frac{\sqrt{\dfrac{K_n W_n}{L_n}}\, V_{Tn} + \sqrt{\dfrac{K_p W_p}{L_p}}\,(V_{DD} + V_{Tp})}{\sqrt{\dfrac{K_n W_n}{L_n}} + \sqrt{\dfrac{K_p W_p}{L_p}}} \tag{13-4}$$

where K_n and K_p are the respective transconductance parameters of the NMOS and PMOS devices, W_n and W_p the gate widths, L_n and L_p the gate

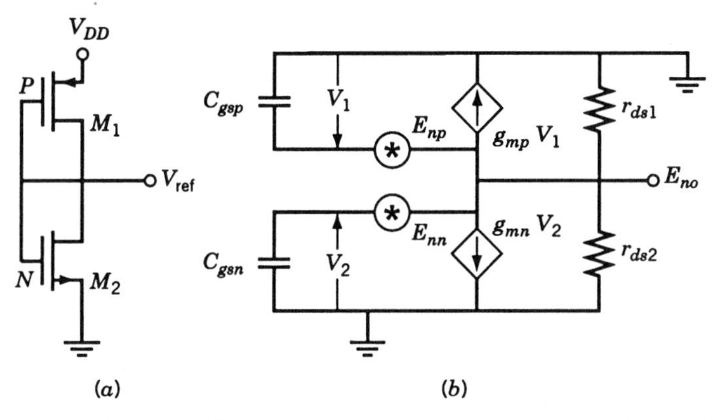

Figure 13-7 (*a*) dc and (*b*) ac equivalent circuits of a CMOS voltage divider.

lengths, and V_{Tn} and V_{Tp} the threshold voltages. (For a detailed explanation of MOSFET behavior and parameters, see Allen and Holberg [3].) Like the resistor divider, this voltage reference has poor supply noise rejection and supply regulation. The threshold voltages of the transistors are also temperature dependent, leading to a variation in V_{ref} with respect to temperature. From the small-signal model in Fig. 13-7, the intrinsic output noise is

$$E_{no} = \sqrt{\frac{\left(\left(g_{mp} + sC_{gsp}\right)E_{np}\right)^2 + \left(\left(g_{mn} + sC_{gsn}\right)E_{nn}\right)^2}{\left(g_{mp} + g_{mn} + \frac{1}{r_{dsp}} + \frac{1}{r_{dsn}} + s\left(C_{gsp} + C_{gsn}\right)\right)^2}} \tag{13-5}$$

where g_{mp} and g_{mn} are the transconductances of the transistors, C_{gsp} and C_{gsn} the gate-to-source capacitances, r_{dsp} and r_{dsn} the drain-to-source resistances, and E_{np} and E_{nn} are the thermal noise voltages of the PMOS and NMOS devices. Assuming that r_{dsp} and r_{dsn} are very large and can be neglected, the noise voltages of the individual MOS transistors will be attenuated by the voltage dividers formed by their transconductances at low frequencies and by the voltage dividers formed by the gate-to-source capacitances at higher frequencies. The thermal and $1/f$ components of E_{np} and E_{nn} can be reduced by increasing the device transconductances and gate areas, respectively.

At low frequencies, assuming $g_{mp} = g_{mn}$ and the drain resistances are very large, Eq. 13-5 reduces to

$$E_{no} = \frac{E_n}{\sqrt{2}} \tag{13-6}$$

when $E_{np} = E_{nn} = E_n$.

13-6-3 Bipolar Diode Reference

Better supply noise rejection than a simple voltage divider can be achieved by using a simple *pn* diode reference as shown in Fig. 13-8. The reference voltage is simply the voltage drop across the diode, or

$$V_{\text{ref}} = V_{BE} = \frac{nkT}{q}\log_n\left(\frac{I_{\text{DC}}}{I_s}\right) = \frac{nkT}{q}\log_n\left(\frac{V_{DD} - V_{\text{ref}}}{I_s R_1}\right) \tag{13-7}$$

where n is a process parameter typically between 1 and 3, I_{DC} is the dc current through the diode, and I_s the diode's reverse saturation current. Because of the logarithmic relationship between diode current and voltage, supply noise rejection and regulation are considerably improved. By replacing the supply resistor R_1 with a current source, these parameters can be improved to the point where the characteristics of the current source will determine the power supply regulation and noise rejection. However, the temperature coefficient of a *pn* diode is about -2 mV/°C, so considerable output voltage variation will occur as the operating temperature varies. From the small-signal model of Fig. 13-8 the intrinsic output noise of a reference with a supply resistor is calculated to be

$$E_{no} = \sqrt{\frac{(r_{d1}E_{R1})^2 + (r_{d1}R_1 I_{d1})^2}{(r_{d1} + R_1)^2}} \tag{13-8}$$

where r_{d1} is the dynamic resistance of the diode, E_{R1} is thermal plus excess noise of resistor R_1, and I_{d1} is the shot noise and excess noise of the diode. If load resistor R_1 is large and diode resistance $r_{d1} \ll R_1$, then Eq. 13-8 reduces to the shot noise voltage of the diode,

$$E_{no} = r_{d1} I_{d1} \tag{13-9}$$

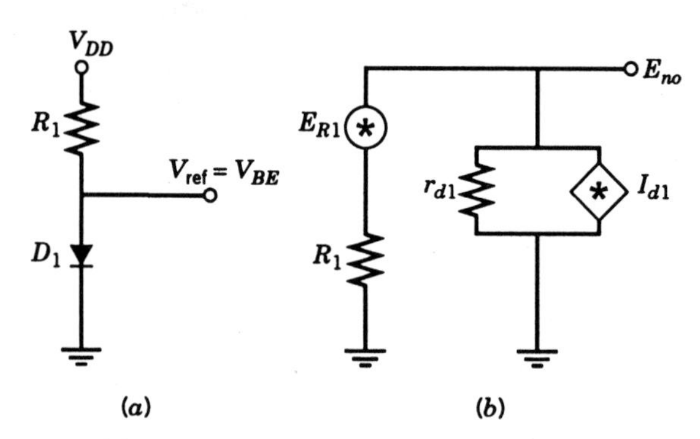

Figure 13-8 (*a*) dc and (*b*) ac equivalent circuits of a diode reference.

Two or more diodes can also be stacked together to obtain higher voltages. If a supply resistor is used, the output noise voltage of a reference with multiple diodes in series is

$$E_{no} = \sqrt{\frac{(nr_d E_{R1})^2 + n(r_d R_1 I_d)^2}{(nr_d + R_1)^2}} \tag{13-10}$$

where n is the number of diodes and r_d is the dynamic resistance of any single diode. Note that while the reference voltage increases linearly with the number of diodes, the output noise voltage due to shot noise roughly increases as the square root of the number of diodes due to rms summing of the individual noise currents. This fact illustrates an interesting principle in low-noise reference design: lower output noise can be achieved by *adding* several reference voltages together rather than by *multiplying* a reference voltage to increase its value. Adding n voltages together results in a $\sqrt{n}$ factor reduction in the value of the total noise voltage as opposed to multiplying a single voltage by a factor of n.

Now consider Eq. 13-8 again. If V_{DD} is much greater than V_{ref}, then R_1 will be much greater than r_{d1}, and the effect of E_{R1} on the output will be greatly attenuated by the resistor divider formed by r_{d1} and R_1. As a result, the noise current of the *pn* diode will be the dominant component of the reference output noise. Ideally, this noise current is due entirely to shot noise, but in practice some amount of excess $1/f$ noise will be present in the diode due to surface and bulk defects. As the diode current increases, this excess noise will increase while the shot noise decreases, and the excess noise will become the dominant noise component at low frequencies. This $1/f$ noise contribution becomes even more apparent if the simple *pn* diode is replaced with an *npn* transistor with its base connected to its collector as shown in Fig. 13-9. In this case the intrinsic output noise voltage can be expressed as

$$E_{no} = \sqrt{\frac{E_{R1}^2 + \left(\frac{R_1(1 + g_m Z_\pi)}{r_x + Z_\pi} E_n\right)^2 + R_1^2 I_n^2}{\left(1 + \frac{R_1(1 + g_m Z_\pi)}{r_x + Z_\pi} + \frac{R_1}{r_o}\right)^2}} \tag{13-11}$$

where E_n and I_n contain $1/f$ noise components as discussed in Chap. 5. In general, higher collector currents in the transistor will result in greater $1/f$ noise and lower shot noise. As with any low-noise bipolar transistor application, a larger β_o will reduce $1/f$ output noise in the reference. If $Z_\pi \gg r_x$, $r_o \gg R_1$, and $g_m Z_\pi \gg 1$, Eq. 13-11 can be simplified to

$$E_{no}^2 = \frac{E_{R1}^2 + (g_m R_1 E_n)^2 + R_1^2 I_n^2}{(1 + g_m R_1)^2} \tag{13-12}$$

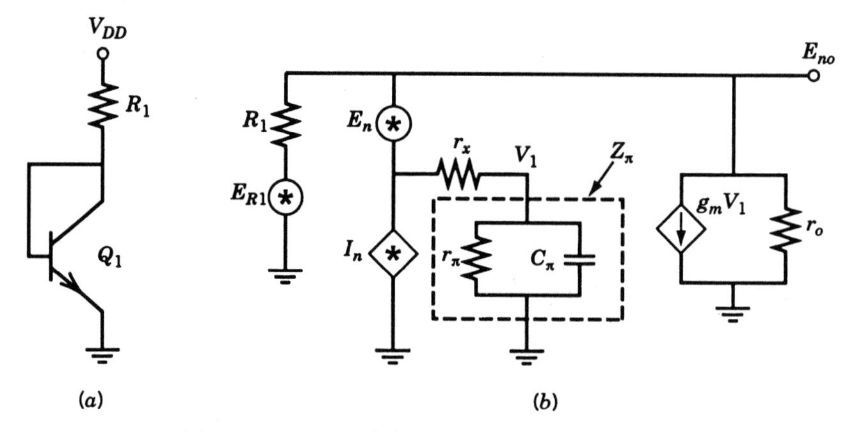

Figure 13-9 (*a*) BJT diode and (*b*) ac equivalent circuit of voltage reference.

13-6-4 Zener Diode References

The simplest Zener diode reference, like a simple *pn* diode reference, consists of the diode in series with a resistor or current source, with similar small signal models. However, the noise output of a Zener diode depends on whether the diode uses the Zener effect or avalanching to achieve its breakdown voltage. Zener breakdown usually occurs at voltages of 5 V or less, while avalanche breakdown dominates at higher voltages. A diode in Zener breakdown will have shot noise and excess noise like that of a *pn* diode, but a Zener diode in avalanche breakdown will generate a much higher level of white noise due to the avalanche generation of carriers in the junction. It is best not to use Zener diodes as a low-noise voltage reference unless breakdown occurs from the Zener effect.

There is a class of Zener diode references (called TC Zeners) which achieve a very low temperature coefficient and low noise in the 0.1 to 10 Hz frequency range by cancelling the negative temperature coefficient of a *pn* diode with the positive temperature coefficient of a Zener diode in avalanche breakdown, resulting in a typical reference voltage of 6.9 to 7 V [7, 8]. Some of these hybrid devices also include a heater to maintain the temperature of the substrate within precise limits. The Zener diodes are created as "buried" components in the substrate by using ion implantation, leading to very little excess noise in the reference from surface effects in the integrated circuit. Since these references are designed for use as discrete components, the additional white noise generated by the avalanche effect can be filtered or bandwidth limited. Two examples of this type of buried Zener reference are the LM199 from National Semiconductor and the LTZ1000 from Linear Technology. The LM199 has 0.7 μV peak-to-peak output noise from a bandwidth of 0.01 to 1 Hz and a typical temperature coefficient of 1 ppm/°C. The LTZ1000, which uses an internal heater, has a maximum of

2 μV peak-to-peak output noise from 0.1 to 10 Hz and a temperature coefficient of 0.05 ppm/°C.

13-6-5 MOS Diode Reference

As shown in Fig. 13-10, a MOS transistor can also be used as a simple diode in a reference circuit by connecting together the drain and gate [3]. Ignoring channel length modulation, the reference voltage is

$$V_{\text{ref}} = V_{GS} = V_T + \sqrt{\frac{2I_{DS}L}{KW}} = V_T + \sqrt{\frac{2(V_{DD} - V_{GS})L}{KWR_1}} \tag{13-13}$$

where I_{DS} is the current flowing through R_1 and M_1. This reference has the advantage of a wider range of output voltages than the bipolar diode reference due to the square-law relationship between gate–source voltage and drain–source current in the MOSFET. However, this increased voltage range also results in poorer supply noise rejection and supply regulation unless R_1 is replaced with a current source. Using the small-signal model of Fig. 13-10, the intrinsic output noise of the reference is

$$E_{no} = \sqrt{\frac{E_{R1}^2 + (g_m R_1 + sC_{gs}R_1)^2 E_n^2}{\left(1 + sC_{gs}R_1 + g_m R_1 + \dfrac{R_1}{r_{ds}}\right)^2}} \tag{13-14}$$

Assuming that r_{ds} is very large and $g_m R_1 \gg 1$, at low frequencies the output

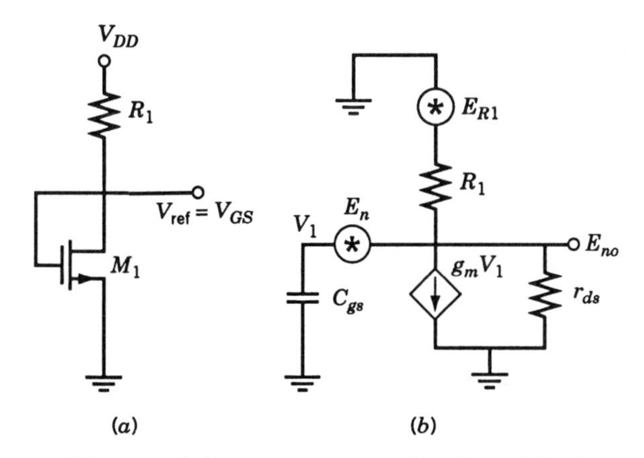

Figure 13-10 (*a*) dc and (*b*) ac equivalent circuits of a MOS voltage reference.

noise can be approximated by

$$E_{no}^2 = \frac{E_{R1}^2}{g_m^2 R_1^2} + E_n^2 \tag{13-15}$$

The $1/f$ equivalent input noise of the MOSFET will be the main component of the intrinsic low-frequency output noise, while the transistor's channel thermal noise dominates at higher frequencies.

13-6-6 V_{BE} Multiplier Reference

A very common method of generating a reference voltage in IC designs is the V_{BE} multiplier circuit of Fig. 13-11. Ignoring the transistor's output impedance and assuming that β_o is very large, the reference voltage is

$$V_{\text{ref}} = \left(1 + \frac{R_2}{R_3}\right) V_{BE} \tag{13-16}$$

Like the bipolar diode reference, the power supply regulation and noise rejection of the V_{BE} multiplier circuit can be greatly improved by replacing R_L with a current source. However, the negative temperature coefficient of the reference voltage will be multiplied by the same factor as the base–emitter voltage with respect to a single diode voltage drop. From the small-signal diagram of Fig. 13-12, thc intrinsic output noise voltage of the V_{BE} multiplier is found to be

$$E_{no}^2 = \left(1 + \frac{R_2}{R_3}\right)^2 \left[\left(\frac{E_{RL}}{g_m R_L}\right)^2 + E_n^2\right] + \left(\frac{R_2}{R_3}\right)^2 E_{R3}^2 + I_n^2 R_2^2 + E_{R2}^2 \tag{13-17}$$

assuming $g_m R_2 \gg 1$, $g_m R_3 \gg 1$, and R_L is very large.

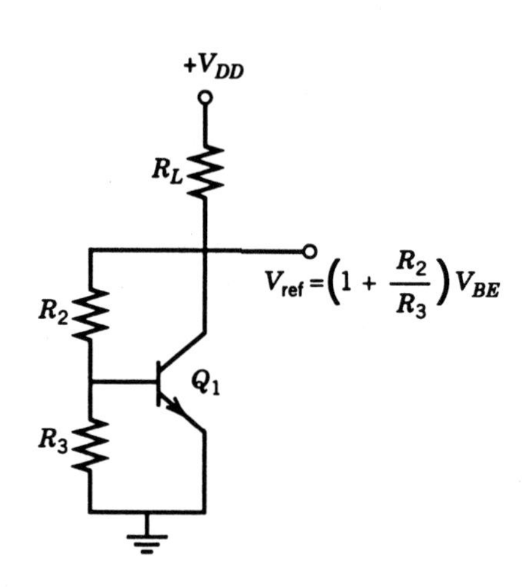

Figure 13-11 The V_{BE} multiplier reference.

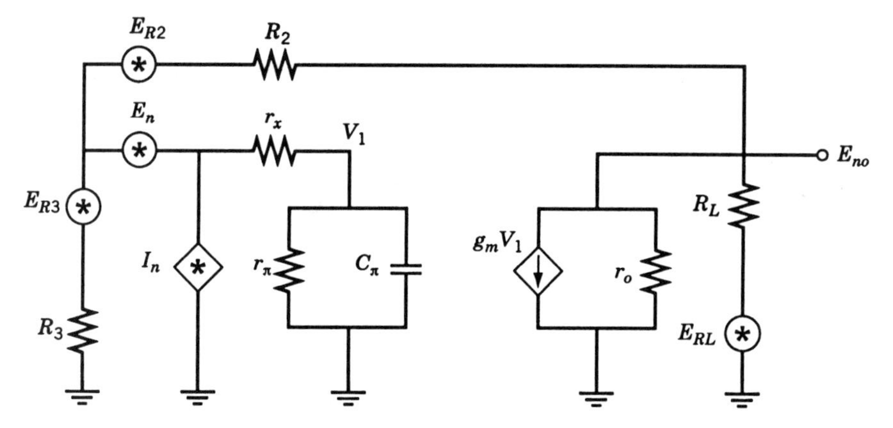

Figure 13-12 ac equivalent circuit for a V_{BE} multiplier reference.

This equation shows that the transistor's equivalent input noise and the noise of resistor R_3 are multiplied by the feedback ratio of R_2/R_3 to produce the output reference noise.

13-6-7 V_{GS} Multiplier Reference

The same feedback principle used in the V_{BE} multiplier can also be applied with an FET to create a V_{GS} multiplier reference (Fig. 13-13) [3]. Neglecting channel length modulation effects, the reference voltage is

$$V_{\text{ref}} = \left(1 + \frac{R_2}{R_3}\right) V_{GS} \tag{13-18}$$

Because of the square-law relationship between drain current and

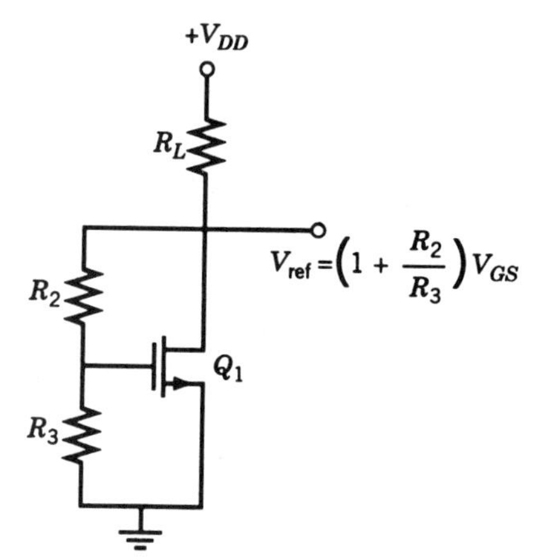

Figure 13-13 The V_{GS} multiplier reference.

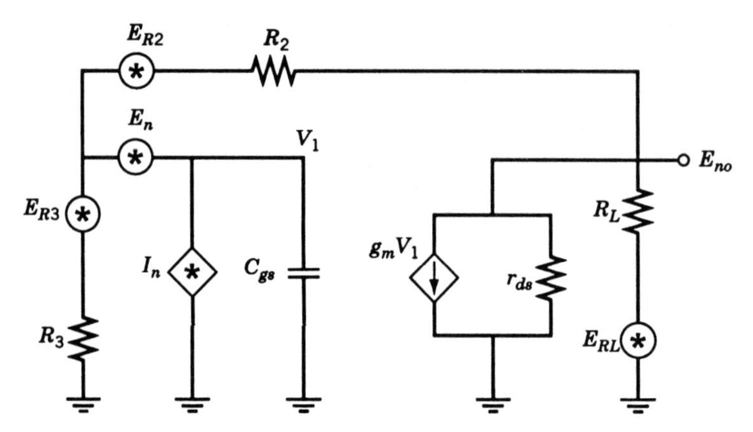

Figure 13-14 ac equivalent circuit of the V_{GS} multiplier reference.

gate–source voltage in a MOSFET, supply noise rejection and supply regulation are not as good as with the V_{BE} multiplier circuit, although they can also be greatly improved by replacing supply resistor R_L with a current source. The gate–source voltage also exhibits a negative temperature coefficient which is multiplied by the resistor feedback ratio. Referring to the small-signal equivalent circuit in Fig. 13-14, the intrinsic output voltage of the V_{GS} multiplier will be

$$E_{no}^2 = E_{RL}^2 + \left(1 + \frac{R_2}{R_3}\right)^2 \left[E_n^2 + E_{R3}^2 + \left(I_n^2 + I_{R2}^2\right)R_3^2\right] \qquad (13\text{-}19)$$

assuming $g_m R_L \gg R_2/R_3$.

Just as with the V_{BE} reference, resistors R_2 and R_3 establish gain ratios which multiply the noise contributions of the transistor's equivalent input noise and resistor R_3 at the reference output.

13-6-8 Noise in Current Sources

One obvious method of generating a reference voltage of any desired value is to source or sink a current through a resistor. The intrinsic output noise of the current source can simply be modeled as a noise current in parallel with the output resistance of the source. However, there is a significant noise mechanism in current sources besides the intrinsic shot or thermal noise of the source transistor. Any high-performance current source requires a bias circuit or current mirror to set the value of the output current as shown in Fig. 13-15. As a result, any noise source present at the base or gate of the current source transistor is amplified by the gain of the common-emitter or common-source stage formed by the current source. If the load resistance is

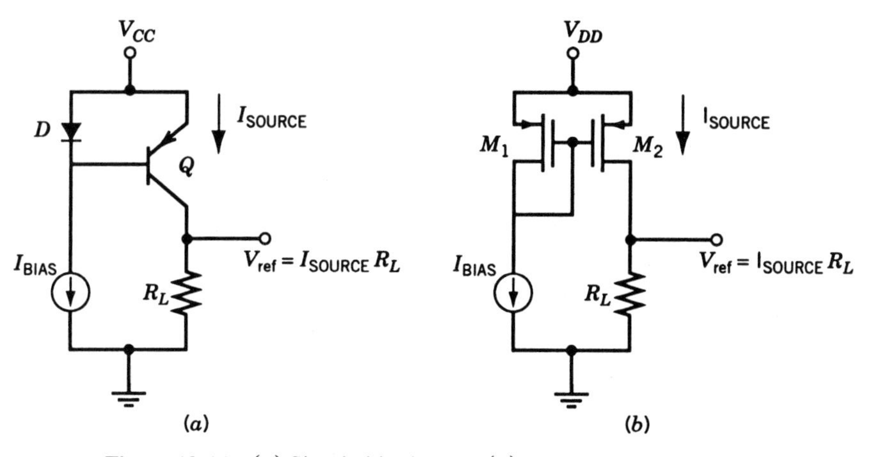

Figure 13-15 (*a*) Simple bipolar and (*b*) MOS current mirrors.

much smaller than the output resistance of the current source, this voltage gain is approximately equal to the load resistance times the transconductance of the transistor. Given the small-signal diagram of a BJT current mirror as shown in Fig. 13-16, the output noise can be expressed as

$$E_{no} = \sqrt{\frac{\left[Z_\pi g_m(R_L \parallel r_{ce})\right]^2\left[E_n^2 + (r_{\text{bias}} \parallel r_d)^2\left(I_n^2 + I_d^2 + I_{\text{bias}}^2\right)\right]}{\left[(r_{\text{bias}} \parallel r_d) + r_b + Z_\pi\right]^2}} \qquad (13\text{-}20)$$

where I_d and r_d are the noise current and dynamic resistance of the current mirroring diode and I_{bias} and r_{bias} are the equivalent noise current and output resistance of the dc bias current circuit I_{BIAS}. Assuming that Z_π is very large with respect to r_d and r_b, and assuming that r_o is much larger than

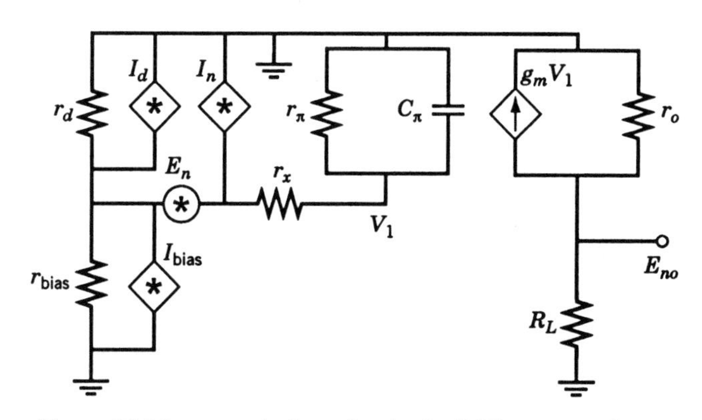

Figure 13-16 ac equivalent circuit of a BJT current mirror.

R_L, the output noise of the BJT current source can be simplified to

$$E_{no} = g_m R_L \sqrt{\left(E_n^2 + (r_{\text{bias}} \parallel r_d)^2 \left(I_n^2 + I_d^2 + I_{\text{bias}}^2\right)\right)} \qquad (13\text{-}21)$$

Since the reference voltage across the load is proportional to the value of I_{SOURCE}, the transconductance of the source transistor cannot be lowered to reduce output noise without reducing the desired reference voltage. However, the small-signal impedance of the load may be reduced by bypassing R_L with a capacitor or by replacing the load resistor with diodes at the cost of increased temperature dependence of the reference voltage. If the dominant noise mechanism of the bias circuit is the shot noise of the mirroring diode, increasing the bias current I_{BIAS} will also reduce the output white noise, although the $1/f$ excess noise level may increase as a result of the higher current. The equivalent input noise voltage and current (E_n and I_n) of the source transistor should also be minimized by choosing a good low-noise device whenever possible.

13-7 BANDGAP VOLTAGE REFERENCE

For many applications, one requires an integrated circuit voltage reference that is independent of both supply voltage and operating temperature. Although several different types of circuits have been developed over the years to meet these requirements, the most popular method remains the bandgap voltage reference [3, 4, 9]. The bandgap reference achieves a very low temperature coefficient by canceling the negative temperature coefficient of a *pn* diode with the positive temperature coefficient of the voltage difference between two *pn* junctions. Bandgap circuits can be implemented in standard bipolar, BiCMOS, and CMOS technologies without the need of special process steps as in the case of ion-implanted Zener diode references. Commercial bandgap references typically have about 50 μV peak-to-peak output noise from 0.1 to 10 Hz with temperature coefficients ranging from 5 to 50 ppm/°C [5, 10].

A standard bandgap voltage reference has an output voltage value that can be expressed as

$$V_{\text{ref}} = V_{BE} + K(\Delta V_{BE}) \qquad (13\text{-}22)$$

where K is a gain constant ranging from 4 to 8 for most circuits. The circuit will achieve minimum temperature coefficient when V_{ref} is approximately 1.2 V, near the bandgap voltage of silicon (hence the name). At this voltage, the negative temperature coefficient of V_{BE} will be canceled by the positive temperature coefficient of $K(\Delta V_{BE})$. In the past several years, high-performance bandgap circuits have been implemented using both linear and switched-capacitor topologies. Switched-capacitor bandgap circuits offer cancelation of the offset errors and $1/f$ noise of the buffer amplifier, but also require a high-speed clock and can only generate the reference voltage for a portion of the clock cycle [9]. Linear bandgap circuits, on the other hand, require no clock and generate a constant output voltage, but are generally subject to offset errors and $1/f$ noise contributions from the amplifier.

13-7-1 Analysis of a Bandgap Voltage Reference

To get a better idea of the relative contributions of each noise source at the output of a typical bandgap voltage reference, consider the circuit in Fig. 13-17. Assuming a noisy operational amplifier with infinite gain, infinite input resistance, and zero output resistance, the reference voltage will be

$$V_{\text{ref}} = V_{BE} + \frac{R_2}{R_3}(\Delta V_{BE}) \tag{13-23}$$

In order to generate the ΔV_{BE} voltage, diode D_2 must have a larger reverse saturation current than diode D_1 (i.e., larger junction area), or the current through D_2 must be smaller than the current through D_1, or both. Note that there is no direct connection to V_{DD} in this circuit. Ideally this circuit would have infinite rejection of power supply noise, although the actual circuit's noise rejection is limited by the power supply rejection ratio of the operational amplifier. In addition, this bandgap circuit will have a second stable state with an output voltage of zero, requiring the use of a start-up circuit to guarantee proper operation.

Because of the relatively low dynamic resistance of a diode for a given dc current, one can safely assume that R_1 and R_2 will be much greater than r_{d1} and r_{d2} if the bandgap output voltage is about 1.2 V. Also assume that $R_2 \gg R_3$ (usually only four to eight times larger, so it is a cautious assumption). Using these assumptions and the small-signal diagram of Fig. 13-18, the

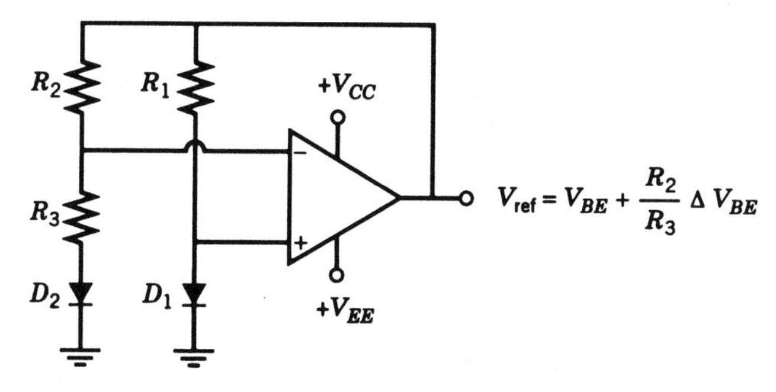

Figure 13-17 A simple bandgap reference.

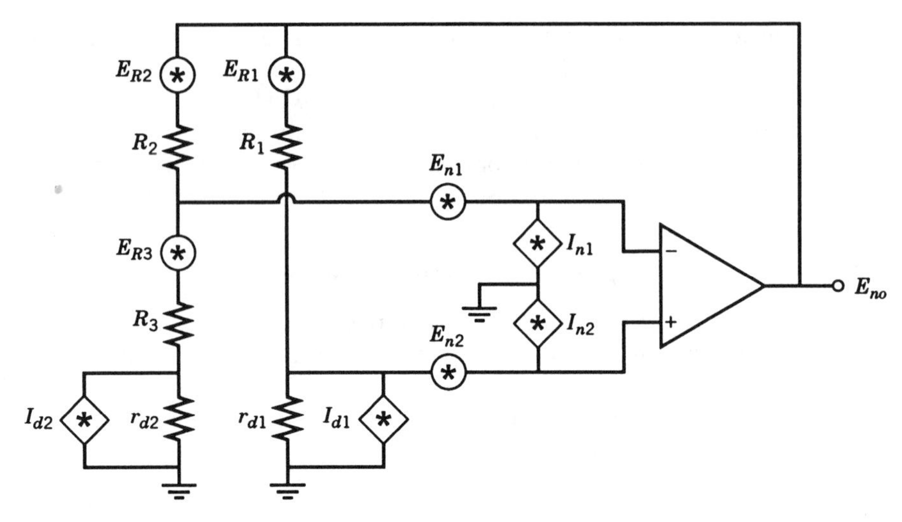

Figure 13-18 ac equivalent circuit of the bandgap reference.

output noise voltage can be expressed as the equivalent input noise voltage times the gain $R_2/(R_3 + r_{d2})$

$$E_{no}^2 = \left(\frac{R_2}{R_3 + r_{d2}}\right)^2 \left[E_{n1}^2 + E_{n2}^2 + (I_{n1}^2 + I_{R2}^2)(R_3 + r_{d2})^2 + I_{n2}^2 r_{d1}^2 + \left(\frac{r_{d1} E_{R1}}{R_1}\right)^2 + E_{R3}^2 + I_{D2}^2 r_{d2}^2 + I_{D1}^2 r_{d1}^2\right] \tag{13-24}$$

As these results show, output noise multiplication in the bandgap circuit is caused by the feedback network formed with R_2 and $(r_{d2} + R_3)$. This ratio can be reduced by increasing the ΔV_{BE} voltage across R_3 by making D_2 much larger than D_1 and/or setting the current through D_2 much smaller than the current through D_1. However, in practice it is difficult to make ΔV_{BE} much larger than 150 mV due to the exponential current–voltage behavior of a bipolar diode, so R_2 will always be considerably larger than $(r_{d2} + R_3)$. A second possibility is to bypass R_2 with a capacitor to reduce the ac gain of the bandgap reference while preserving the dc characteristics. The noise gain will be attenuated above the frequency set by the pole of R_2 and the capacitor, although area considerations for most circuits would limit the size of an integrated capacitor. An external discrete capacitor is a more practical solution in this case.

Besides reducing the gain of the feedback network, the intrinsic noise sources themselves can be reduced by using a good low-noise operational amplifier and by reducing the values of R_1 through R_3 while preserving the

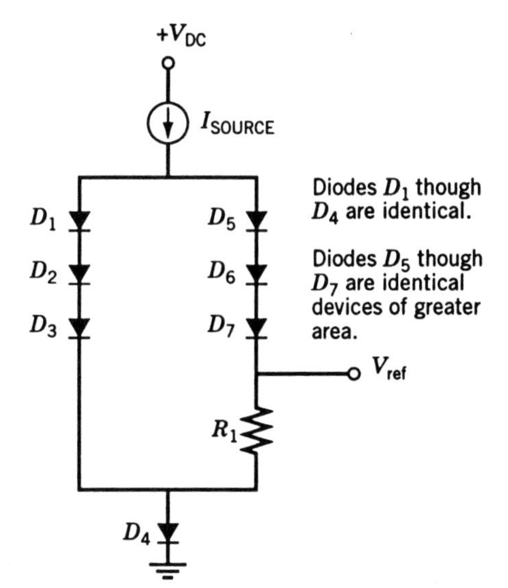

Figure 13-19 A simple low-noise bandgap reference.

required ΔV_{BE} gain. The trade-off in this case will be higher dc currents for lower noise, although excess noise in the diodes will also increase as the currents increase. The final limit of noise is the shot noise currents I_{D1} and I_{D2} of the two reference diode strings and the thermal noise of R_3.

13-7-2 Low-Noise Bandgap Reference Topology

In the previous noise analysis for the bandgap reference, it was shown that the multiplication of the ΔV_{BE} voltage leads to an undesirable increase in noise at the output of the bandgap circuit. By generating the full ΔV_{BE} voltage directly, the output noise of the bandgap circuit could be significantly reduced. Figure 13-19 shows one possible method of direct bandgap voltage generation. By stacking diodes D_1 through D_7, the ΔV_{BE} voltage drops across R_1 are effectively summed together. Since most of the current from I_{SOURCE} flows through the D_1 to D_4 path, the current source has a very low ac load impedance, which minimizes the contribution of the bias circuit noise. If a unity-gain buffer amplifier is required at the output of this circuit, the noise contributions of the buffer will not be multiplied by a gain factor. The disadvantages of this circuit include larger circuit area, higher current consumption, and a larger minimum operating voltage than conventional designs. Furthermore, this design can only be implemented in integrated circuit processes where a floating *pn* diode is available.

SUMMARY

a. To eliminate stray common-mode coupling, as well as other sources of pickup, transformer shielding is recommended.

b. A power supply that provides good ripple filtering also discriminates against noise reaching the electronic system through the power supply.
c. The ripple clipper circuit uses negative feedback from the load to control conduction in a series-pass transistor.
d. A power supply regulator will remove much of the power supply noise.
e. The regulator is essentially a voltage reference followed by a buffer amplifier.
f. Noise of a voltage regulator comes primarily from the noise sources in the voltage reference with contributions from the buffer amplifier.
g. A low-noise regulator must start with a low-noise reference.

PROBLEMS

13-1. Given a CMOS voltage divider as in Fig. 13-7*a* with $V_{DD} = 5$ V which uses the CMOS devices whose parameters are for the *n*-channel device $K_n = 41.8\ \mu\text{A/V}^2$, $V_T = 0.79$ V, and $\lambda = 0.01\ \text{V}^{-1}$. For the *p*-channel device $K_p = 15.5\ \mu\text{A/V}^2$, $V_T = -0.93$ V, and $\lambda = 0.01\ \text{V}^{-1}$. Design the divider circuit by determining reasonable W_n/L_n and W_p/L_p aspect ratios so that $V_{\text{ref}} = 2.5$ V.

13-2. Compare the output noise voltages for a diode and base-collector shorted *npn* transistor with each device carrying 1 mA of bias current and $V_{DD} = 9$ V. For the transistor assume $\beta = 200$, $r_x = 100\ \Omega$, and the Early voltage $E_A = 200$ V.

13-3. Find the aspect ratio required for M_1 in Fig. 13-10*a* if we desire a $V_{\text{ref}} = 2.5$ V and $V_{DD} = 5$ V. Assume that $V_T = 1$ V and that the dc drain current is 200 μA.

13-4. Verify Eq. 13-16 for the V_{BE} multiplier voltage reference.

13-5. Show that ΔV_{BE} in Eq. 13-23 is given by

$$\Delta V_{BE} = \frac{kT}{q} \ln \frac{I_{D1} I_{S2}}{I_{D2} I_{S1}}$$

where I_D represents the diode current, and I_S is the reverse saturation current.

13-6. For Fig. 13-19 show that

$$V_{\text{ref}} = \frac{kT}{q}\left(4 \ln \frac{I_{D1}}{I_{S1}} - 3 \ln \frac{I_{D5}}{I_{S5}}\right)$$

where I_{D1} and I_{D5} are the currents through the left and right diode strings and I_{S1} and I_{S5} are the left and right reverse saturation currents.

13-7. Verify that the dc output voltage of the active capacity multiplier of Fig. 13-3 is KV_{BB} and the filter time constant is KR_1C_1, where

$$K = \frac{R_2}{R_1 + R_2}$$

REFERENCES

1. Carr, J. J., Elektor Electronics USA, July 1992, pp. 20–23.
2. Holman, W. T., Georgia Institute of Technology, Ph.D. Qualifying Examination, (unpublished) 1991.
3. Allen, P. E., and D. R. Holberg, *CMOS Analog Circuit Design*, Holt, Rinehart and Winston, New York, 1987.
4. Gray, P. R., and M. G. Meyer, *Analysis and Design of Analog Integrated Circuits*, Wiley, New York, 1984.
5. *Linear Circuits Data Book*, Vol. 3, Texas Instruments, 1989.
6. *1992 Amplifier Applications Guide*, Analog Devices, Inc.
7. Knapp, R., "Selection Criteria Assist in Choice of Optimum Reference," *EDN*, (February 18, 1988), 183–192.
8. Goodenough, F., "IC Voltage References: Better Than Ever," *Electronic Design*, (September 22, 1988); 83–89.
9. Song, B., and P. R. Gray, "A Precision Curvature-Compensated CMOS Bandgap Reference," *IEEE J. Solid-State Circuits*, **SC-18** (December 1983); 634–643.
10. *Linear ICs for Commercial Applications*, Harris Semiconductor, 1990.

CHAPTER 14

LOW-NOISE AMPLIFIER DESIGN EXAMPLES

It would be a poor design manual that did not provide a few practical low-noise circuits as examples. After all, we want to help you get started in your low-noise application. Technology moves so rapid that it is impossible to provide the ultimate circuit. Also, various source impedances require different amplifiers for optimum noise so we will show some examples of amplifiers we have designed and some selections of good designs from the literature. These designs employ the methodologies of the previous chapters to illustrate their application.

14-1 CASCADE AMPLIFIER

To illustrate the design trade-offs, the basic designs for a cascaded CE–CE stage and a cascode-connected CE–CB amplifier are shown. These are designed with discrete transistors and use a single power supply. A dual power supply will simplify the biasing. Both these circuits can be integrated using active loads and current sources. For higher impedances, FETs can be substituted for BJTs.

As the first example consider the complementary cascade circuit shown in Fig. 14-1. The voltage gain is determined by the $g_m R_C$ of each stage of the CE–CE pair, Q_1 and Q_2. This stage and its biasing were analyzed in Chap. 10. The output stage Q_3 is an emitter follower to provide a low output impedance and to buffer the amplifier from loading.

Direct coupling is employed between stages. Capacitor C_{C1} is used at the input port to isolate the amplifier biasing from the signal source. A value for

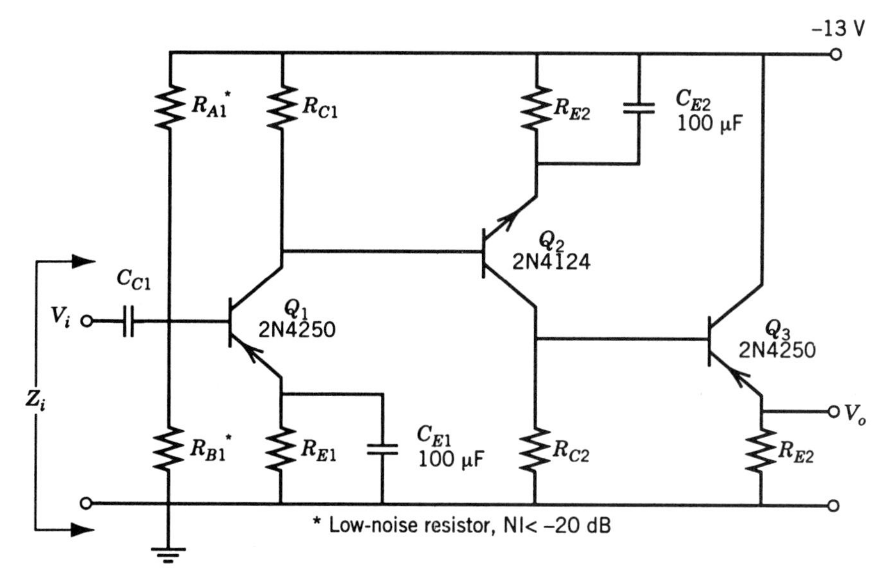

Figure 14-1 Complementary cascade example.

TABLE 14-1 Performance Data on Complementary Cascade Amplifier of Fig. 14-1[a]

I_{C1}	1 mA	100 μA	10 μA	1 μA
I_{C2}	3 mA	300 μA	30 μA	10 μA
R_{A1}	470 kΩ	4.7 MΩ	30 MΩ	30 MΩ
R_{B1}	220 kΩ	2.2 MΩ	22 MΩ	22 MΩ
R_{E1}	3 kΩ	30 kΩ	300 kΩ	3 MΩ
R_{C1}	3.6 kΩ	36 kΩ	360 kΩ	3 MΩ
R_{E2}	820 Ω	8.2 kΩ	82 kΩ	240 kΩ
R_{C2}	1.8 kΩ	18 kΩ	180 kΩ	500 kΩ
R_{E3}	1 kΩ	1 kΩ	100 kΩ	100 kΩ
A_v	6700	5250	2900	600
f_2	240 kHz	36.5 kHz	7.5 kHz	4.0 kHz
R_i (1 kHz)	11 kΩ	90 kΩ	825 kΩ	3 MΩ
C_i (1 kHz)	310 pF	280 pF	160 pF	62 pF
E_{nT} (10 Hz)	2.0 nV	2.5 nV	5.4 nV	14 nV
E_{nT} (10 kHz)	1.5 nV	2 nV	5 nV	14 nV
I_{nT} (10 Hz)	6.0 pA	1.2 pA	0.4 pA	0.1 pA
I_{nT} (10 kHz)	0.9 pA	0.3 pA	0.1 pA	0.06 pA
R_o (10 Hz)	330 Ω	2.4 kΩ	13.5 kΩ	140 kΩ
R_o (10 kHz)	1.7 kΩ	6.7 kΩ	50 kΩ	230 kΩ

[a]Noise values are given on a hertz$^{-1/2}$ basis.

the coupling capacitor is determined by the product of the source resistance and the noise current I_n of the first stage at the lowest frequency of interest. To assure that I_n does not contribute noise, it is necessary that

$$I_n|R_s - jX_C| \le E_n \tag{14-1}$$

at the lowest frequency of interest.

In order to show the characteristics of this amplifier, performance data for various values of I_C are tabulated in Table 14-1. The four columns represent four values of quiescent I_C ranging from milliamperes to microamperes. To achieve these operating points, the values of the biasing elements must be changed. For a reduction of 10:1 in I_C, R_E and R_C increase by a factor of 10 to maintain the designed bias and gain. E_{nT} and I_{nT} are the total noise voltage and noise current of the total amplifier at the input. Usually this is the same as the first stage E_n and I_n.

Certain trends are clear from the data. The voltage-gain upper-cutoff frequency f_2 declines with decreased I_C, as does the input capacitance C_i and the voltage gain. Input resistance, as expected, increases.

From the listing of noise parameters, we find that I_{nT} is reduced by a factor of 10 and E_{nT} is increased by 10 when I_C is lowered. The optimum source resistance R_o increases by more than 100:1.

14-2 CASCODE AMPLIFIER

A discrete cascode amplifier is shown in Fig. 14-2. This circuit was also described in Chap. 10. Although the components and noise are similar to the

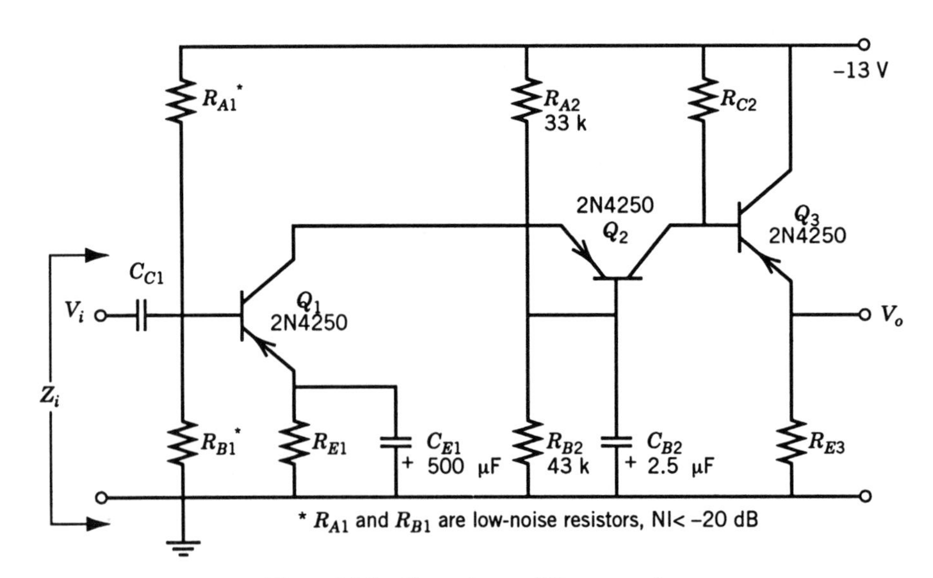

Figure 14-2 Cascode amplifier example.

TABLE 14-2 Performance Data on Cascode Amplifier on Fig. 14-2[a]

I_{C1}	1 mA	100 μA	10 μA	1 μA
I_{C2}	1 mA	100 μA	10 μA	1 μA
R_{A1}	300 kΩ	3 MΩ	20 MΩ	20 MΩ
R_{B1}	150 kΩ	1.5 MΩ	10 MΩ	10 MΩ
R_{E1}	3 kΩ	30 kΩ	300 kΩ	3 MΩ
R_{C2}	3 kΩ	30 kΩ	300 kΩ	3 MΩ
R_{E3}	10 kΩ	100 kΩ	1 MΩ	1 MΩ
A_v	97	115	120	145
f_2	5 MHz	500 kHz	47 kHz	5.3 kHz
R_i (1 kHz)	11.1 kΩ	75 kΩ	730 kΩ	4.0 MΩ
C_i (1 kHz)	130 pF	34 pF	35 pF	27 pF
E_{nT} (10 Hz)	2 nV	2.5 nV	5.4 nV	14 nV
E_{nT} (10 kHz)	1.5 nV	2 nV	5 nV	14 nV
I_{nT} (10 Hz)	6 pA	1.2 pA	0.4 pA	0.1 pA
I_{nT} (10 kHz)	0.9 pA	0.3 pA	0.1 pA	0.06 pA
R_o (10 Hz)	330 Ω	2.4 kΩ	13.5 kΩ	140 kΩ
R_o (10 kHz)	1.7 kΩ	6.7 kΩ	50 kΩ	230 kΩ

[a]Noise values are given on a hertz$^{-1/2}$ basis.

cascade amplifier, the frequency response and impedance are different. This illustrates that similar noise performance can be obtained with quite different signal performance. This is similar to using overall feedback to have low noise and still optimize signal performance.

A performance summary is given in Table 14-2 for the cascode amplifier. When compared with the cascade data, the total voltage gain of the cascode circuit is lower than the CE–CE. The noise parameters are about equal. A major reduction in C_i for the cascode is achieved and the voltage-gain upper-cutoff frequency is extended.

14-3 GENERAL-PURPOSE LABORATORY AMPLIFIER

When making low-noise measurements, it is handy to have a low-noise amplifier to quietly amplify the input signals. This amplifier should have low-noise voltage E_n to operate from low source resistances, low $1/f$ noise to operate at low frequencies, and high input impedance so it will not load the test circuit. The AD745 is a BIFET op amp with high input impedance and low noise voltage. With the 20-MHz gain–bandwidth, it can be used into the megahertz region. Most of the noise measurements for this book have been made with the circuit shown in Fig. 14-3. This circuit cascades two amplifiers for a gain of 1000 × and a frequency response of 500 kHz. If the gain of each stage is reduced to 10 × for an overall gain of 100 ×, the bandwidth increases to 1 MHz. The input impedance is set by the 10-MΩ input bias resistance. The input is: $E_{nT} = 2.5$ nV/Hz$^{1/2}$, $1/f$ corner = 50 Hz, $I_{nT} =$ 10 fA/Hz$^{1/2}$.

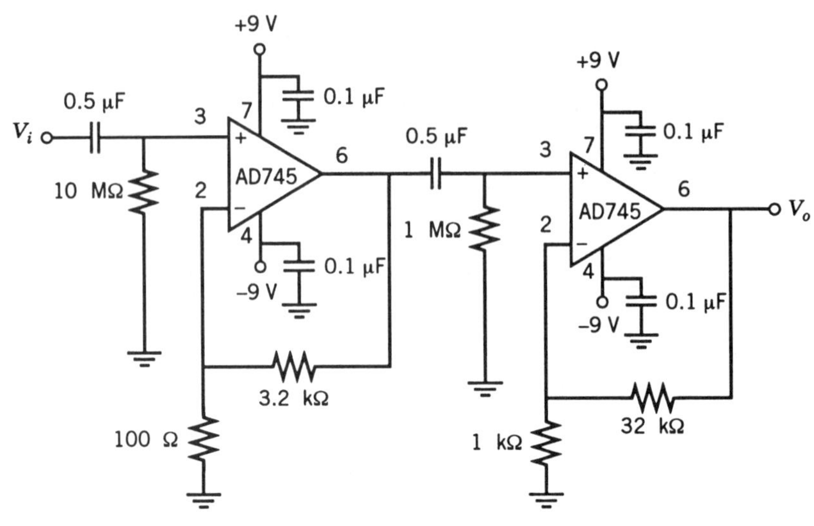

Figure 14-3 Low-noise lab amplifier. $E_n = 2.5\ \text{nV/Hz}^{1/2}$, $A_v = 1000$, $f_2 = 500$ kHz.

To avoid ground loops and power supply noise, use two 9-V batteries for power. The maximum load resistance is 600 Ω, without adding a unity-gain buffer stage.

14-4 IC AMPLIFIER WITH DISCRETE PARALLEL INPUT STAGES

When the noise of an integrated amplifier is too high, it is possible to reduce the input noise by adding a discrete transistor to the input of the IC. This can be a special transistor with a very low base resistance or the designer can use several transistors in parallel as shown in the novel circuit of Fig. 14-4 from D. Bowers of Analog Devices [1, 2]. This circuit starts with a low-noise OP-27 or one of the later lower-noise versions and adds additional gain stages ahead of the op amp. The three MAT-02 transistor pairs operate at a collector current of 1 mA each and give a reported noise voltage of only 0.5 $\text{nV/Hz}^{1/2}$.

For a MAT-02 with a noise of 0.9 $\text{nV/Hz}^{1/2}$ for each transistor, a pair would have a noise of $\sqrt{2} \times 0.9 = 1.27\ \text{nV/Hz}^{1/2}$. By paralleling the three pairs, the noise is reduced by $\sqrt{3}$ so the input noise would now be $1.27/\sqrt{3} = 0.73\ \text{nV/Hz}^{1/2}$. Additional stages would decrease the noise voltage E_n proportional to the square root of the number of stages and increase the noise current I_n proportional to the square root of the stages as derived in Chap. 10.

Another novel feature of this circuit is the use of the red LED to bias the 2N2905 transistor providing the 6-mA emitter bias for the MAT-02 transistors. The LED has about 1 V more forward voltage drop than the transistor,

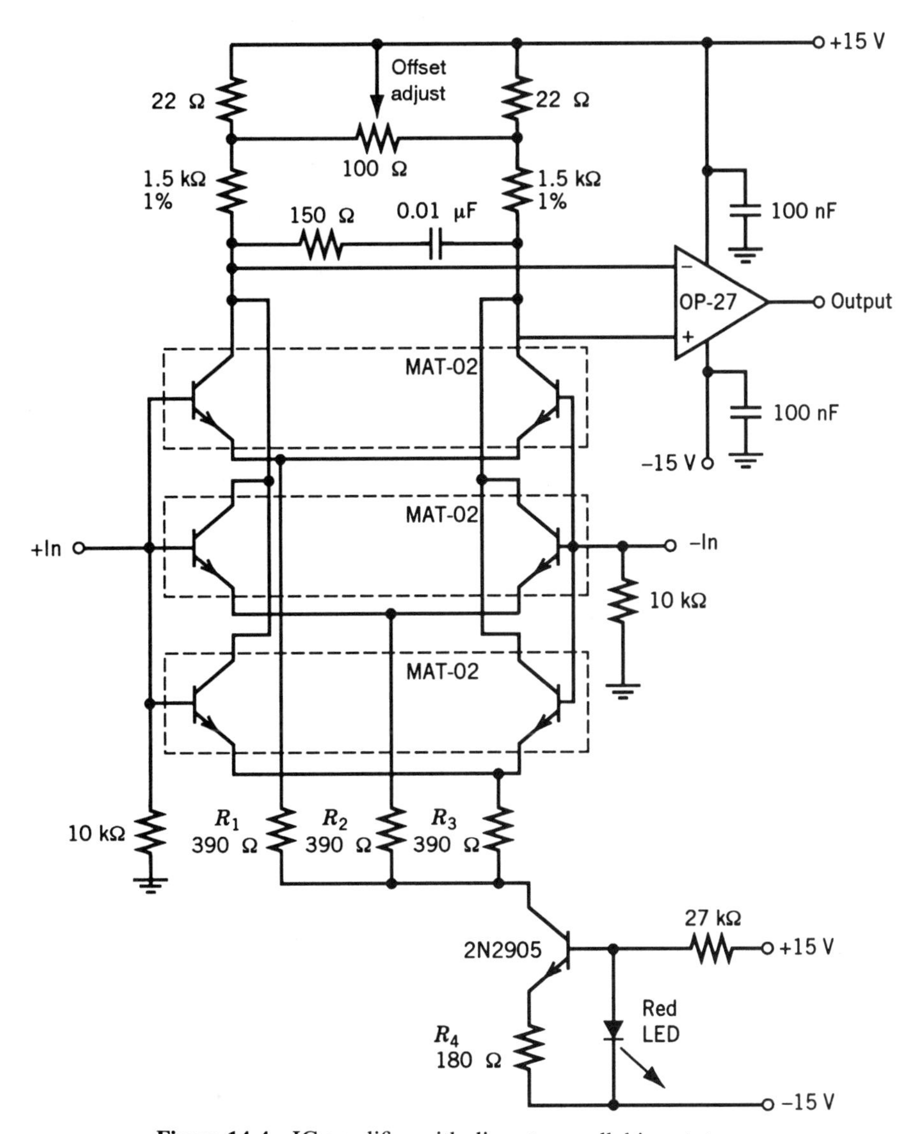

Figure 14-4 IC amplifier with discrete parallel input stages.

but about the same temperature coefficient. This generates a stable voltage across the 180-Ω emitter resistor R_4. The 390-Ω emitter resistors R_1 through R_3 provide equal current splitting between the MAT-02 transistors.

14-5 DIRECT-COUPLED SINGLE-ENDED AMPLIFIER

Another example of an amplifier that uses both discrete and integrated stages is shown in Fig. 14-5 [3]. This amplifier has a single-ended input with

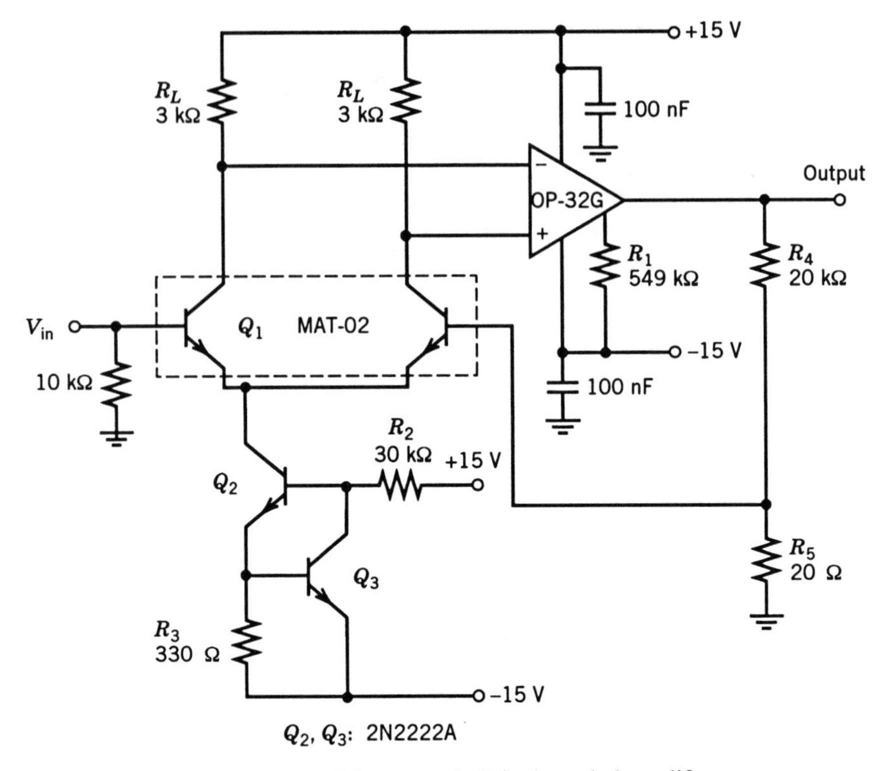

Figure 14-5 Direct-coupled single-ended amplifier.

overall feedback to set the gain at 1000 ×. This circuit offers the advantage of both low noise and direct coupling. There are no large capacitors. The noise voltage E_{nT} is determined by the input differential stage Q_1 as $\sqrt{2}$ times the noise of one transistor. E_{nT} is then the noise of a single transistor, $(0.9\ \text{nV/Hz}^{1/2}) \times 1.414 = 1.27\ \text{nV/Hz}^{1/2}$. The noise current I_{nT} is the same as the noise current of one single transistor of the MAT-02. The gain is set by the ratio of R_4 to R_5.

One advantage of using a matched dual monolithic transistor is the improved rejection of the noise of the bias circuit consisting of Q_2 and Q_3. The noise generated in the current bias circuit is rejected in proportion to the common-mode rejection ratio. This is significant since the noise of the bias is greater than the amplifier itself. The load resistors R_L should be matched or 1% tolerance.

14-6 ac-COUPLED SINGLE-ENDED AMPLIFIER

Additional design flexibility is gained by capacitively coupling the input stage to the IC amplifier stage as shown in Fig. 14-6. Although this circuit uses a matched pair for low E_{nT}, there are other arrays of transistors such as the

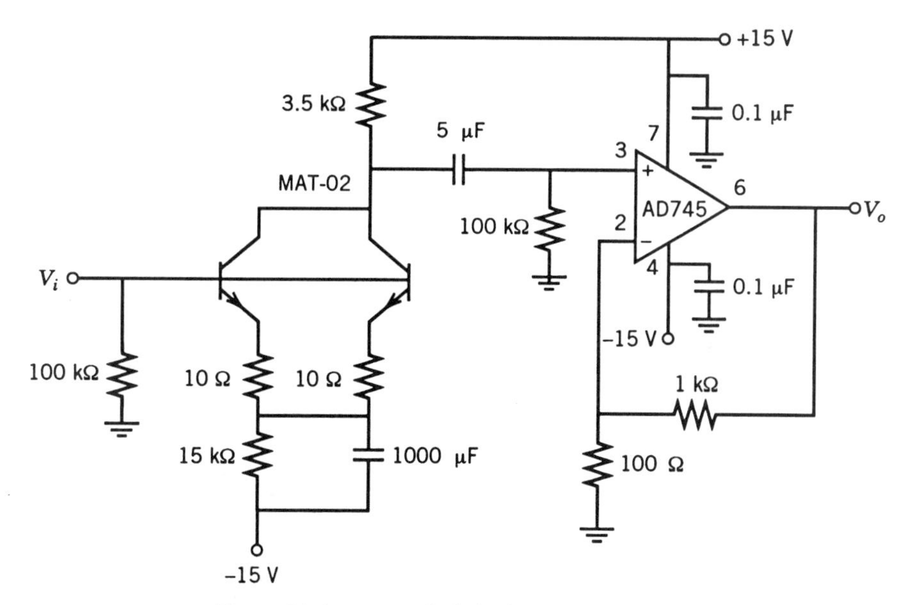

Figure 14-6 ac-coupled single-ended amplifier.

HA 3046 from Harris Semiconductor. Also, discrete or pairs of JFETs can be used to obtain higher impedance and isolation while maintaining the lowest available noise.

The overall gain of the amplifier is 1000 × with 100 × in the first stage and 10 × in the second. E_{nT} is 0.7 nV/Hz$^{1/2}$ and the frequency response ranges from 10 Hz to 1 MHz.

SUMMARY

a. The input noise of a cascaded BJT amplifier stage can be adjusted by selecting the operating point of the devices.
b. Using a cascode connection allows a trade-off of frequency response versus gain.
c. The addition of parallel stages to the input of an op amp will reduce the input noise voltage by the square root of the number of parallel stages.
d. The use of ac coupling with a single-ended input will give the lowest total equivalent input noise.

REFERENCES

1. 1992 Amplifier Applications Guide, Analog Devices Inc., 1992, p. X-5.
2. Jenkins, A., and D. F. Bowers, "*NPN* Pairs Yield Ultra-Low Noise Op-Amp," *EDN* (May 3, 1984).
3. MAT-02 Application Notes, PMI Analog Integrated Circuits Data Book, Vol. 10, 1990.

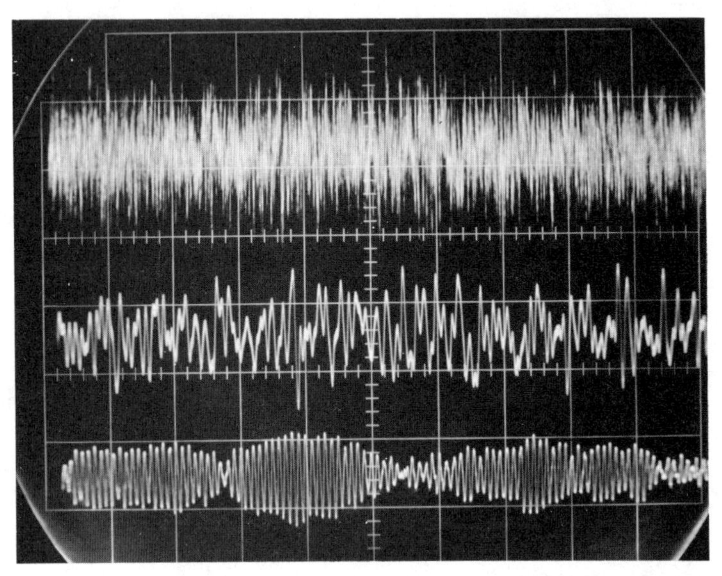

Broadband white noise is shown in the top trace. When passed through a 2-kHz crystal filter with $\Delta f = 50$ Hz, we obtain the middle trace. In the bottom trace, the same noise is passed through a 2-kHz *RLC* high-Q filter with $\Delta f = 50$ Hz. Note the "ringing" effect in the bottom trace. Horizontal sensitivity is 5 ms/cm.

CHAPTER 15

NOISE MEASUREMENT

When designing low-noise electronic systems, we measure the system noise to evaluate performance or to compare alternate designs. Noise is measured in much the same manner as other electrical quantities; the most significant difference is the voltage level. Several specific noise parameters have been derived such as equivalent input noise voltage E_n, equivalent input noise current I_n, and noise figure NF. These are useful unambiguous quantities capable of being measured and serving as comparison indexes among different amplifying systems.

Since noise voltages are often in the nanovolt (10^{-9} V) region, it is not possible to measure noise directly at its source. We cannot put a sensitive voltmeter at the input of the amplifier and say, "Here is the noise." Often, noise generation is not physically located at the input but is distributed throughout the system. The total noise is the sum of contributions from all noise generators. In any case, the signal-to-noise ratio at the output is the main concern, for that is where the response, actuator, relay, meter, display, or other output device is located. Noise is measured at the output port where the level is highest and then referred to the input for reference to the signal source.

15-1 TWO METHODS FOR NOISE MEASUREMENT

Two general techniques for noise measurement are the *sine wave method* and the *noise generator method*. In the sine wave method we measure the rms noise at the output of the amplifier, measure the transfer voltage gain with a sine wave signal, and, finally, divide the output noise by the gain to obtain the equivalent input noise. In this way, both the noise and the gain can be measured at high levels.

For the noise generator method we use a calibrated broadband noise generator placed at the amplifier input. With the noise generator set to zero noise, the total noise power at the output of the amplifier is measured; then the calibrated noise voltage is increased until the output noise *power* is doubled. This means the noise generator voltage is now equal to the equivalent input noise of the amplifier.

Sine wave and noise generator methods each have certain areas of application, as well as specific limitations. The choice between methods depends on the frequency range and the equipment available. The sine wave method requires more measurements, but it uses common laboratory instruments and is more applicable at low frequencies; the noise generator method is usually simpler and more applicable at high frequencies. These methods are contrasted in Sec. 15-5.

15-2 SINE WAVE METHOD

When considering the noise of a component such as a resistor, we are often concerned with a single-noise mechanism. With a sensor–amplifier system we seek not only the noise of an amplifier composed of many noise sources, but the signal-to-noise ratio of the entire system. Since the signal is located at the input of the amplifier, it is logical to sum all of the amplifier and input network noise into an equivalent input noise parameter.

We have defined the equivalent input noise voltage E_{ni} as a Thevenin equivalent noise voltage generator located in series with the sensor impedance and equal to the sum of the sensor and amplifier noise as in Fig. 15-1. This places all of the amplifier and input noise in series with the signal. Since the signal and noise generators are located at the same point and the same transfer function applies, the equivalent input noise is inversely proportional to the signal-to-noise ratio.

The measurement of equivalent input noise is basic to both the determination of NF and the characterization of the amplifier noise voltage and noise current parameters. Either the sine wave or the noise generator method can be used to measure E_{ni}.

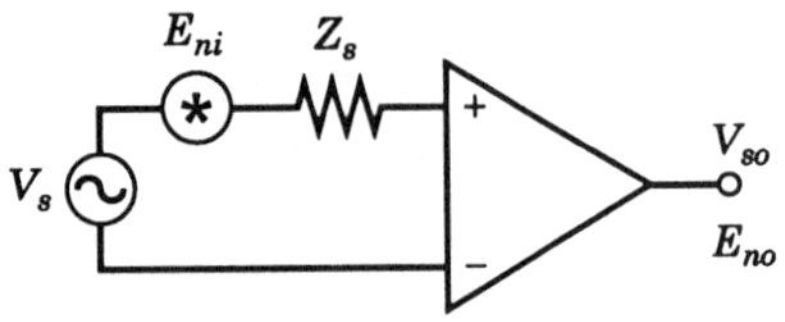

E_{ni} Z_s V_s V_{so} E_{no}

Figure 15-1 Measurement of equivalent input noise using the sine wave method.

The sine wave method requires measurement of both output noise E_{no} and transfer voltage gain K_t. The procedure for measuring equivalent input noise for a voltage amplifier is

1. Measure the transfer voltage gain K_t.
2. Measure the total output noise E_{no}.
3. Calculate the equivalent input noise E_{ni} by dividing the output noise by the transfer voltage gain.

Referring to Fig. 15-1, the transfer voltage gain K_t can be defined as

$$K_t = \frac{V_{so}}{V_s} \tag{15-1}$$

where V_s is the input sine wave signal and V_{so} is the output sine wave signal. The equivalent input noise E_{ni} is then the total output noise divided by the gain:

$$E_{ni} = \frac{E_{no}}{K_t} \tag{15-2}$$

where E_{no} is the output noise of the amplifier.

Now, following the preceding three steps to measure the equivalent input noise, we can measure the transfer voltage gain K_t by inserting a voltage generator V_s in series with the source impedance Z_s and measuring the signal V_o. The transfer voltage gain K_t is the ratio of V_{so} to V_s. *Since this gain must be measured at a signal level higher than the noise level, ensure that the amplifier is not saturating* by doubling and halving the input signal; the output signal should double and halve proportionately. It is important to note that this transfer voltage gain K_t is dependent on the source impedance and amplifier input impedance while the typical voltage gain A_v is not. By using a transfer voltage gain K_t, the input noise is *independent* of the input impedance!

The transfer voltage gain must be measured using a generator impedance equal to the impedance of the signal source. Do not use the voltage gain of the amplifier. If the equivalent input noise is to be measured at various frequencies or source impedances, both the transfer voltage gain K_t and the output noise E_{no} must be remeasured *each* time with *each* source impedance at *each* frequency.

The next step in calculating E_{ni} is measurement of the total output noise E_{no}. Remove the signal generator and replace it with a shorting plug. Do not remove the source impedance Z_s. The output noise E_{no} is now measured with a rms voltmeter as described in Sec. 15-3-7. The equivalent input noise E_{ni} is the ratio E_{no}/K_t.

The signal generator must be removed from the noise test circuit before measuring the noise. This is necessary whether the signal generator is ac or battery operated. In either case, its capacitance to ground can result in noise pickup. Also, if it is plugged into the line or grounded, there is a possibility of ground loop pickup.

If the output noise is measured on an average responding rather than a rms meter, multiply by 1.13 to get the rms. This is discussed in more detail in Sec. 15-3-7. To obtain the noise spectral density, divide the equivalent input noise by the square root of the noise bandwidth Δf.

15-2-1 Measurement of E_n and I_n

The amplifier noise voltage E_n and noise current I_n parameters are calculated from the equivalent input noise for two source resistance values. As defined previously, the equivalent input noise is

$$E_{ni}^2 = E_t^2 + E_n^2 + I_n^2 R_s^2 + 2CE_n I_n R_s \tag{15-3}$$

Measurement gives the total equivalent input noise E_{ni}. To determine each of the three quantities, E_n, I_n, and E_t, make one term dominant or subtract the effects of the other two. In general, the correlation coefficient C is zero and can be neglected. Measurement of C is discussed later in this section.

To measure the noise voltage E_n, measure the equivalent input noise with a small value of source resistance. When the source resistance is zero, the thermal noise of the source E_t is zero and the noise current term $I_n R_s$ is also zero; therefore, *the total equivalent input noise is the noise voltage* E_n. How small should R_s be for E_n measurement? The thermal noise E_t of R_s should be much less than E_n. Usually 5 Ω is adequate, 50 Ω can add some noise, and 500 Ω is too large.

To measure the noise current I_n, remeasure E_{ni} with a very large source resistance. Measure the output noise and transfer voltage gain again to calculate the equivalent input noise E_{ni}. Assuming the $I_n R_s$ term to be dominant, I_n *is simply the equivalent input noise* E_{ni} *divided by the source resistance* R_s. If R_s is large enough, the $I_n R_s$ term dominates the E_n term, and it also dominates the thermal noise since the thermal noise voltage E_t increases as the square root of the resistance, whereas the $I_n R_s$ term increases linearly with resistance. When the $I_n R_s$ term cannot be made dominant, the thermal noise voltage $[E_t = (4kTR\,\Delta f)^{1/2}]$ can be subtracted from the equivalent input noise. Since this is a rms subtraction, a thermal noise of one-third the noise current term only adds 10% to the equivalent input noise.

The source resistor for measuring I_n can be calculated in a general way for any amplifier since the noise current source is physically located at the input of the amplifier. The limiting noise current I_n is caused by the shot noise of the input dc gate or base bias current I_B:

$$I_n \cong \sqrt{2qI_B \, \Delta f} \tag{15-4}$$

To make the noise current term dominate the thermal noise of the measuring resistor, we have

$$I_n R_s = R_s \sqrt{2qI_B \, \Delta f} \geq 3\sqrt{4kTR_s \, \Delta f} \tag{15-5}$$

Solving for source resistance R_s,

$$R_s \geq \frac{18kT}{qI_B} = \frac{0.45}{I_B} \tag{15-6}$$

This defines the minimum value of a series source resistor for I_n measurement.

Example 15-1 What are the minimum source resistances for a BJT amplifier with 1 μA bias current, a JFET amplifier with 0.2 nA bias current, and a MOSFET amplifier with 1 pA bias current?

Solution For the BJT amplifier, $R_s = 450$ kΩ; for the JFET amplifier, $R_s = 2.25$ GΩ; and for the MOSFET amplifier, $R_s = 450$ GΩ or 4.5×10^{11} Ω.

Amplifier gain and bias requirements, however, may make it difficult or impossible to achieve these values since each of the amplifiers in Example 15-1 will have an input offset of at least 0.45 V with an unbalanced input. From Eq. 15-6, $I_B R_s = 0.45$ V. For an amplifier with a balanced input and equal resistors in each input, the voltage input offset will be determined by the unbalance in input currents which is about 10% of I_{off} as shown by typical IC specifications. With the same value of source resistance in each input, the offset will be reduced to about 50 mV. It may be necessary to use additional feedback to reduce the overall amplifier gain to avoid excessive offset at the amplifier output. For the amplifier with an unbalanced input, it is usually necessary to use a gain of less than 10× to stay within linear operation.

For an accurate I_n measurement, the amplifier input bias resistors must be much larger than the source resistance R_s or you will only be measuring the noise current of the shunting biasing resistors. It is best to use the measuring source resistor as the bias current path.

The source resistor should be shielded to prevent pickup of stray signals. The resistor used for R_s need not be low noise as long as there is no dc voltage drop across it. When the source resistance is also used for biasing, a resistor with a low noise index may be needed.

Another method of obtaining a high source impedance for I_n measurement is with a reactive source. A low-loss 47-pF mica capacitor can be used for R_s. Now, the $I_n X_c$ term is large and since the reactive impedance has no thermal noise, the equivalent input noise is

$$E_{ni}^2 = I_n^2 X_c^2 + E_n^2 \tag{15-7}$$

where X_c is the reactance of the source impedance at the measurement frequency. I_n is calculated from

$$I_n^2 X_c^2 = E_{ni}^2 - E_n^2 \tag{15-8}$$

The only unknown in Eq. 15-8 is I_n. This method is most useful at frequencies below 100 Hz. One difficulty with a capacitive source is biasing. To provide a path for the input bias current or the offset current, it is necessary to parallel the amplifier input terminals with a large resistance. This bias resistance R_B generates thermal noise current $I_t = (4kT\Delta f/R_B)^{1/2}$ which can easily dominate the amplifier input shot noise current. A FET input amplifier often has a noise current of less than 10 fA/Hz$^{1/2}$. This is equivalent to the thermal noise of a 160-MΩ resistor. Thus the bias resistance *for all* FET I_n measurements must be in the GΩ (10^9) range.

The correlation coefficient C can be measured after the other three quantities, E_n, I_n, and E_t, have been determined. Select the optimum source resistance value, $R_s = R_o$ such that $I_n R_o$ is equal to E_n. For this resistance

TABLE 15-1 Table for Recording Measured Noise Values

	Frequency			
	f_1	f_2	f_3	f_4
I_C (dc collector current or operating point)				
V_g (ac generator terminal voltage)				
V_s (V_g/attenuation)				
V_{so} (output signal)				
$K_t = (V_{so}/V_s)$				
R_s (source resistance)				
E_{no} (total output noise)				
E_{na} (postamplifier noise at measurement frequency)				
$E'_{no} = (E_{no}^2 - E_{na}^2)^{1/2}$				
$E_{ni} = (E'_{no}/K_t)$				
Δf (noise bandwidth)				
M (meter correction factor)				
$E_{ni}/\text{Hz}^{1/2} = ME_{ni}/\text{Hz}^{1/2} = E_n$, if R_s is 0				
E_t (source resistance thermal noise)				
$I_n^2 R_s^2 = (E_{ni}/\text{Hz}^{1/2})^2 - (E_t/\text{Hz}^{1/2})^2 - E_n^2$				
$I_n = I_n R_s / R_s$				

the correlation term $2CE_nI_nR_s$ has a maximum effect. To determine C, measure the equivalent input noise a third time with the optimum source resistance R_o. The correlation coefficient C can be calculated from Eq. 15-3 by subtracting the contributions of E_n, I_n, and E_t. This is a difficult measurement of E_{ni} since the correlation term can increase E_{ni} at most by 40% when $R_s = R_o$ and $C = 1$.

A table for systematically recording measured noise data and calculating E_n and I_n is shown in Table 15-1. This table serves as a reminder to include the various correction factors.

15-2-2 Noise Figure Measurement

An amplifier's noise figure (NF) must be measured with a specific source resistance. The determination of NF is straightforward once you have the equivalent input noise E_{ni} for that value of source resistance. The noise figure is the ratio of the total amplifier and sensor noise to the thermal noise of the sensor alone:

$$\mathrm{NF} = 10 \log \frac{E_{ni}^2}{E_t^2} \tag{15-9}$$

Another definition of NF is the degradation in the signal-to-noise power ratio when the signal is passed through a network. This definition in equation form is

$$\mathrm{NF} = 10 \log \frac{S_i/N_i}{S_o/N_o} = 10 \log \frac{S_i}{N_i} - 10 \log \frac{S_o}{N_o} \tag{15-10}$$

where S_i and N_i are the signal and noise powers at the input, and S_o and N_o are the signal and noise at the output. In terms of voltage, Eq. 15-10 is

$$\mathrm{NF} = 20 \log \frac{V_s}{E_t} - 20 \log \frac{V_{so}}{E_{no}} \tag{15-11}$$

The NF of a transistor can be *measured* using the definition in Eq. 15-11. Make the input signal level equal 100 times the thermal noise. Then Eq. 15-11 becomes

$$\mathrm{NF} = 40\text{ dB} - 20 \log V_{so} + 20 \log E_{no} \tag{15-12}$$

Using a meter that is *calibrated in decibels* as well as voltage, adjust the gain of the amplifier with an attenuator or post amplifier until the output meter reads OdB on some convenient range, while the 100X input signal is connected. Now remove the input signal and increase the meter sensitivity by 40 dB, reduce the meter range by 100X. Substituting 40 dB for the V_{so} term in Eq. 15-12 gives

$$\mathrm{NF} = 20 \log E_{no} \tag{15-13}$$

The meter reading, after removing the input signal generator, is equal to the NF in decibels.

If the amplifier considered in the preceding discussion was nearly noiseless, then the input and output signal-to-noise ratios would be 40 dB. On removal of the input signal, the output meter would decrease by 40 dB to 0 dB on the new scale. In a practical case, the amplifier is not noiseless and the output does not decrease by 40 dB, but by some lesser value such as 35 dB. The NF is, therefore, 40 minus 35 or 5 dB. Thus the output meter reading is easily converted to the NF for the specific source resistance and noise bandwidth of the system under study.

15-3 NOISE MEASUREMENT EQUIPMENT

The sine wave generator method of noise measurement requires measurement of total output noise and transfer voltage gain. The instruments needed are a sine wave oscillator, attenuator, test circuit, postamplifier, spectrum analyzer or bandpass filter, and rms voltmeter. With the exception of the spectrum analyzer, there is no special equipment required. Most engineering labs have this equipment available for general measurements.

A diagram of the noise measurement instrumentation is shown in Fig. 15-2. We will now discuss the system shown in the diagram stage by stage.

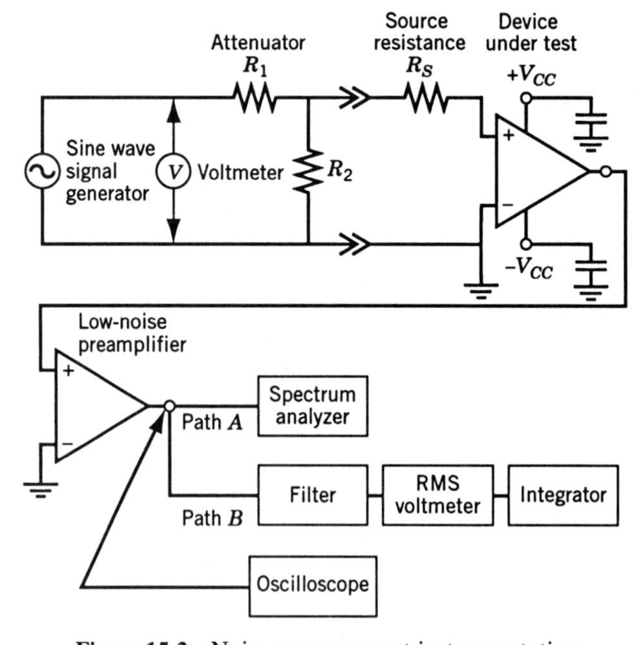

Figure 15-2 Noise measurement instrumentation.

15-3-1 Signal Generator

The input signal generator is a sine wave oscillator. A sine wave generator is preferred over a function generator because some of the function generators radiate too much signal and the grounding is poor. An oscillator with a built-in voltmeter and attenuator is convenient but be careful of using attenuation over 40 dB because of radiated signal and ground loops. It is better to use an external attenuator as shown in Fig. 15-2. If the meter is not built in, measure the signal level before attenuation with a broadband ac voltmeter or better yet, a spectrum analyzer.

Caution! Do not try to read the low-level sine wave source signal directly at the input terminals of the amplifier with a DVM or an ac voltmeter. At these low levels, the input noise of the DVM and pickup in the leads add to the sine wave signal and will give an erroneously high reading of the input. This will make the amplifier gain calculation too small and will make your calculation of equivalent input noise too large. Measure the input voltage ahead of the attenuator or read the signal with the spectrum analyzer which rejects more of the stray pickup interference.

The oscillator can be line or battery operated since it is disconnected before making any noise measurements. One precaution: Some of the function generators have excessive radiated signals that can yield poor results because of pickup.

15-3.2 Attenuator

Since low-noise amplifiers often have high gain, it is necessary to reduce the oscillator signal below its minimum setting to avoid overdriving the amplifier. This is accomplished with an attenuator that serves two functions: It reduces the oscillator signal by a known factor and it provides a low-impedance voltage source for measuring the amplifier gain.

If a good low-impedance attenuator is not available, a calibrated attenuator can be built as shown in Fig. 15-3. The attenuation is $R_2/(R_1 + R_2)$ at low frequencies. Resistor R_3 provides the matching load for the signal generator, if needed. Metal film resistors with their lower capacitance and higher stability are recommended.

For operation above a few kilohertz, the divider must be frequency compensated. Resistors such as R_1 have a shunt capacitance C_1 of 0.1 to 1 pF. At high frequencies, the attenuation ratio approaches $C_1/(C_1 + C_2)$. To compensate for a flat frequency response, connect a capacitor C_2 across R_2 so that $C_1/(C_1 + C_2) = R_2/(R_1 + R_2)$ to make $R_1C_1 = R_2C_2$. Now the attenuation is equal for all frequencies. To test for frequency compensation, insert a square wave at V_g and look for overshoot at V_s. Attenuations of more than 1000 to 1 are difficult to attain in a single stage because stray capacitances cause feedthrough at high frequencies.

Although the attenuator is usually removed before measuring noise, it should be shielded because we are concerned with low-level signals and

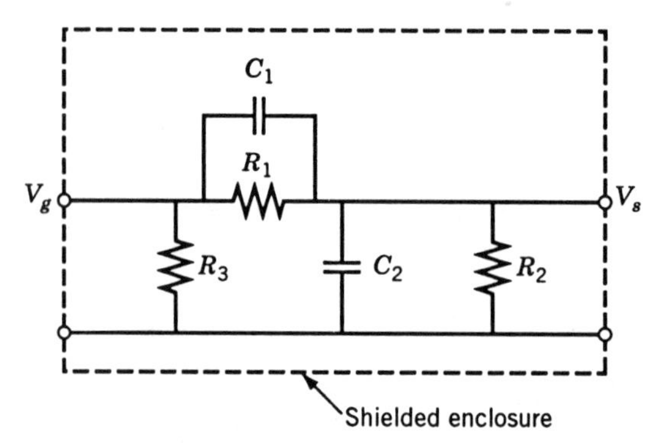

Figure 15-3 Attenuator.

high-gain amplifiers. Construct the attenuator in a small shielded package that can be plugged into the amplifier input terminals.

15-3-3 Source Impedance

To measure an amplifier noise current or NF, a resistor R_s or impedance Z_s equivalent to the source impedance is required. Following the attenuator in Fig. 15-2 is a low-capacitance resistor R_s that simulates the source resistance. It is not necessary for R_s to be a low-noise resistor if there is no dc voltage across it. Since the source resistor is often in the high megohm range, it should be included in the shielded enclosure.

Any capacitance shunting R_s causes two kinds of measurement errors. An apparent increase in gain K_t at high frequencies results, and a reduction in thermal noise is apparent. Such capacitance increases the transfer voltage gain by causing a decrease in the series source impedance. In addition, shunt capacitance across R_s shorts out part of the thermal noise voltage and therefore reduces the noise. Both of these effects cause the equivalent input noise E_{ni} to appear too low. (This may be good if you have trouble meeting a specification, but it is a headache when trying to predict performance.)

For noise measurement the attenuator can be replaced by a shielded resistor of value equal to the attenuator output resistance. If R_s is built into the test circuit, the attenuator can be replaced by a shorting plug; thus either the oscillator signal or a short circuit is connected.

15-3-4 Device Under Test

The fourth stage in the diagram of Fig. 15-2 represents the device or circuit being tested. The sine wave measurement method applies equally well for measuring the noise of a single transistor, an IC, or a complete amplifier.

When testing an integrated op amp circuit, use overall negative feedback to set the gain of the op amp system to some nominal value such as 10 or 100. This does not change the basic noise voltage and current measurements as long as the feedback resistors do not add appreciable additional noise. Be careful of noise from power supplies; they must be very well filtered. It is good practice to use batteries mounted inside the shielded enclosure. This avoids both pickup and ground loops.

Be careful of ground loops in your instrumentation. It is best to have a single ground at the amplifier input. There is a very good reference on the subject of grounding and shielding by Ott [1].

When measuring the noise of a discrete transistor, the noise is the same for all three configurations; but it is normally measured in the common-emitter or common-source configuration. The biasing network must not contribute additional noise. When measuring noise over a wide range of collector currents and frequencies, biasing becomes a problem. It is *best not* to use variable resistors at these low levels.

Careful packaging of the test circuit is important. If the test circuit is constructed as a breadboard and operated on a bench, significant low-frequency pickup is likely. As a rule of thumb, construct the test circuitry to be as compact as possible and place it in a small shielded box. There is a positive correlation between the size of the box and pickup. In fact, the pickup seems to increase exponentially with increasing package size. We have found that small cast aluminum boxes such as those made by the Pomona Electronic Company do a very good job of shielding. It does require extra effort to construct the test circuitry carefully and package it compactly, but this pays off later in ease of measurement and freedom from pickup. At most frequencies the circuit under test can be constructed on a small universal test board. This allows for easy circuit changes and source impedance adjustments.

15-3-5 Preamplifier

Frequently, the output of the test stage in Fig. 15-2 is very low. Since spectrum analyzers and voltmeters have their own noise level, the addition of a low-noise decade amplifier with a gain of 10 or 100 after the test stage may be needed. These amplifiers are available from several companies or better, you can build an additional stage on your test board.

Ideally, the noise of a preamplifier does not contribute to the noise of the device under test. To test for added noise, terminate the amplifier input in the output impedance of the device under test and turn the test stage "off." The remaining output noise comes from the amplifier and wave analyzer. If the amplifier is contributing noise, it can be subtracted as the difference of the squares of the voltages as noted in Table 15-1. Similarly, any noise contribution of the spectrum analyzer can also be removed. These noise contributions should be checked at each test frequency.

15-3-6 Spectrum Analyzer Techniques

For spectral or spot noise measurement, bandwidth limiting is required. The rms noise is measured in a specified noise bandwidth Δf. Two methods of measuring noise versus frequency are illustrated in Fig. 15-2 as paths A and B. Path A uses a spectrum analyzer and path B uses a filter.

Path A illustrates the simplest method for making spectral noise measurements over a wide range of frequencies. Since noise is composed of all frequencies, the spectrum analyzer will display the noise spectrum. Some problems arise because the spectrum analyzers were designed to measure and display continuous signals and not noise. Three issues must be considered when making spectrum analyzer readings of noise: analyzer bandwidth, detection method, and interfering signals.

The analyzer bandwidth must be considered because noise is composed of many frequencies occurring randomly. The bandwidth of the analyzer changes with the scan range so the noise reading will increase with increasing bandwidth. In effect, if the bandwidth is increased, more noise frequency components are allowed to be measured. Noise measurements cannot be compared directly because more noise will be measured with higher bandwidths. The correct value is obtained by dividing each reading by the square root of the noise bandwidth Δf. This brings up another problem: The analyzers are calibrated in terms of a resolution bandwidth and not the ideal rectangular noise bandwidth. Although the flat-topped filters used in the analyzer are close, they may be 10% to 50% higher than the noise bandwidth. The best approach is to determine the equivalent noise bandwidth of the analyzer filters from the instrument manufacturer and multiply the output reading by 1.1 to 1.5 as needed.

Most spectrum analyzers are not true rms meters. They are average reading meters calibrated to display the rms of a sine wave. In this case, it is necessary to multiply the reading by 1.13 to correct the noise reading to rms. The analyzer instruction manual should indicate the type of meter being used.

Often, noise measurements will have low-level signals buried in noise because of pickup and interference. These signals will raise the apparent noise floor. Since the low-level signals are usually at specific frequencies, they can be detected by using a narrowband scan. This narrow scan is slow but will resolve the discrete signals and show the true noise floor around the signals. When trying to read $1/f$ noise at low frequencies, the 60-Hz signals and harmonics may look like noise. Reducing the scan bandwidth will resolve the difference between the pickup and the noise.

Some spectrum analyzers have a feature called a "noise marker" [2]. A series of noise measurements are made at a frequency indicated by the marker, the readings averaged and normalized to a 1-Hz bandwidth. Check with the instrument manufacturer to see if your spectrum analyzer has this convenient feature.

Path *B* of Fig. 15-2 illustrates an alternate system to be used for testing a number of devices for a specific application. For manufacturing testing, a repeated acceptance test is performed. A simplified noise measurement setup can be constructed to replace the slower spectrum analyzer. A bandpass filter such as a passive *RLC* or active filter, determines the noise bandwidth. The filter of the system being designed can often be used. The signal is read on a rms or averaging meter with a bandwidth wider than the filter. If the readings are narrowband, use averaging and smoothing as described in the next section.

15-3-7 Averaging Methods

Although the long-term rms value of noise is a constant, the instantaneous amplitude is totally random, and therefore the meter jitters. For an accurate noise instrument we can smooth the meter fluctuations by averaging over a long period of time. Three principal methods of smoothing are

1. Use a long *RC* time constant.
2. Integrate the signal over a period of time.
3. Record the signal and average its value.

Filtering or averaging with a long *RC* time constant is the most commonly used method. The time constant is increased by placing a large capacitor across the meter terminals. Accuracy is inversely proportional to the square root of the time-constant bandwidth as defined in Sec. 15-6. In addition to the theoretical averaging time, wait several time constants for the capacitor to charge. If the bandwidth is greater than 100 Hz, the meter response provides adequate damping. Usually, it is only necessary to add an external capacitor. The charge and discharge time constants of the capacitor should be equal. If the charge time constant is shorter than the discharge time constant, a peak responding instrument results and the ratio of average to rms no longer holds.

A true integration technique digitizes the noise signal with a voltage-to-frequency converter and sums the total number of cycles on a digital counter. The total number of cycles divided by the number of seconds gives the average frequency and the average noise voltage. Integration methods are fast and give unambiguous readings, but may have the disadvantages of poor stray noise rejection and higher cost.

For a simple laboratory method of averaging, record the rectified noise output on a storage scope and read the average value. The speed and time constant of the trace are adjusted to see the fluctuations in signal amplitude. In theory, the *RC* method produces the same accuracy as the recorded scope trace, but in practice the recorder has an advantage. In a lab environment there are sporadic noise spikes. An integrating or averaging circuit sums these with the circuit noise. With the recorder method, noise spikes or a sudden change in noise level show on the record. These spikes can often be traced to an interference source, or a record for a longer period of time can be made until the signal becomes more typical. Essentially, this is using the operator's brain as an additional selective filter. The recorder and *RC* methods are laboratory techniques since they use operator judgment. For production testing the true integration method is both easier and faster.

For an analyzer with a 5-Hz noise bandwidth, a 2-min recording time is used to obtain 5% repeated ability. This is higher accuracy than necessary for most engineering noise work. It may be needed when measuring the noise contribution of an amplifier that is dominated by a large thermal noise component. As a rule of thumb, 10% is good accuracy for noise measurements, considering that the excess noise of a transistor type may vary by as much as 100% to 300%.

15-3-8 Oscilloscope

It is difficult to read noise levels with an oscilloscope because of the random nature of noise. Total broadband noise can be estimated by taking the peak-to-peak 'scope reading and dividing by about 6 ($\pm 3\sigma$). This allows for the "grass" effect of the noise.

An oscilloscope is more useful as a monitoring device. As illustrated in Fig. 15-2, a 'scope is temporarily connected to the output of the amplifier to monitor the character of the output signal. Any clipping, distortion, or interfering signals will often be obvious to the eye. This is a course check but it can keep you from being fooled by erroneous measurements. Remove the 'scope after making the reading to avoid introducing another ground loop into the system.

An audio amplifier can also be temporarily connected to the output as a monitor. One author, at one time, had a high noise level in some sensor measurements. When he connected a speaker to the output he heard music from a local "rock" radio station which was being picked up and demodulated by the unshielded sensor. While such music might be considered "noise," it was not the fundamental noise which the author was trying to measure.

15-3-9 Meters

Variations in noise readings can frequently be traced to the use of different types of meters to measure the output noise power. To measure noise accurately requires that the meter respond to the rms noise voltage and have an adequate crest factor and sufficient bandwidth.

The meter must provide a rms response proportional to the power or voltage squared. A typical meter movement is designed to measure the average value of constant-amplitude, repetitive waveforms. It is calibrated to

TABLE 15-2 Crest Factors for Gaussian Noise [3]

Percentage of Time Peak Is Exceeded	$\frac{\text{Peak}}{\text{rms}}$	Peak Factor in $\text{dB} = 20\log_{10}\frac{\text{Peak}}{\text{rms}}$
10.0	1.645	4.32
1.0	2.576	8.22
0.1	3.291	10.35
0.01	3.890	11.80
0.001	4.417	12.90
0.0001	4.892	13.79

indicate the rms amplitude of a sine wave, but does not usually measure the rms value. When such a voltmeter measures noise whose waveform is neither sinusoidal nor of constant amplitude, we do not get an accurate reading.

Random noise has a well-defined average value when rectified, but its instantaneous value cannot be defined except in terms of probability (see Sec. 1-2). The percentage of time that Gaussian noise exceeds a certain peak level is shown in Table 15-2. *The crest or peak factor is defined as the ratio of the peak value to the rms value of a waveform.* As shown in Table 15-2, noise has a crest factor of 3 for 1% of the time, but it reaches 4 less than 0.01% of the time. This means, to measure noise a meter must not overload or clip on a signal that is three times the full-scale reading.

The meter must respond to the total signal. Its bandwidth must be much greater than the bandwidth of the amplifier system being measured. Consider the case where the meter and the amplifier system each have a single-time-constant high-frequency roll-off. The transfer functions would be

$$T_m(jf) = \frac{1}{1 + jf/f_m} \quad \text{and} \quad T_s(jf) = \frac{1}{1 + jf/f_s} \tag{15-14}$$

where f_m is the -3-dB cutoff frequency of the meter and f_s is the -3-dB cutoff frequency of the system. If $f_m \to \infty$ and $\Delta f = \pi f_s/2$, the total measured noise would be

$$E_{no}^2 = 4kTR\,\Delta f = \frac{4kTRf_s\pi}{2} = 2\pi kTRf_s \tag{15-15}$$

Now let f_m be finite. The noise bandwidth Δf of the measurement system and amplifier is

$$\Delta f = \int_0^\infty \frac{df}{\left[1 + \left(\frac{f}{f_m}\right)^2\right]\left[1 + \left(\frac{f}{f_s}\right)^2\right]} \tag{15-16}$$

TABLE 15-3 Output Meter Bandwidth Measurement Error

Meter Bandwidth (BW)	Relative Reading	Percentage Error
BW = Δf	0.707	−29.3
BW = 2 Δf	0.818	−18.2
BW = 5 Δf	0.915	−8.5
BW = 10 Δf	0.956	−4.4
BW = 20 Δf	0.979	−2.1
BW = 50 Δf	0.993	−0.7
BW = 100 Δf	0.998	−0.2
BW = 1000 Δf	1.000	0.0

Solving gives

$$\Delta f = \frac{\pi f_s f_m}{2(f_m + f_s)} \tag{15-17}$$

The relative reading for a meter with bandwidth f_m is

$$\text{Relative reading} = \sqrt{\frac{f_m}{f_m + f_s}} \tag{15-18}$$

As an example, when measuring broadband white noise, if the meter 3-dB bandwidth is equal to the 3-dB bandwidth of the system, the noise reading will be 0.707 of the correct reading (see Table 15-3). If the meter 3-dB bandwidth is 10 times the system bandwidth, the error will only be 0.956 of the value. It is clear that the output power meter must have adequate frequency response.

Three general requirements of the output noise meter, then, are that it must respond to the rms value of the signal, have a crest factor greater than 3, and have a bandwidth greater than 10 times the noise bandwidth. Consider, now, the three common types of meters: true rms, average responding, and peak responding.

15-3-9-1 True rms Instruments A rms responding voltmeter gives true rms indication of the noise if the voltmeter bandwidth is greater than the noise bandwidth of the system and the noise peaks do not exceed the maximum crest factor of the voltmeter. The crest factor of 4 to 5 on most rms voltmeters is adequate except for a system with unusually high noise spikes such as popcorn noise. To check for clipping, monitor the noise meter signal with a scope or change the range switch and observe whether the reading is identical.

Two common types of rms responding meters are the quadratic devices and those that respond to heat. Most rms meters use a form of squaring circuit so that the meter averages the square of the instantaneous value of the signal. The meter scale is calibrated in terms of the square root of the quantity, thus giving an indication of the rms value.

Thermal responding meters use the signal power to heat a resistor linearly and detect the temperature rise with a thermistor or thermocouple. This follows the basic definition that the rms value of an ac current has the same heating power as an equal dc current. Thermal-type meters are usually slower in response and tend to be subject to burnout on overload.

15-3-9-2 Average Responding Instruments *To measure the rms value of a signal, use a rms responding meter*. This seems obvious, yet it is frequently ignored. Most ac voltmeters and spectrum analyzers are average responding instruments. They typically use a diode rectifier and a dc meter that responds to the average value of one or both halves of the rectified waveform. The average value of a full-wave rectified sine wave is 0.636 of the peak amplitude; the rms is 0.707 of the peak. To indicate the rms value of a sine wave, the meter has a multiplying *scale correction factor* of $0.707/0.636 = 1.11$. In other words, 1.11 times the average value gives the rms of a sine wave.

A problem arises when the input signal is noise, and therefore not a sine wave. Gaussian noise has an average value of 0.798 times the rms value. Since an average responding voltmeter indicates 1.11 times higher, it reads $1.11 \times 0.798 = 0.885$ of the true rms value. The averaging meter reads too low. *When reading noise on an average responding meter, multiply the reading by* 1.13 *or add* 1 *dB to obtain the correct value*. If you do not apply this correction, you may try to claim that the noise of a resistor is less than the thermal noise, a phenomenon that is difficult to explain.

Average responding meters may be suitable for noise measurements with proper precautions, but it is desirable to verify the reading with a calibrated noise source. The accuracy of the output meter is particularly important for the sine wave noise measurement method since there is no noise reference signal. To measure noise with a average responding meter, two more characteristics must be considered: the bandwidth and crest factor of the meter. As pointed out in Table 15-3, the meter bandwidth must be much greater than the noise bandwidth of the system.

The crest factor limitation is a special problem. To protect the indicator, some average responding voltmeters are designed to saturate on greater than full-scale readings corresponding to crest factors of only 1.4 to 2. As shown in Table 15-2 a crest factor of 3 to 5 is desirable for Gaussian noise. There are two possible solutions: Use a meter with a high crest factor, or simply read at less than half-scale. For half-scale reading, place a 6-dB attenuator ahead of the meter and multiply all readings by 2. This increases the crest factor from 2 to 4. The attenuator must have a flat frequency response.

The addition of a shunt capacitor to increase the averaging time can convert the meter to a peak detecting instrument. This problem arises when the meter is connected directly to the rectifier. The averaging circuit should have equal charge and discharge time constants.

15-3-9-3 Peak Responding Instruments A peak reading voltmeter responds to the peak or peak-to-peak value of a signal. It can be calibrated to indicate the peak or the true value of a sine wave. This works well when the peak value is constant, but as shown in Table 15-2, the peak value of noise is a variable.

The reading of noise on a peak responding meter depends on the charge versus discharge time constants. Peak voltmeters of different types produce different results. In general, *noise should not be measured with a peak responding meter* unless the peak amplitude of the noise has some significance such as for an application that involves level detection.

15-4 NOISE GENERATOR METHOD

Amplifier noise can be measured by comparison using a calibrated noise source E_{ns} and a rms output noise meter located at E_{no} in Fig. 15-4. The unknown noise level of the amplifier is compared with the known amplitude of the noise generator. Accuracy is determined primarily by calibration of the noise generator. It must have a uniform noise spectral density over the bandwidth of measurement (white noise). Although this criteria may be difficult to meet at low frequencies where $1/f$ noise dominates, at higher frequencies the noise generator method is the easiest form of noise measurement.

The calibrated noise source is shown in Fig. 15-4 as a noise voltage generator in series with the sensor resistance R_s. The system equivalent input noise is summed at E_{ni}. The amplifier and generator noise are measured as E_{no}. Alternately, a high-impedance noise current generator can be connected in parallel with the source impedance to measure the equivalent input noise current I_{ni}.

The purpose of the noise measurement is to determine the value of the equivalent input noise E_{ni} or I_{ni}. The procedure is to measure the noise at the output twice as E_{no1} and E_{no2}. Output noise with the generator E_{ng} connected is E_{no1}, and output noise with the noise generator disconnected is E_{no2}. These measurements are described by two equations:

$$E_{no1}^2 = K_t^2\left(E_{ng}^2 + E_{ni}^2\right) \tag{15-19}$$

and

$$E_{no2}^2 = K_t^2 E_{ni}^2 \tag{15-20}$$

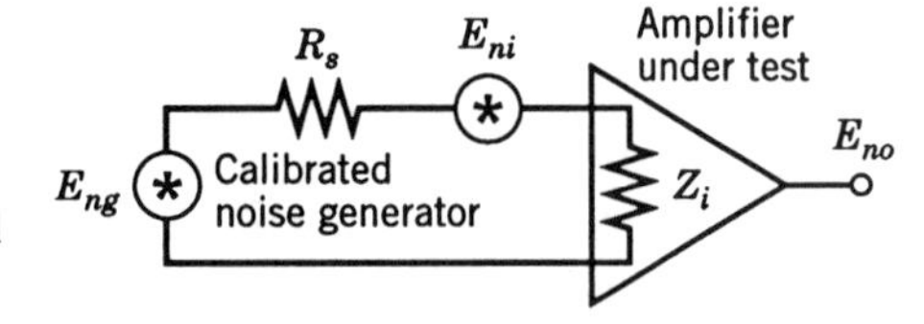

Figure 15-4 Noise generator method for noise measurement.

where K_t is the transfer voltage gain as defined in Eq. 15-1. For good accuracy, E_{no1} should be much larger than E_{no2}. From these two noise measurements and the known value of the noise generator E_{ng}, the transfer voltage gain K_t can be calculated:

$$K_t^2 = \frac{E_{no1}^2 - E_{no2}^2}{E_{ng}^2} \tag{15-21}$$

The equivalent input noise E_{ni} can be calculated for any known noise generator level as follows:

$$E_{ni}^2 = \frac{E_{no2}^2}{K_t^2} = \frac{E_{no2}^2 E_{ng}^2}{E_{no1}^2 - E_{no2}^2} \tag{15-22}$$

The technique most commonly employed is to increase E_{ng} to double the output noise power. So

$$E_{no1}^2 = 2E_{no2}^2 \tag{15-23}$$

By substituting this relation into the general expression for equivalent input noise, Eq. 15-22, we find

$$E_{ni}^2 = \frac{E_{no2}^2 E_{ng}^2}{2E_{no2}^2 - E_{no2}^2} = E_{ng}^2 \tag{15-24}$$

Therefore, *the noise voltage of the noise generator necessary to double the output noise power is equal to the equivalent input noise of the amplifier*.

In summary, the measurement procedure is

1. Measure the total output noise.
2. Insert a calibrated noise signal at the input to increase the output noise power by 3 dB.
3. The noise generator signal is now equal to the amplifier's equivalent input noise.

An uncalibrated power meter can be used to measure the noise. First, the noise of the amplifier is measured, then the amplifier gain is attenuated by 3 dB, and, finally, the noise generator is increased until the output power meter returns to its original level. The added noise is equal to the amplifier equivalent input noise.

15-4-1 Noise Measurement Equipment

There are several types of dispersed signal generators. Although most are noise generators, some produce a nonrandom swept-frequency sine wave.

Either type serves as a broadband noise source. There is one basic criterion: For noise measurement the generator must have a flat spectral noise density. Random noise signals are generated by temperature-limited vacuum diodes, Zener diodes, and amplifiers.

15-4-1-1 Temperature-Limited Vacuum Diode One of the original noise sources is the temperature-limited vacuum diode commonly called a noise diode. As a noise diode, it is operated with an anode voltage large enough to collect all the electrons emitted by the cathode; hence the name temperature-limited diode. When there is no space-charge region around the filament to smooth out the electron emission, the anode current shows full shot noise I_{sh}, as given in Chap. 1:

$$I_{sh}^2 = 2qI_{dc}\,\Delta f \tag{15-25}$$

where q is the electron charge, I_{dc} is the direct anode current, and Δf is the noise bandwidth. The noise is Gaussian and independent of frequency from a few kilohertz to several hundred megahertz. At low frequencies, $1/f$ noise dominates. In the high-frequency region, the electron transit time can become significant and the output noise decreases. The anode current is controlled by varying the temperature of the filament. Since the emission current is an exponential function of the filament temperature, it is important to control the filament current very accurately. Use a regulated supply.

A circuit diagram of the vacuum noise diode used as a noise current generator is shown in Fig. 15-5. The filament temperature and, therefore, the anode current is controlled by a variable resistor R_1. The ac load impedance is set by the tuned circuit C_1L_1. Using a resonant load minimizes the effects of anode and wiring capacitance. The noise current I_{sh} is coupled through capacitor C_C. Capacitors C_B are bypass capacitors.

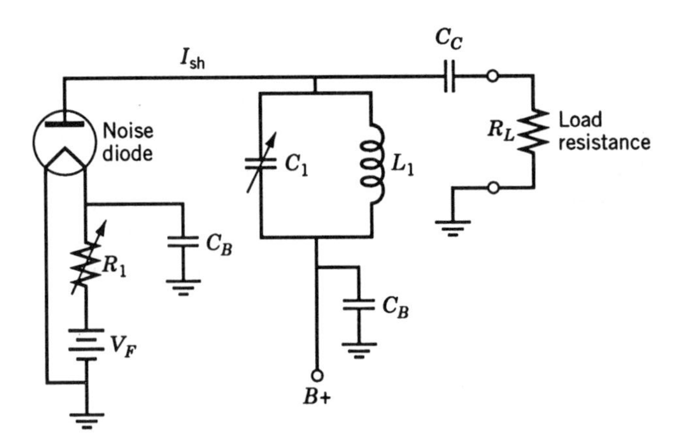

Figure 15-5 Circuit diagram of vacuum noise diode.

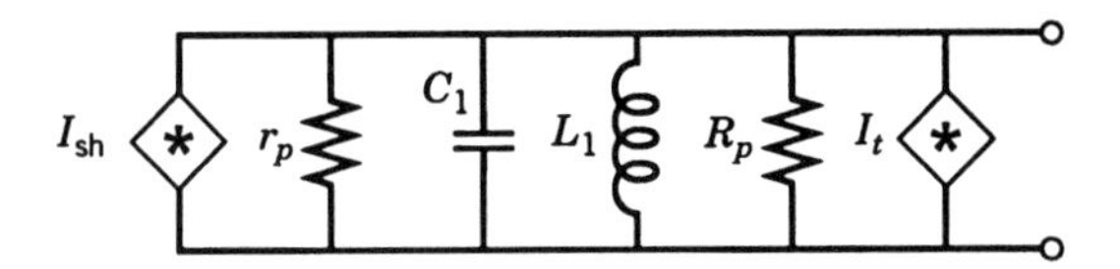

Figure 15-6 Noise equivalent circuit for noise diode.

Refer to the noise equivalent circuit in Fig. 15-6. The noise generator is a current source I_{sh} shunted by the plate resistance r_p. The plate resistance of a noise diode is typically from 25,000 Ω to 1 MΩ. The tuned circuit is represented by C_1 and L_1, and R_p is the effective parallel resistance of the tuned circuit. R_p has a parallel thermal noise current generator I_t. The total noise current is the sum of the shot noise and thermal noise components:

$$I_{tot}^2 = I_{sh}^2 + I_t^2 \tag{15-26}$$

For a calibrated noise source the shot noise term must dominate the thermal noise. This requires a large R_p or high circuit Q.

If a noise voltage source with a specific output resistance is desired, a resistor is placed in parallel with the noise diode output. This resistor must be much smaller than the plate resistance of the diode. The output noise voltage is the shot noise current I_{sh} times the shunt resistance R_L plus the thermal noise voltage of the resistor. When using a noise generator the output cable capacitance must not seriously degrade the noise signal.

When a resistive load R_L is used in place of L_1C_1, the output noise has a flat frequency distribution. Although required for broadband measurements, there are two disadvantages of a resistive load: The thermal noise of R_L is likely to be significant, and the shunt capacitance of the tube and wiring causes a roll-off in noise at high frequencies. In general, a reactive load is useful to keep the losses low and the impedance high.

15-4-1-2 Zener Noise Diode A more common calibrated noise generator is the low-voltage Zener diode. The Zener mechanism is shot noise limited. Selected Zener diodes can have a $1/f$ noise corner as low as a few hertz. The higher-voltage avalanche diodes, often called Zener diodes, may exhibit excess noise and must be calibrated for use as noise sources.

A circuit diagram of a Zener diode noise generator is shown in Fig. 15-7. The dc supply V_{BB} must be much larger than the Zener voltage so that the bias resistor R_B can be large. The thermal noise current of R_B parallels the output as shown in the equivalent circuit of Fig. 15-8. The capacitor C_B bypasses the battery or power supply noise and C_C decouples the dc output level. The load resistor R_L provides the desired generator impedance to the amplifier input, R_B and r_{ze} must be much larger than R_L. R_L will generate

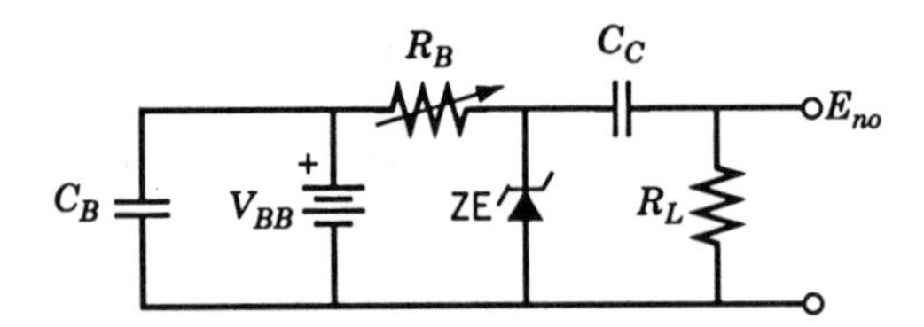

Figure 15-7 Zener diode noise generator.

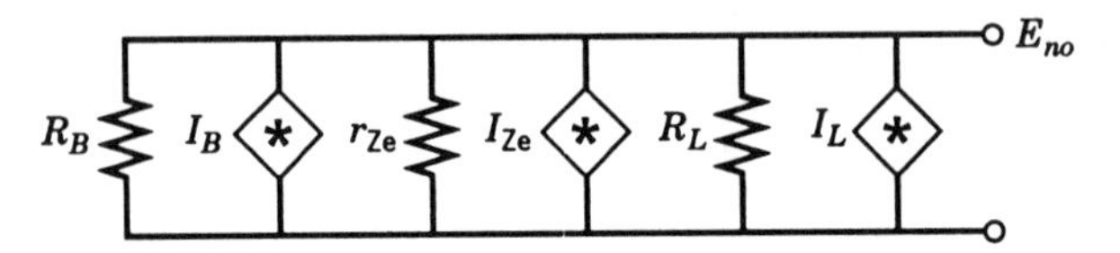

Figure 15-8 Noise equivalent circuit for Zener diode.

thermal noise I_L and R_B generates thermal noise I_B (Fig. 15-8). To reduce the noise contribution of R_L and provide the maximum output impedance, replace it with an LC circuit as discussed in the previous section.

15-4-1-3 Low-Noise Amplifier as a Noise Generator A forward-biased semiconductor diode can be used as a calibrated noise voltage generator. The diode shunt resistance is 25 Ω divided by its forward current in milliamperes. The noise voltage E_{sh} is equal to the shot noise current times the forward resistance of the diode as discussed in Chap. 1. Often the base–emitter junction of a low-noise transistor can be used as a forward-biased noise diode. The value of the shot noise voltage of a semiconductor diode is

$$E_{sh}^2 = \frac{(0.025)^2 2q\,\Delta f}{I_{dc}} = \frac{2 \times 10^{-22}\,\Delta f}{I_{dc}} \tag{15-27}$$

A low-noise amplifier operating with a shorted input and a first-stage collector current of less than 10 μA may be a good calibrated noise generator. In this case, the E_n noise is limited by the shot noise voltage of the first transistor so that the noise is constant if the collector current is constant. The amplifier gain can be adjusted to provide the desired output noise level. This may sound like an anomaly, using a low-noise amplifier as a calibrated noise source, but it meets the basic criteria. The amplifier noise is set by known fundamental noise mechanisms. This feature is valuable to circuit designers because they can use the same op amp package for both the amplification and calibrated noise generator.

15-5 COMPARISON OF METHODS

Comparing the sine wave and noise generator methods of noise measurement, the main advantage of the noise generator method is its ease of measurement. The main advantage of the sine wave method is its applicability at low frequencies and the availability of equipment.

The noise generator method is straightforward because we simply connect a noise generator to the amplifier input and adjust the generator to double the output noise power of the amplifier. In a wideband system this is done very quickly. Another advantage is the availability of low-cost calibrated noise diodes. However, these noise diodes may exhibit $1/f$ noise below a few hundred hertz. Since the noise generator is a broadband source, the system noise bandwidth Δf is not required. Some form of bandwidth limiting may be necessary to measure the spot noise. This could be a narrowband filter, a one-third octave filter, or a spectrum analyzer.

When using the noise generator method, the noise source is connected to the amplifier while measuring output noise, and the probability of pickup is increased. With the sine wave method, only a small shielded resistor is connected to the input terminals. Another disadvantage of the noise generator method is the long measurement time at low frequencies. For the instrumentation and control fields, noise measurements extend down to a few hertz or less. Spot noise measurements require a bandwidth of less than 1 octave. As pointed out in Sec. 15-6, a noise bandwidth of 5 Hz requires an averaging time constant of 10 seconds for 10% accuracy. Also, the noise generator method requires two or more measurements of output noise, the amplifier alone and the amplifier plus the noise generator.

A general rule of thumb: Use the sine wave method for low and medium frequencies and the noise generator method for high frequencies and RF.

15-6 EFFECT OF MEASURING TIME ON ACCURACY

Although it is impossible to predict the instantaneous value of noise, the long-term average of rectified noise can be determined statistically.

Let us examine the fluctuations of the pointer of a dc meter when rectified noise is being indicated. It is desired that the meter indicate the average value of the waveform. However, if the meter circuit has a short time constant, the pointer will try to respond to the instantaneous value of the noise, instead of its average value. It is of course impossible to read a wildly fluctuating meter.

A relation between noise bandwidth, output meter time constant τ, and meter relative error ε is

$$\varepsilon = \frac{1}{\sqrt{2\tau \Delta f}} \tag{15-28}$$

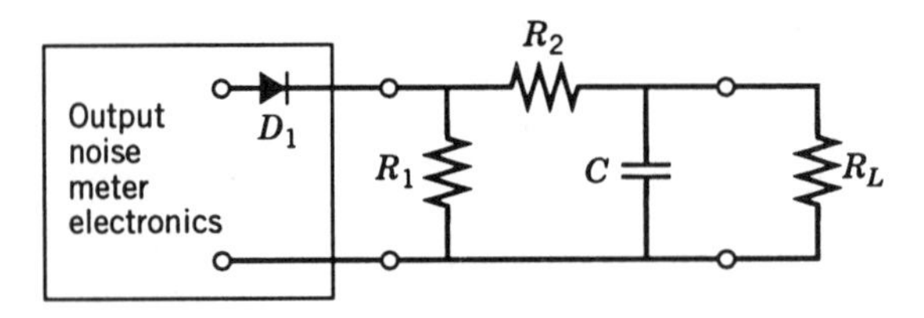

Figure 15-9 Averaging circuit.

where ε is the ratio of the rms value of meter fluctuations to the average meter reading. For the same level of accuracy, narrowband measurements require a longer averaging time than do wideband measurements. For bandwidths of 1000 Hz or more, the meter time constant is usually sufficient. On the other hand, a 5-Hz bandwidth requires a 40-s time constant for 5% accuracy, whereas 1% accuracy requires a 1000-s time constant. This illustrates one type of difficulty encountered when making accurate noise measurements.

As a rule of thumb, use the widest possible bandwidth for the shortest reading times and greatest accuracy.

Several methods of obtaining a long time constant were described in Sec. 15-3-7. The most common way is to parallel the meter with a capacitor. The capacitor must not interfere with the calibration of the meter. Use a low-leakage capacitor, such as a tantalum electrolytic, so that the capacitor does not shunt the dc signal. The capacitor also may overload the output of the amplifier by requiring larger peak currents. Test the accuracy with a sine wave by noting whether the presence of the capacitor changes the meter reading. In addition to the accuracy requirements, allow several time constants for the capacitor to charge.

If the output meter is connected directly to the rectifier D_1, adding a capacitor C makes it peak responding. By connecting a small shunt resistor R_1 and a large series resistor R_2 as shown in Fig. 15-9, the meter reads correctly; although the gain is reduced to $R_L/(R_2 + R_L)$, where the meter resistance is R_L.

15-7 BANDWIDTH ERRORS IN SPOT NOISE MEASUREMENTS

Spot noise or noise spectral density is the noise in a 1-Hz noise bandwidth. This implies that the noise was measured with a 1-Hz bandwidth filter. In actual practice, the noise is measured with a noise meter or spectrum analyzer, and the bandwidth may be significantly greater than 1 Hz. To obtain spot noise, the reading is divided by the square root of the noise bandwidth.

When the noise is white, bandwidth poses no problem. The total noise voltage in a wideband is equal to the sum of identical noise contributions from each frequency that make up the total bandwidth. Frequently, however, noise waveforms contain a significant amount of $1/f$ noise power. The question arises: *How much error is introduced when measuring the noise spectral density with a wide bandwidth and dividing by the square root of bandwidth?*

Now compare the spot noise with the integrated noise in a bandwidth, divided by the bandwidth, for an amplifier with a $1/f$ noise power distribution. It was shown in Chap. 1 that the mean square value of $1/f$ noise voltage is given by

$$E_f^2(\Delta f) = K \ln \frac{f_h}{f_l} \tag{15-29}$$

where f_h and f_l are the upper- and lower-cutoff frequencies in the band being considered. This relation can be rearranged in the form

$$E_f^2(\Delta f) = K \ln \frac{f_l + \Delta f}{f_l} \tag{15-30}$$

and written as

$$E_f^2(\Delta f) = K \ln\left(1 + \frac{\Delta f}{f_l}\right) \tag{15-31}$$

For purposes of analysis we define a center frequency f_o and consider that Δf is geometrically centered about f_o. Then $f_o = (f_1 f_h)^{1/2}$.

The bandwidth error can be calculated. Using Eq. 15-31 to integrate over a bandwidth and dividing by the square root of the bandwidth, we compare this calculated value with the true value of spot noise at the center frequency f_o as shown in Table 15-4.

TABLE 15-4 Errors in Spot Noise Measurements of a $1/f$ Noise Dominated System

Δf (as percentage of f_o)	Value of Wideband Noise Divided by Δf	Value of Spot Noise at f_o	Percentage Error
0.1	1	1	0
1	1	1	0
10	0.9998	1	−0.02
30	0.9978	1	−0.22
100	0.9809	1	−1.91

Even with a $1/f$ noise power distribution, a bandwidth of one-third the frequency gives a negligible error. If part of the noise is white, the error is even smaller. This is convenient because it allows us to use the readily available one-third octave filters. One-third of an octave is approximately equal to one-third of the frequency. In general, it is possible to use a bandwidth equal to the frequency, geometrically centered around the frequency, and still have negligible error. In some cases, such as with non-Gaussian "popcorn" noise, the noise *power* spectral density is proportional to $1/f^2$. The mean square noise voltage is

$$E_p^2 = \frac{K_2}{f^2} \tag{15-32}$$

and the total noise in the bandwidth from f_1 to f_h is

$$E_p^2(\Delta f) = K_2 \int_{f_l}^{f_h} \frac{df}{f^2} \tag{15-33}$$

Therefore,

$$E_p^2(\Delta f) = K_2 \left(\frac{f_h - f_l}{f_l \times f_h} \right) \tag{15-34}$$

Substituting in the noise expression of Eq. 15-34 gives

$$E_p^2(\Delta f) = K_2 \left(\frac{\Delta f}{f_o^2} \right) \tag{15-35}$$

Equation 15-35 shows that the total noise voltage of a $1/f^2$ power distribution is proportional to the square root of the bandwidth Δf just as for white noise. The measurement accuracy is independent of the measurement bandwidth.

15-8 NOISE BANDWIDTH

As discussed in Chap. 1, the noise bandwidth is not equal to 3-dB bandwidth. The noise bandwidth is always greater than the 3-dB bandwidth. A definition of noise bandwidth was given:

$$\Delta f = \frac{1}{A_{vo}^2} \int_0^\infty |A_v(f)|^2 \, df \tag{1-9}$$

Here we add a few comments to the discussion given earlier. If the power gain curve is not symmetrical about the center frequency as shown in Fig.

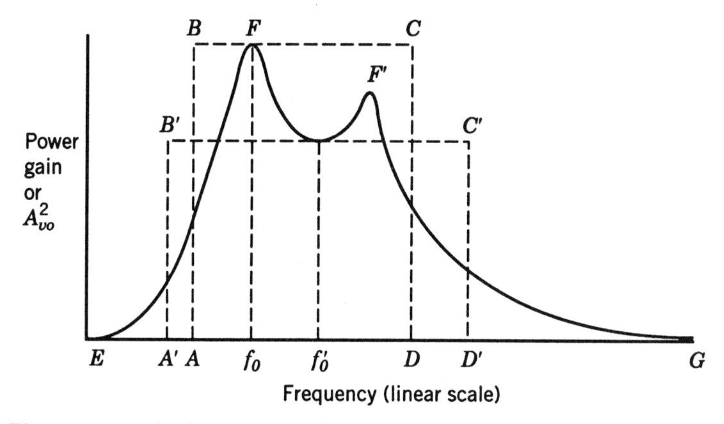

Figure 15-10 Noise bandwidth of a nonsymmetrical transfer function.

15-10, the selection of f_o is arbitrary. Consider the noise bandwidth of curve $EFF'G$ which has two peaks at F and F'. Two reference frequencies f_o and f_o' are defined in the figure. These produce rectangular bandwidths of $ABCD$ and $A'B'C'D'$, respectively. Each of these rectangles has the same gain–bandwidth product, but the peak gains differ; we see that the noise bandwidth AD is smaller than $A'D'$. Either determination of the noise bandwidth can be correct, although the preferred is AD.

The noise bandwidth can be measured with a calibrated noise generator. The noise generator must be the dominant noise in the system. Measure the transfer voltage gain K_t at f_o. Then, with the noise generator connected to the input terminals, measure the total output noise E_{no}. The output is

$$E_{no}^2 = K_t^2 E_i^2 \, \Delta f \tag{15-36}$$

where K_t is the transfer voltage gain, E_i is the input noise signal in volts per hertz$^{1/2}$, and Δf is the noise bandwidth. The unknown Δf is

$$\Delta f = \left(\frac{E_{no}}{K_t E_i} \right)^2 \tag{15-37}$$

SUMMARY

a. Two widely used techniques for noise measurement are the sine wave and the noise generator methods.
b. The sine wave method requires measurement of the total output noise E_{no} and the transfer gain K_t. Then, $E_{ni} = E_{no}/K_t$.
c. For $R_s \to 0$, $E_{ni} = E_n$. For $R_s \to \infty$, $E_{ni} = I_n R_s$.

d. To eliminate thermal noise from R_s, a capacitor can sometimes be substituted for the source resistance.

e. A spectrum analyzer is convenient for making spectral density measurements.

f. The noise voltage of a noise generator necessary to double the output noise power is equivalent to the E_{ni} of the system under test. Thus the generator must be calibrated.

g. The shot noise of a calibrated noise source is used for noise measurements.

h. The requirements for an output noise meter are: (1) it must respond to noise power; (2) it must have a crest factor greater than 3; and (3) it must have a bandwidth 10 or more times the system noise bandwidth. A true rms reading meter is best; an average responding meter must be corrected by a 1.13 multiplier; and a peak responding meter is valueless.

PROBLEMS

15-1. Design an attenuator of the Fig. 15-3 type to match a 600-Ω signal generator. Consider that $R_2 = 10\ \Omega$. The output V_s should be 1% of V_g. Specify values for R_1 and R_3.

15-2. The spot noise figure is to be determined, but we do not have a measuring instrument calibrated in decibels; we just have a true value reading voltmeter. The transfer voltage gain K_t is measured to be 10,000. We adjust the level of the input signal so that at the output port $V_{so} = 100E_{no}$. We measure the input signal to be 1 μV at the source resistance is 1000 Ω. Find the NF.

15-3. Design a test setup to measure the excess noise of resistors. Develop a block diagram and specify the gain necessary and any special requirements for the equipment.

15-4. The probability density function (pdf) of a Gaussian distribution was given in Chap. 1 by Eq. 1-1 as

$$f(x) = \frac{1}{\sigma\sqrt{2\pi}} \exp - \left[\frac{(x-\mu)^2}{2\sigma^2} \right]$$

The expected value of any pdf corresponds to its average value (dc value) and can be found by

$$E(x) = \int_{-\infty}^{+\infty} xf(x)\,dx$$

Now consider that a regular ac voltmeter (not a true rms voltmeter) is to be used to measure a Gaussian noise voltage. This ac voltmeter uses a full-wave rectifier to determine the dc value of any ac applied waveform. Then the dc reading is "scaled" to a corresponding "rms" value by a gain constant. Show that the expected value of the absolute value of the Gaussian pdf is given by

$$E(|x|) = \frac{2\sigma}{\sqrt{2\pi}} = 0.798\sigma$$

Assume that $\mu = 0$ in the pdf expression. *Hint*: Make a change-of-variable substitution before performing the integration by letting $y = x/\sigma$.

REFERENCES

1. Ott, H. W., *Noise Reduction Techniques in Electronic Systems*, Wiley-Interscience, New York, 1988.
2. Wilkie, S., "Tips on Spectrum Analyzer Noise Measurements," *EE-Evaluation Engineering*, (February 1992), 54–60.
3. Bennett, W. R., *Electrical Noise*, McGraw-Hill, New York, 1960, p. 44.

APPENDIX A

MEASURED NOISE CHARACTERISTICS OF INTEGRATED AMPLIFIERS

This appendix contains the measured data from a small sample of IC amplifiers that represent the current noise technology. Since books change slowly and technology advances rapidly, these data arc only an indication of the state of the art.

A variety of process technologies are included to contrast their performance for various source impedances and frequency ranges. Unfortunately, there is no single low-noise amplifier that will fill all requirements. The eight devices shown are:

OP-27/OP-37 bipolar op amp. Gain-bandwidth = 63 MHz (OP-37).
TL061 JFET input op amp. Gain-bandwidth not specified.
TL071 JFET input op amp. Gain-bandwidth = 3 MHz.
OP-97 super β bipolar. Gain-bandwidth = 0.9 MHz.
μA 741 general purpose bipolar op amp. Gain-bandwidth = 1 MHz.
AD745 BiFET op amp. Gain-bandwidth = 20 MHz.
TLE2037 bipolar op amp. Gain-bandwith = 80 MHz.
TLC2201 advanced Lin CMOS^{Tm} op amp. Gain-bandwidth = 2 MHz.

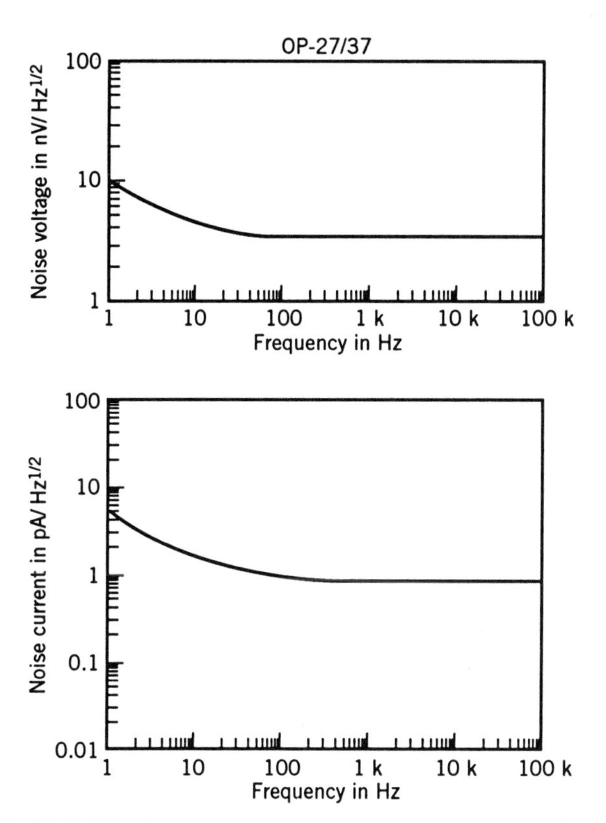

OP-27/OP-37 bipolar op amp. Gain–bandwidth = 63 MHz (OP-37).

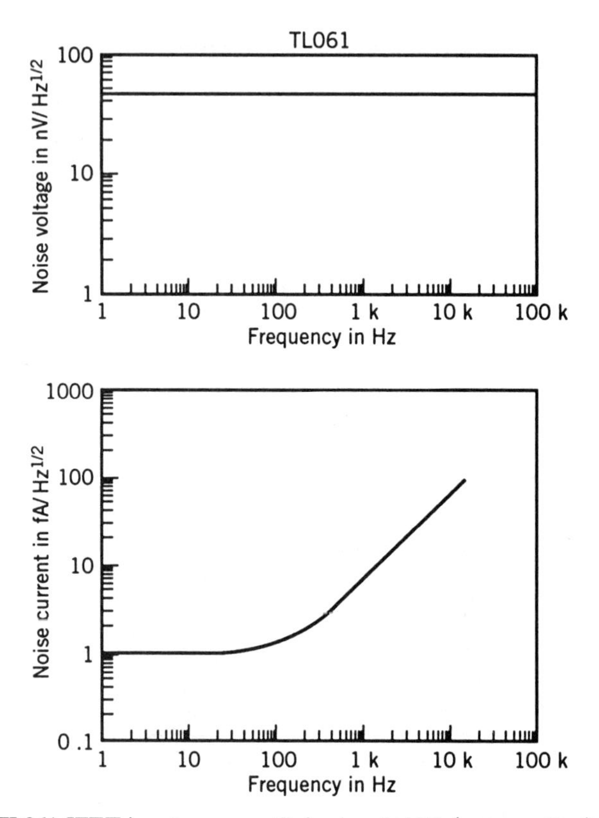

TL061 JFET input op amp. Gain–bandwidth (not specified).

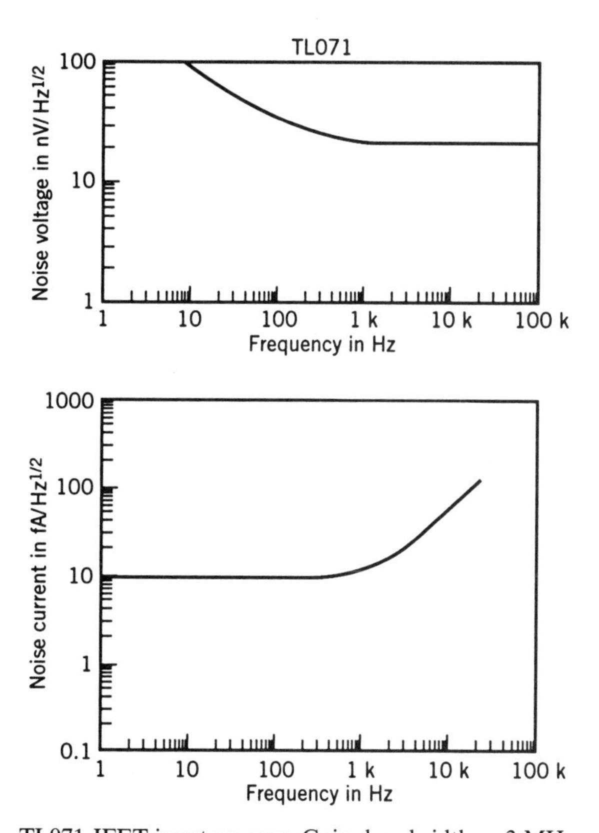

TL071 JFET input op amp. Gain–bandwidth = 3 MHz.

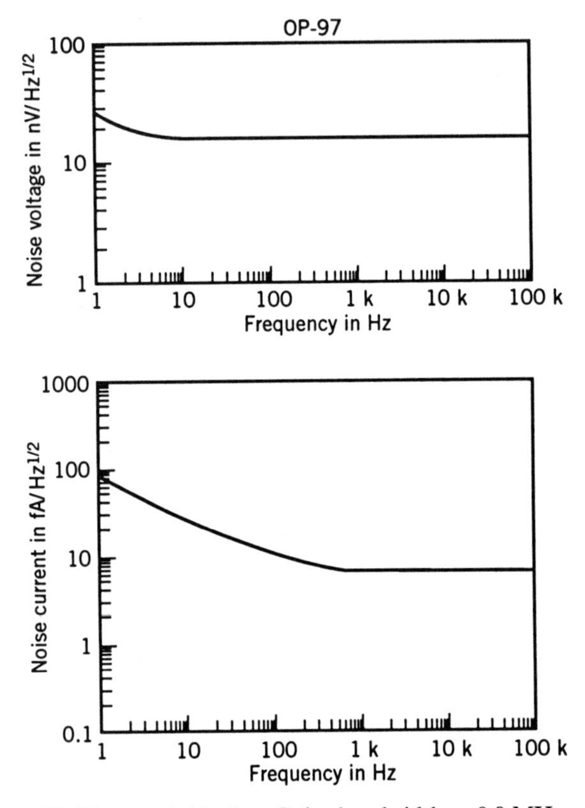

OP-97 super β bipolar. Gain–bandwidth = 0.9 MHz.

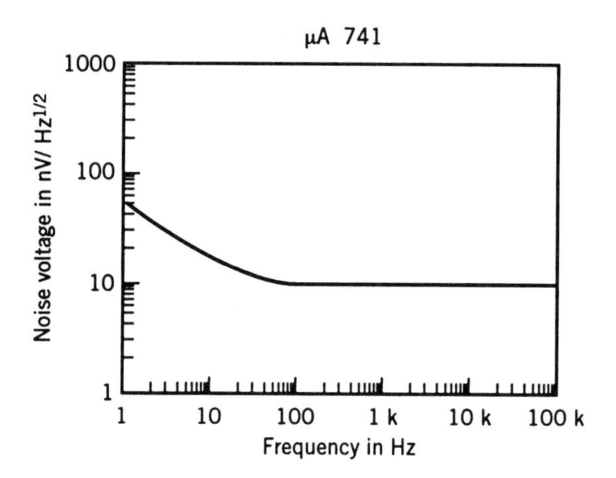

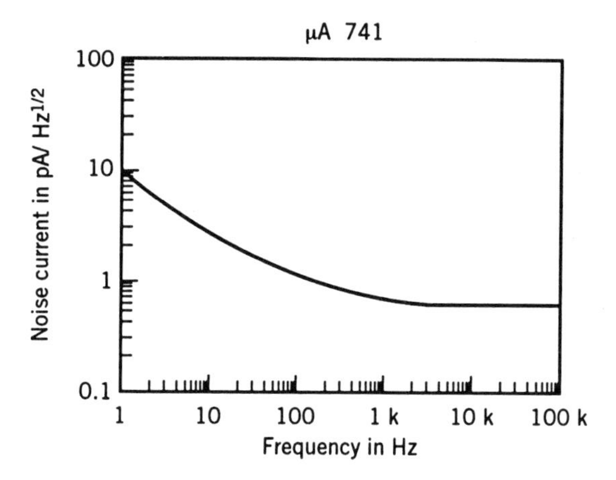

μA 741 general purpose bipolar op amp. Gain–bandwidth = 1 MHz.

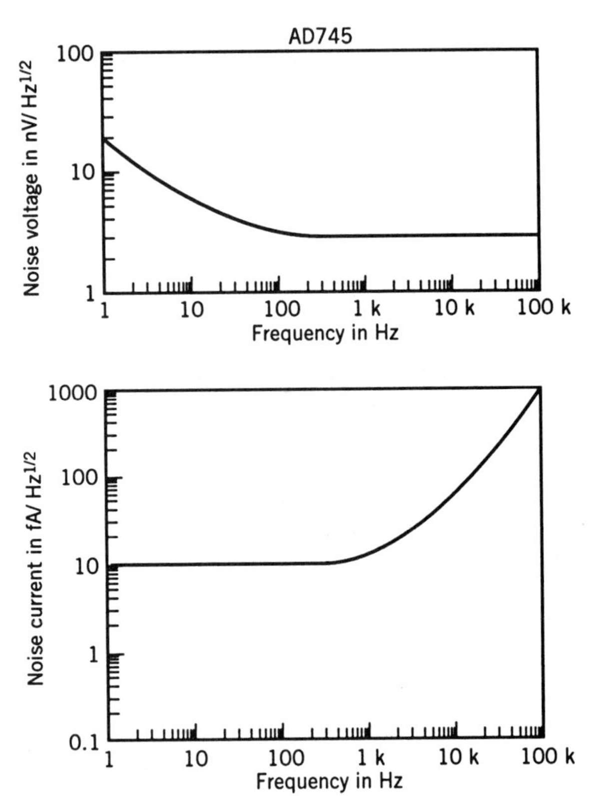

AD745 BIFET op amp. Gain–bandwidth = 20 MHz.

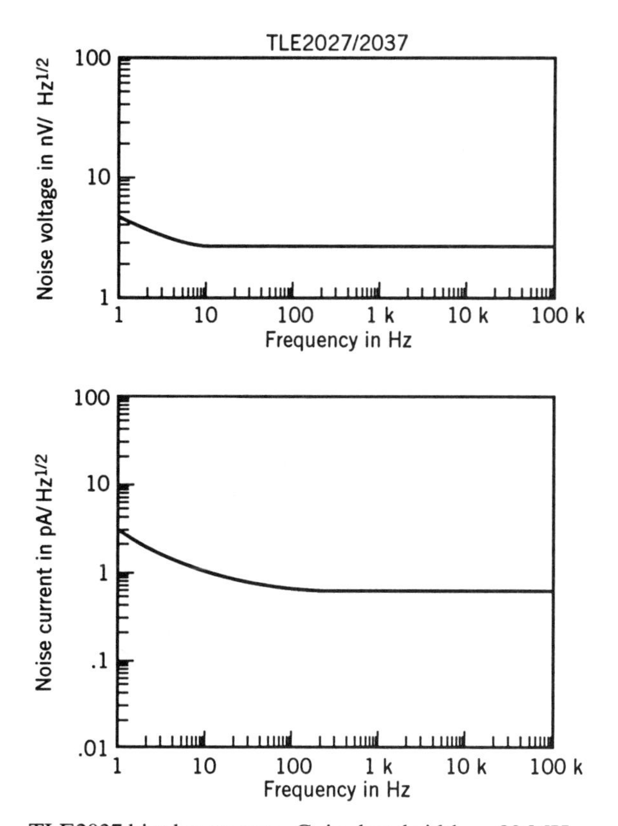

TLE2037 bipolar op amp. Gain–bandwidth = 80 MHz.

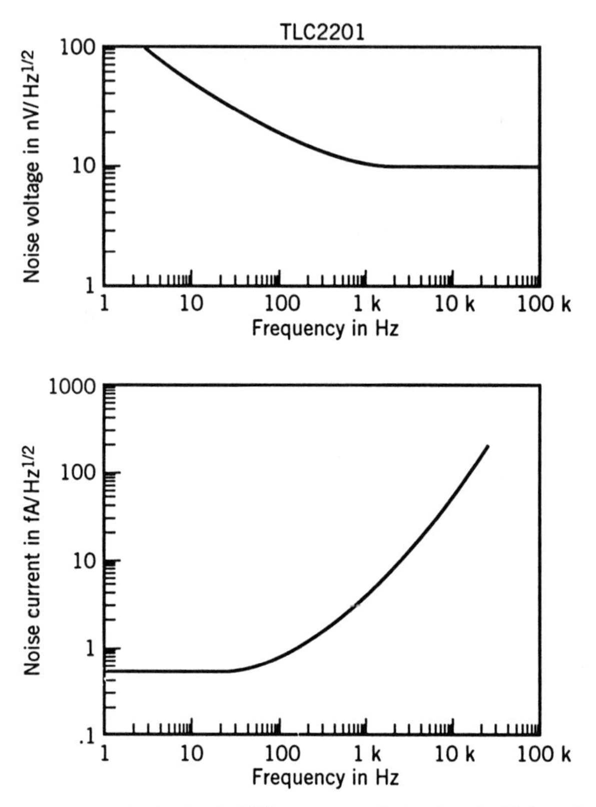

TLC2201 advanced LinCMOS™ op amp. Gain–bandwidth = 2 MHz.

APPENDIX B

MEASURED NOISE CHARACTERISTICS OF FIELD EFFECT TRANSISTORS

This appendix contains measured data on some discrete FETs to illustrate noise performance. Compare the discrete devices with the newer data on JFET input op amps in App. A.

Dashed lines in the plots indicate extrapolated data based on extension of the low- and high-frequency data. The seven devices shown are:

2N2609 *n*-channel JFET.
2N3460 *n*-channel JFET.
2N3684 *n*-channel JFET.
2N3821 *n*-channel JFET.
2N4221A *n*-channel JFET.
2N4416 *n*-channel JFET.
2N5116 *p*-channel JFET.

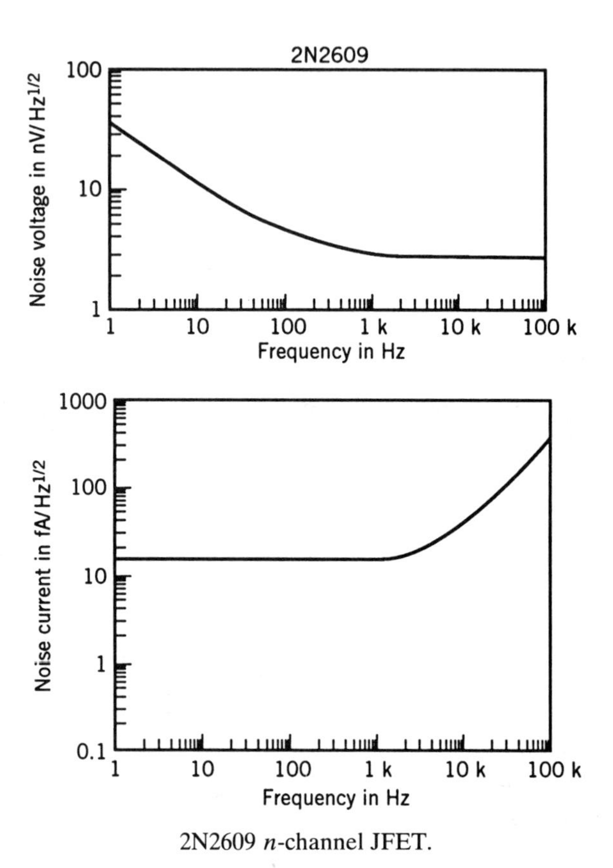

2N2609 *n*-channel JFET.

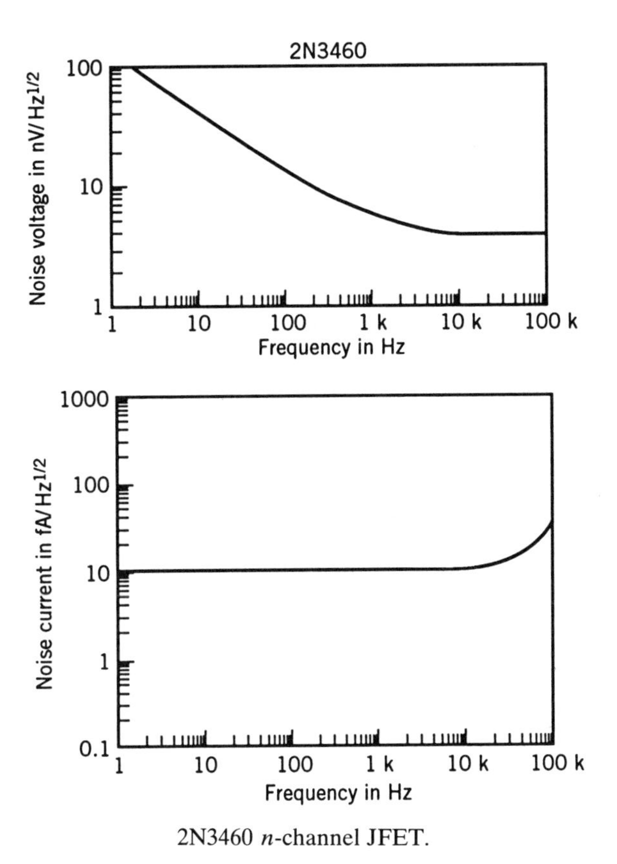

2N3460 *n*-channel JFET.

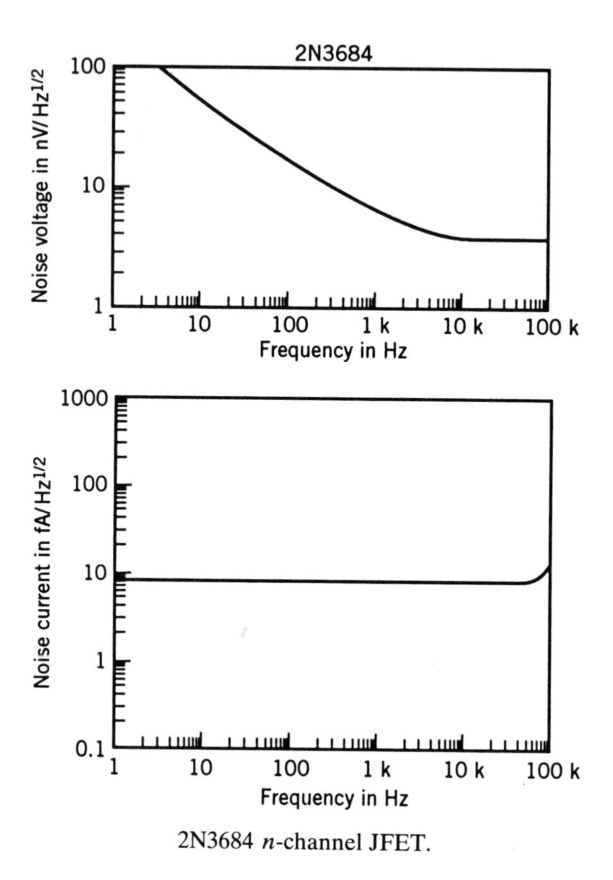

2N3684 *n*-channel JFET.

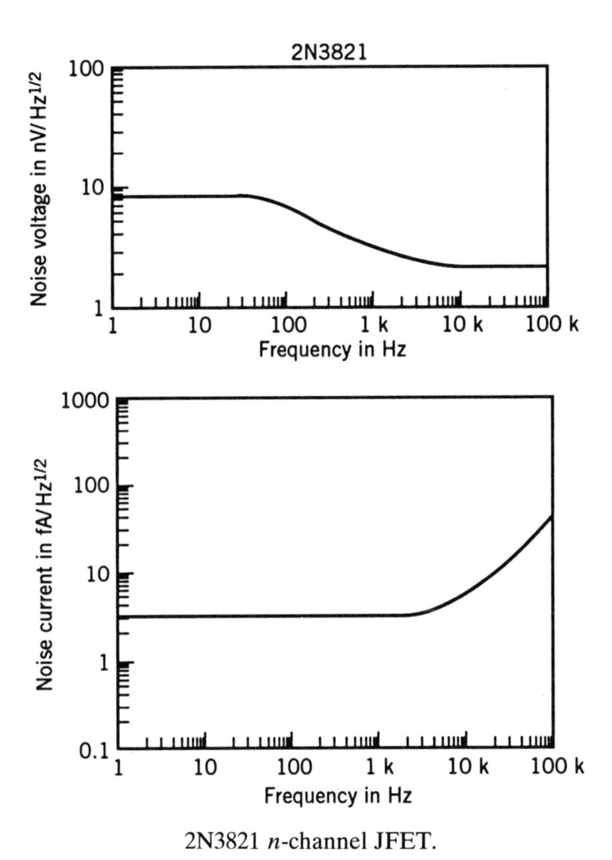

2N3821 *n*-channel JFET.

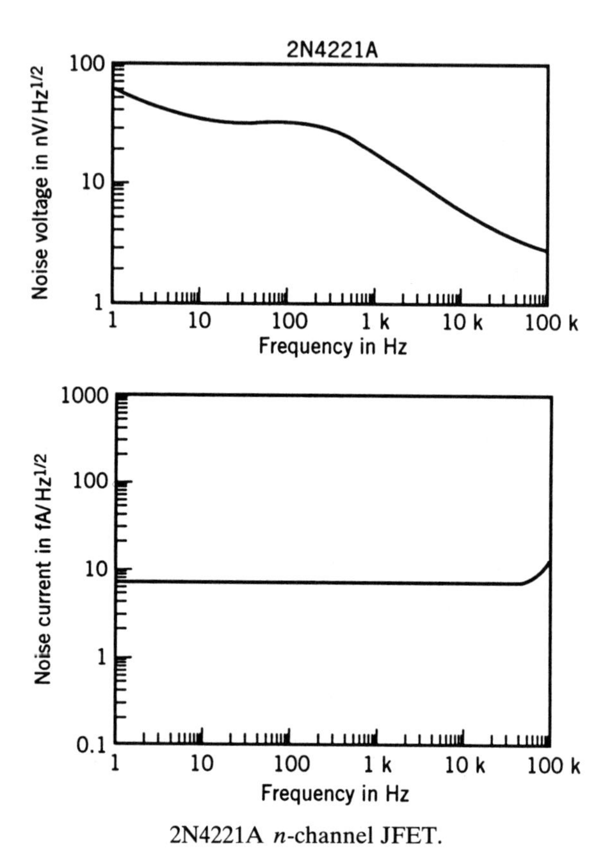

2N4221A *n*-channel JFET.

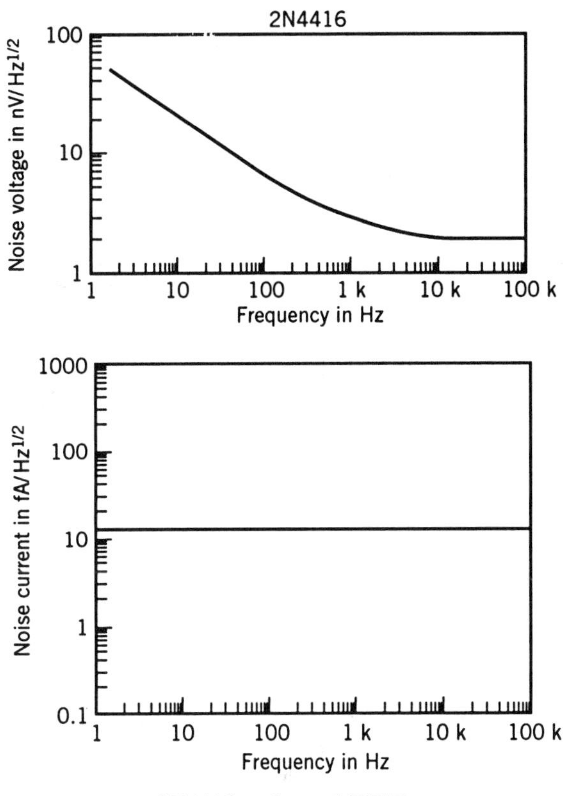

2N4416 *n*-channel JFET.

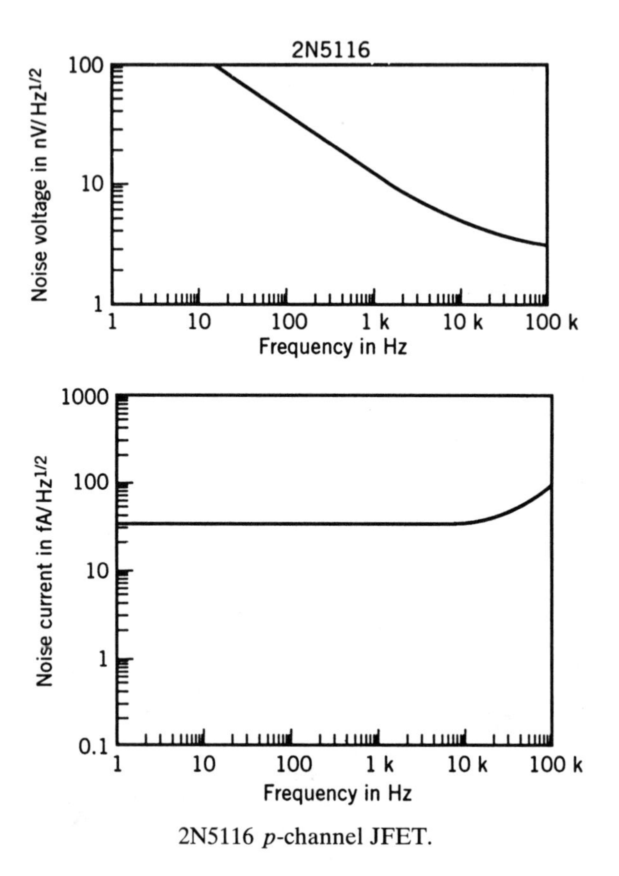

2N5116 *p*-channel JFET.

APPENDIX C

MEASURED NOISE CHARACTERISTICS OF BIPOLAR JUNCTION TRANSISTORS

Examples of the noise characteristics of BJTs are shown in this appendix. Noise constants for these transistors are shown in Table 6-1. Noise measurements were made on a small sample of each device to show the general characteristics of discrete BJTs and may not represent current device technology. The eight devices shown are:

2N930	*npn*
2N3964*	*pnp*
2N4124	*npn*
2N4125	*pnp*
2N4250*	*pnp*
2N4403	*pnp*
2N5138	*pnp*
MPS–A18	*npn*

*Same curves for these two devices.

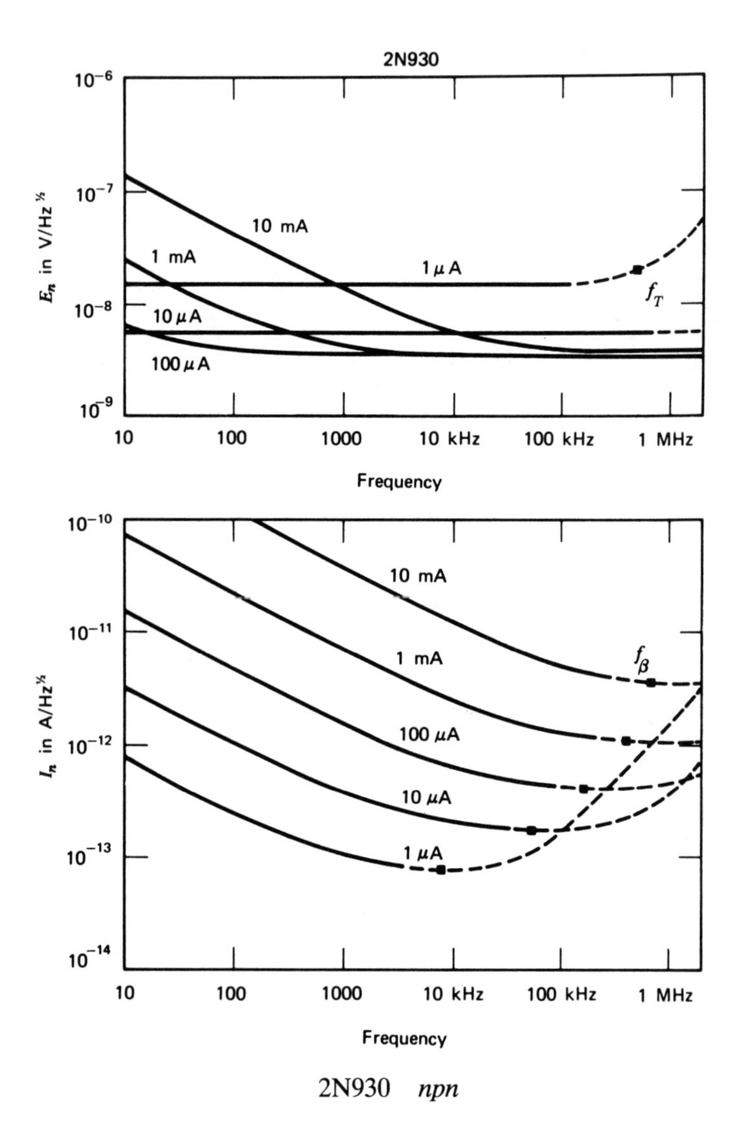
2N930
10^{-6}
10^{-7}
10^{-8}
10^{-9}
E_n in V/Hz$^{1/2}$
10 mA
1 mA
1 μA
10 μA
100 μA
f_T
10
100
1000
10 kHz
100 kHz
1 MHz
Frequency
10^{-10}
10^{-11}
10^{-12}
10^{-13}
10^{-14}
I_n in A/Hz$^{1/2}$
10 mA
1 mA
100 μA
10 μA
1 μA
f_β
10
100
1000
10 kHz
100 kHz
1 MHz
Frequency

2N930 *npn*

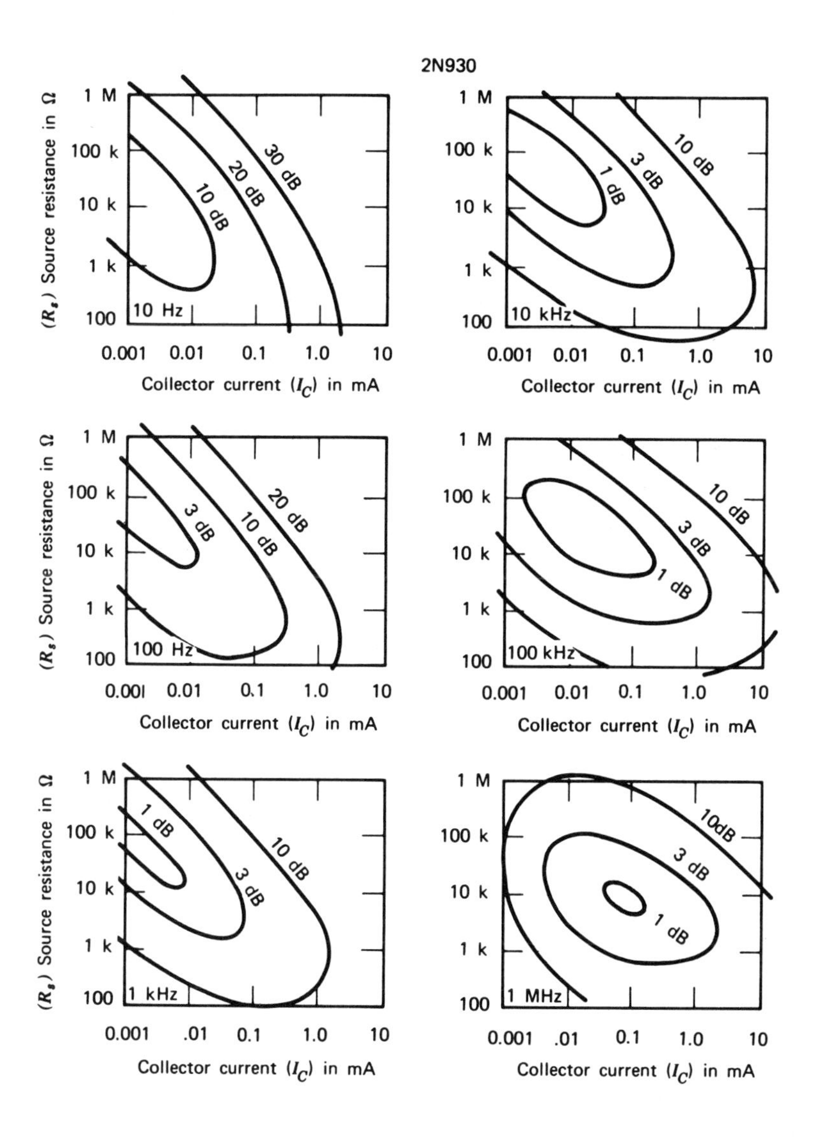
2N930
1 M
100 k
10 k
1 k
100
0.001
0.01
0.1
1.0
10
10 dB
20 dB
30 dB
10 Hz
1 dB
3 dB
10 dB
10 kHz
3 dB
10 dB
20 dB
100 Hz
1 dB
3 dB
10 dB
100 kHz
1 dB
3 dB
10 dB
1 kHz
1 dB
3 dB
10dB
1 MHz
.01
Collector current (I_C) in mA
(R_s) Source resistance in Ω

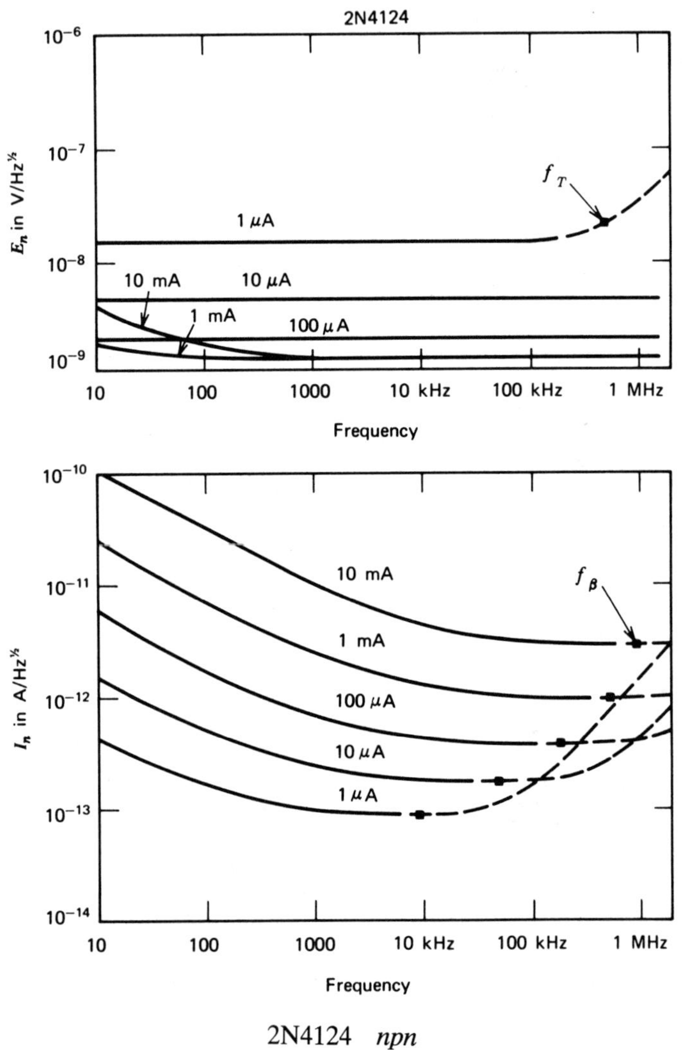
2N4124
10⁻⁶
10⁻⁷
10⁻⁸
10⁻⁹
En in V/Hz½
f T
1 μA
10 mA
10 μA
1 mA
100 μA
10
100
1000
10 kHz
100 kHz
1 MHz
Frequency
10⁻¹⁰
10⁻¹¹
10⁻¹²
10⁻¹³
10⁻¹⁴
In in A/Hz½
10 mA
f β
1 mA
100 μA
10 μA
1 μA
10
100
1000
10 kHz
100 kHz
1 MHz
Frequency

2N4124 *npn*

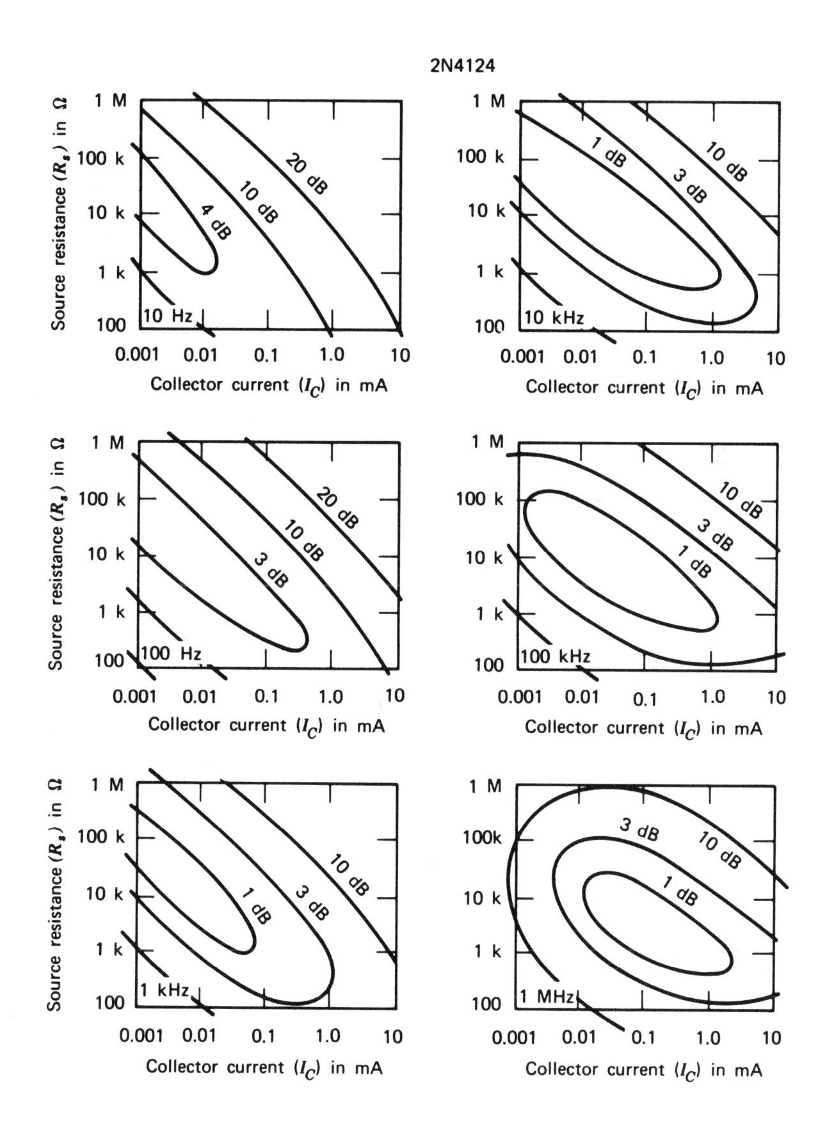
2N4124
Source resistance (R_s) in Ω
1 M
100 k
10 k
1 k
100
0.001
0.01
0.1
1.0
10
Collector current (I_C) in mA
10 Hz
4 dB
10 dB
20 dB
10 kHz
1 dB
3 dB
10 dB
100 Hz
3 dB
10 dB
20 dB
100 kHz
1 dB
3 dB
10 dB
1 kHz
1 dB
3 dB
10 dB
1 MHz
1 dB
3 dB
10 dB

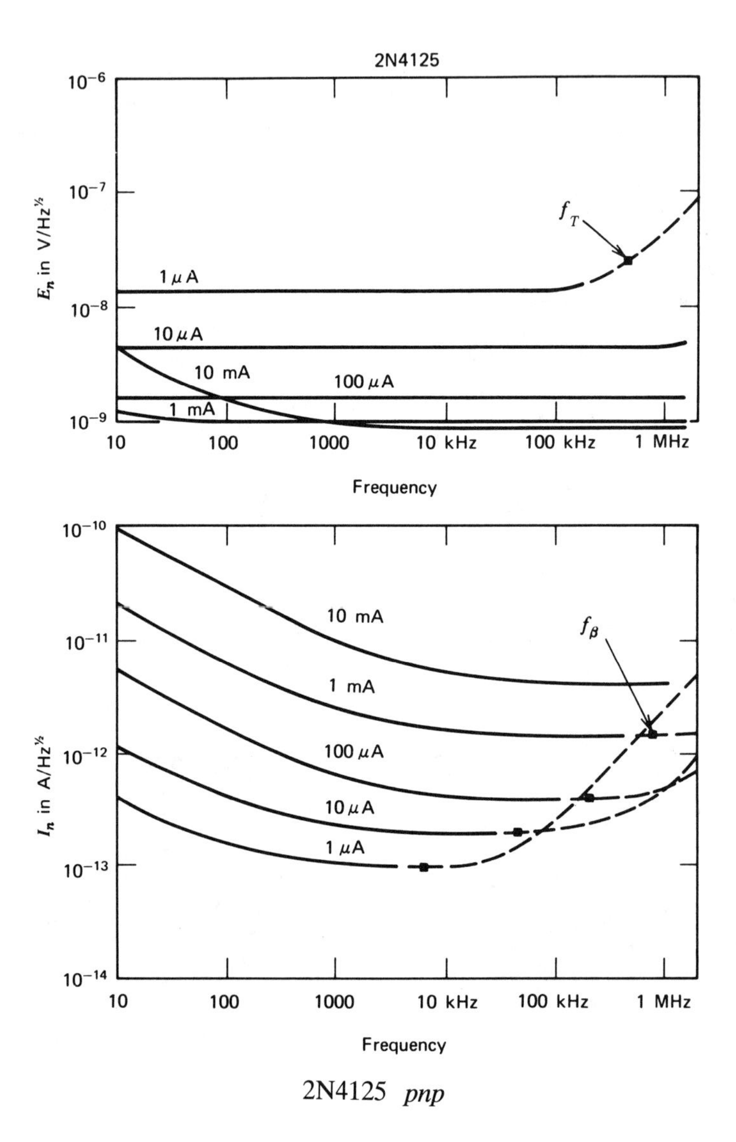
2N4125
10^{-6}
10^{-7}
10^{-8}
10^{-9}
E_n in V/Hz$^{1/2}$
f_T
1 μA
10 μA
10 mA
100 μA
1 mA
10
100
1000
10 kHz
100 kHz
1 MHz
Frequency
10^{-10}
10^{-11}
10^{-12}
10^{-13}
10^{-14}
I_n in A/Hz$^{1/2}$
10 mA
f_β
1 mA
100 μA
10 μA
1 μA
10
100
1000
10 kHz
100 kHz
1 MHz
Frequency

2N4125 *pnp*

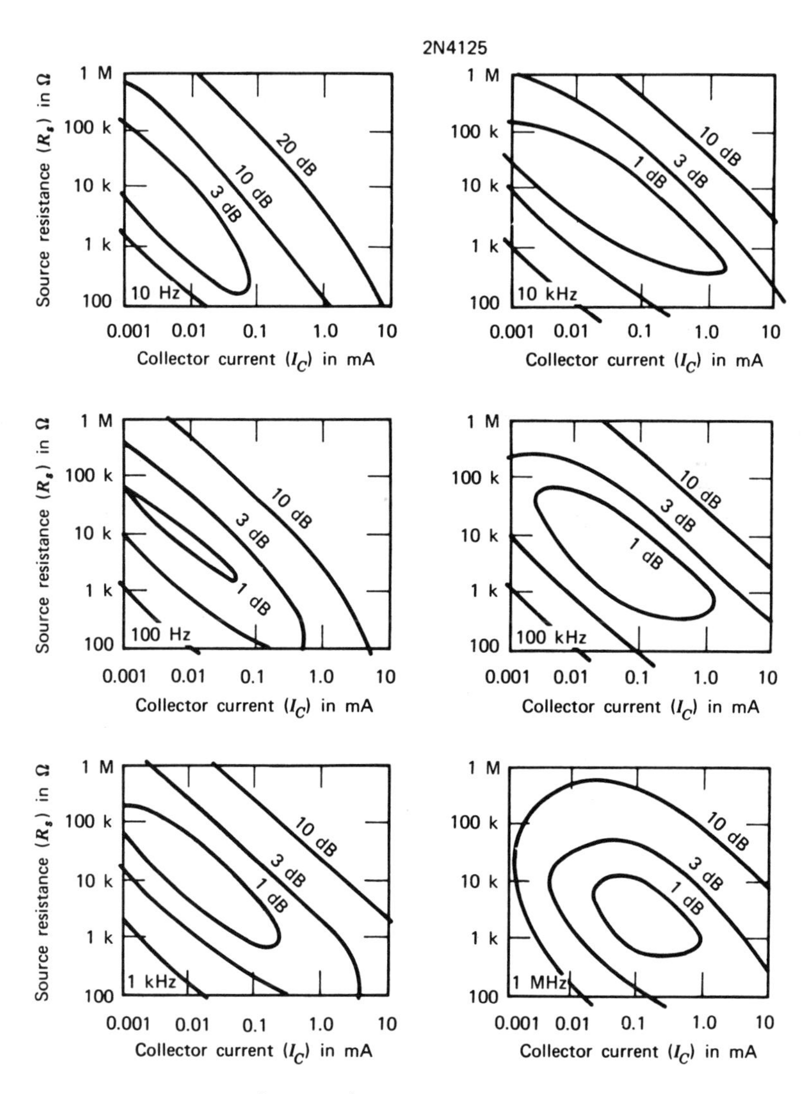

Contours of constant narrowband noise figure

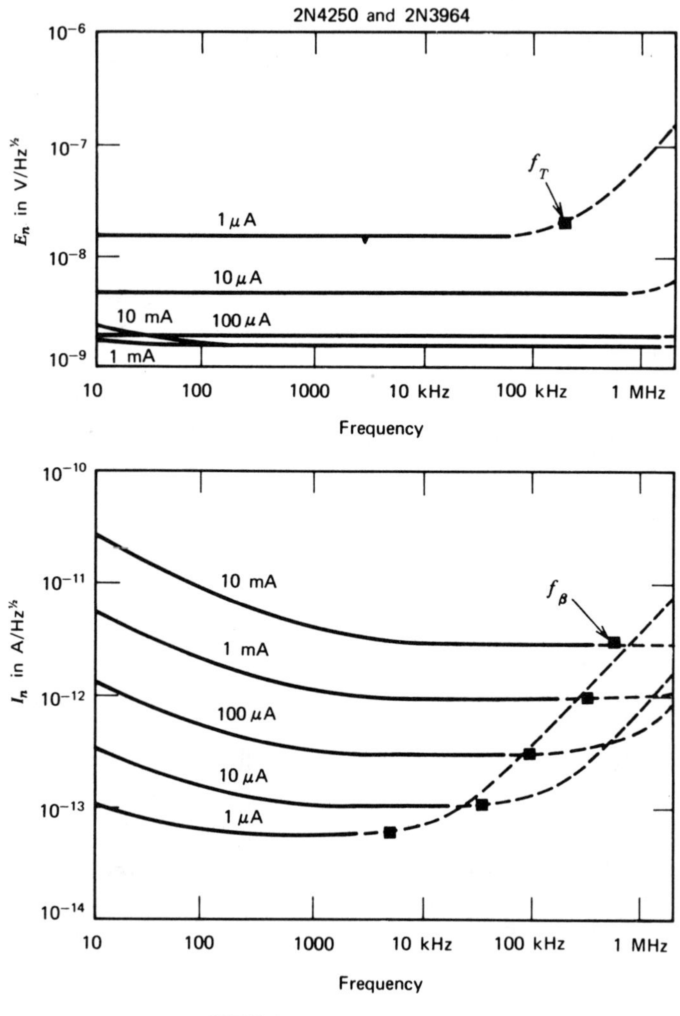

2N4250 and 2N3964 *pnp*

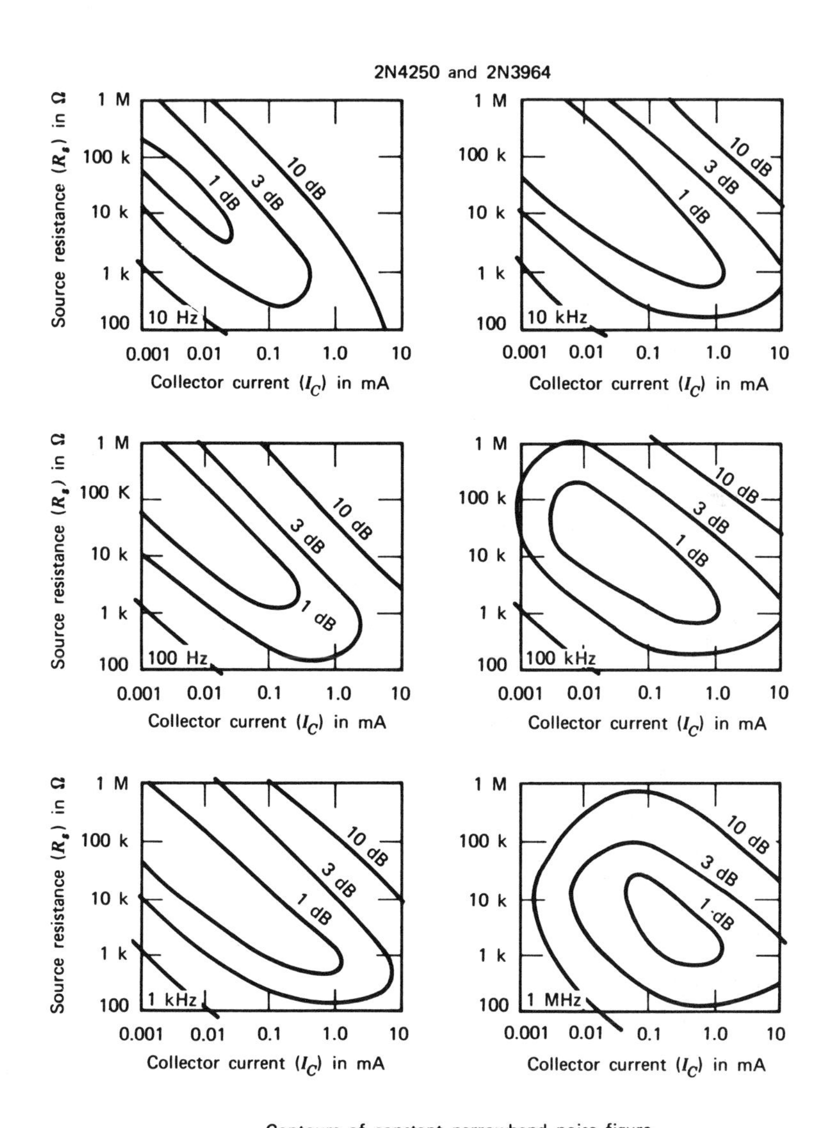

Contours of constant narrowband noise figure

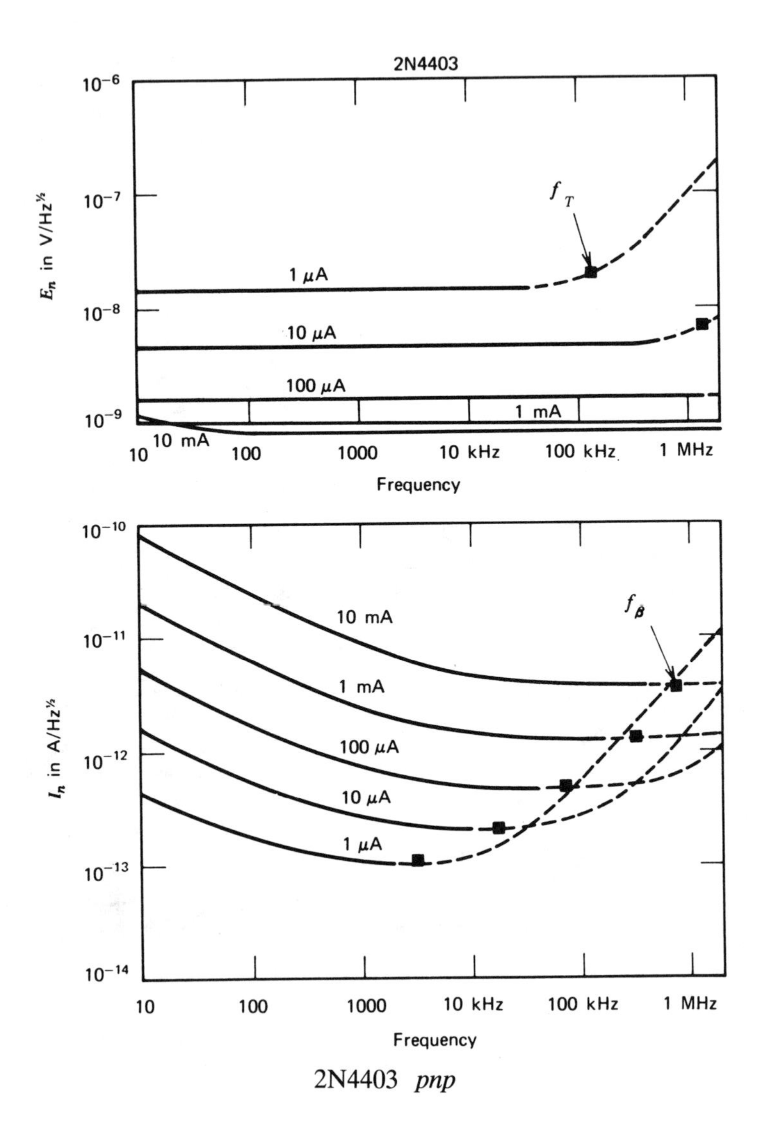
2N4403
10^{-6}
10^{-7}
10^{-8}
10^{-9}
E_n in V/Hz½
f_T
1 μA
10 μA
100 μA
1 mA
10 mA
10
100
1000
10 kHz
100 kHz
1 MHz
Frequency
10^{-10}
10^{-11}
10^{-12}
10^{-13}
10^{-14}
I_n in A/Hz½
f_β
10 mA
1 mA
100 μA
10 μA
1 μA
10
100
1000
10 kHz
100 kHz
1 MHz
Frequency

2N4403 *pnp*

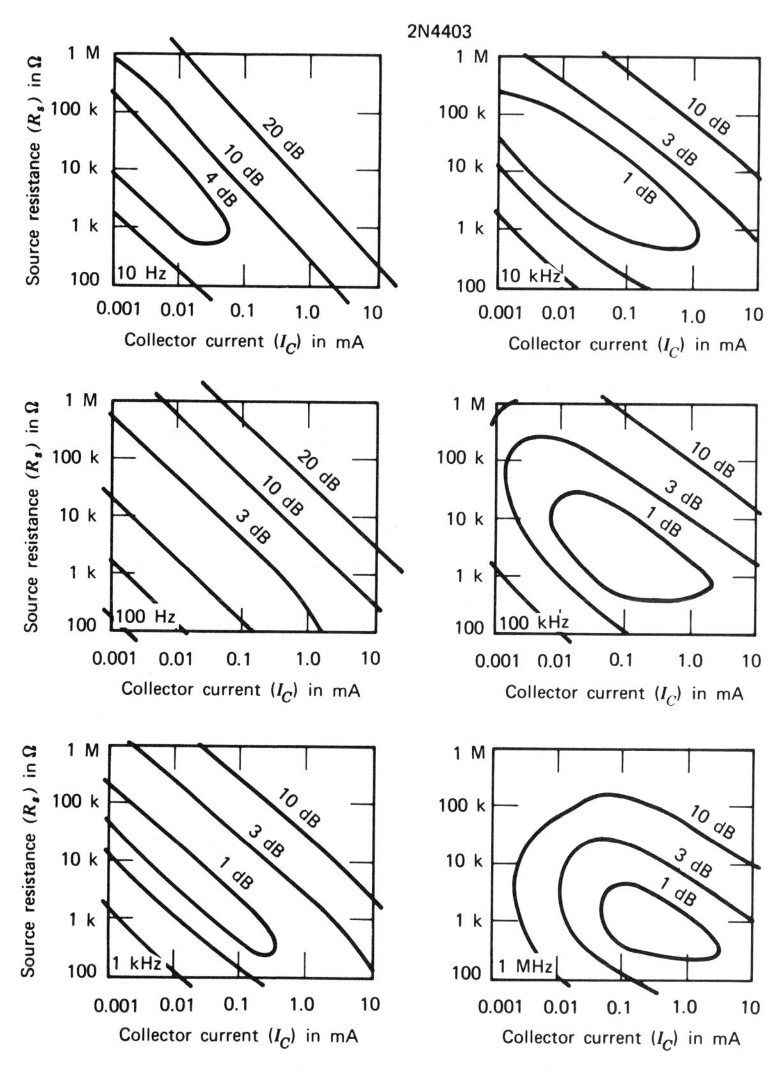

Contours of constant narrowband noise figure

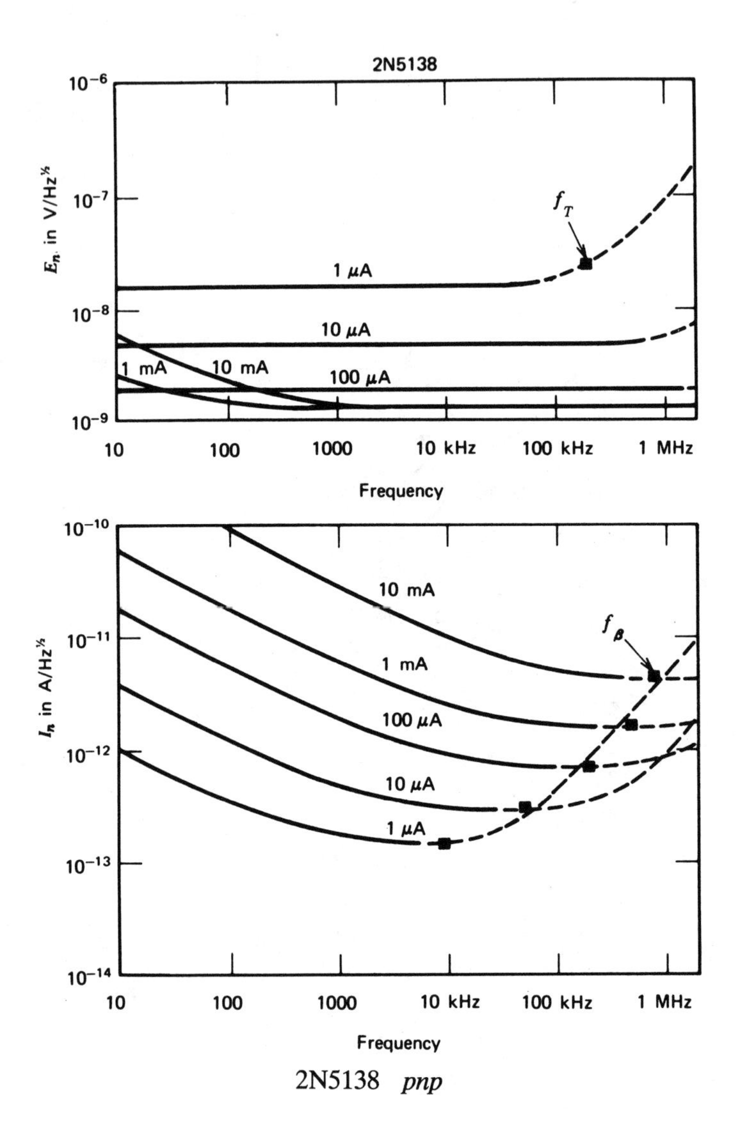
2N5138
10^{-6}
10^{-7}
10^{-8}
10^{-9}
E_n in V/Hz½
f_T
1 μA
10 μA
1 mA
10 mA
100 μA
10
100
1000
10 kHz
100 kHz
1 MHz
Frequency
10^{-10}
10^{-11}
10^{-12}
10^{-13}
10^{-14}
I_n in A/Hz½
10 mA
f_β
1 mA
100 μA
10 μA
1 μA
10
100
1000
10 kHz
100 kHz
1 MHz
Frequency

2N5138 *pnp*

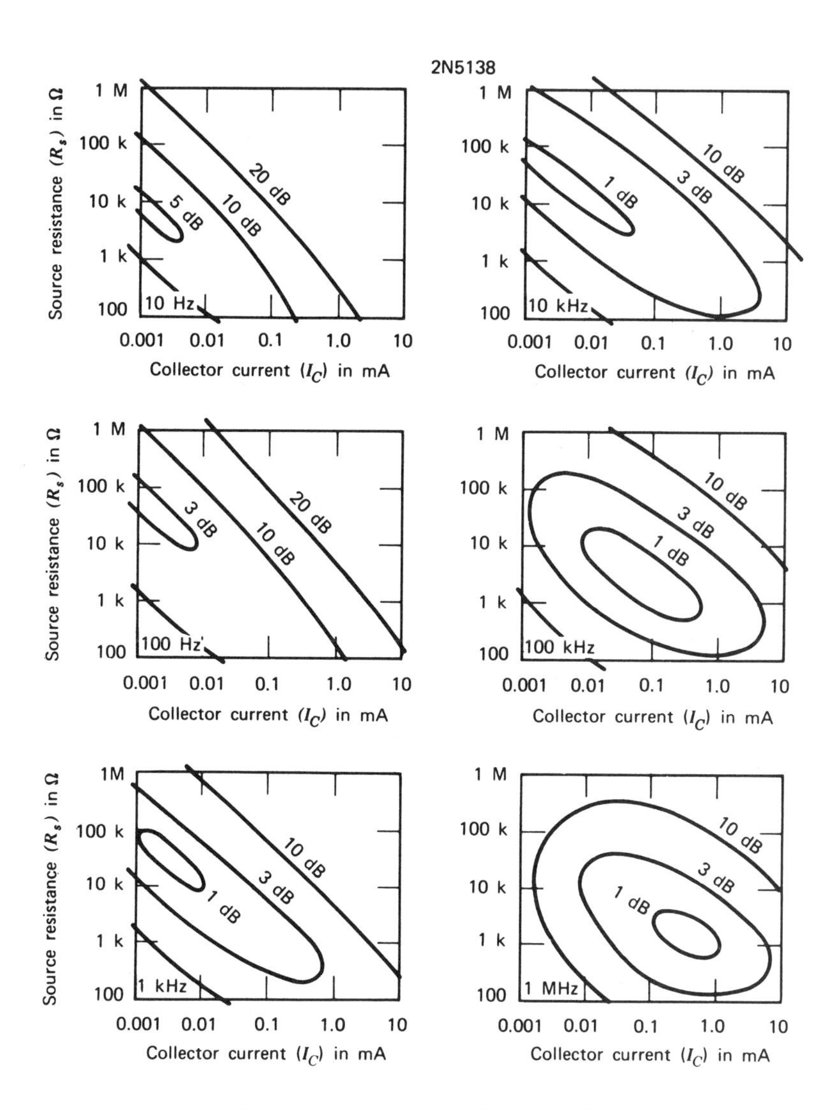

Contours of constant narrowband noise figure

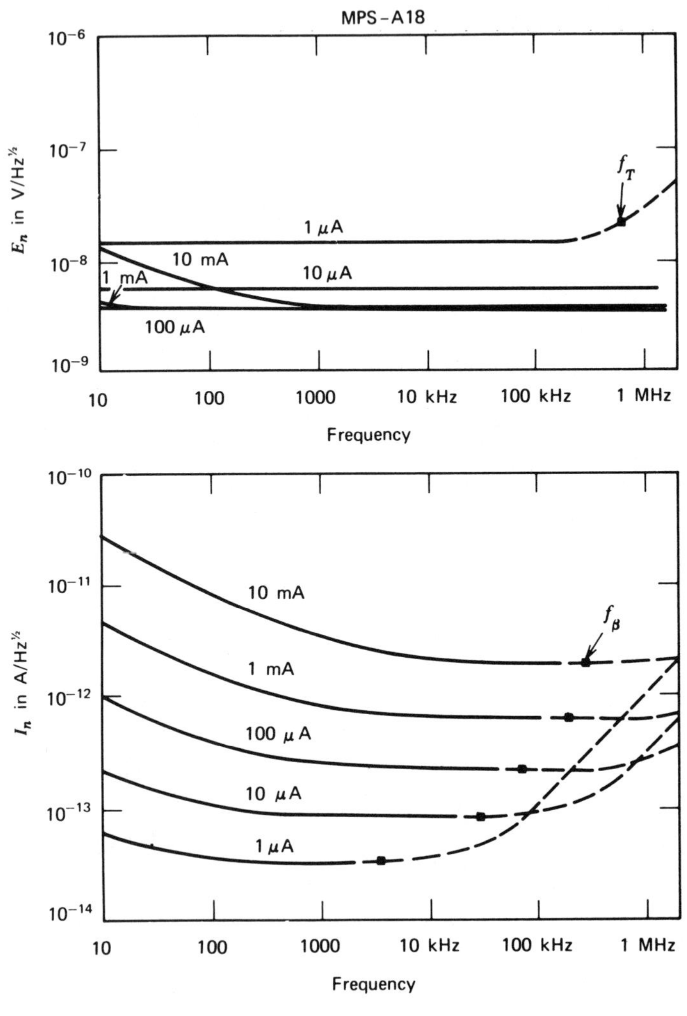
MPS-A18
10⁻⁶
10⁻⁷
10⁻⁸
10⁻⁹
E_n in V/Hz½
f_T
1 μA
10 mA
1 mA
10 μA
100 μA
10
100
1000
10 kHz
100 kHz
1 MHz
Frequency
10⁻¹⁰
10⁻¹¹
10⁻¹²
10⁻¹³
10⁻¹⁴
I_n in A/Hz½
10 mA
f_β
1 mA
100 μA
10 μA
1 μA
10
100
1000
10 kHz
100 kHz
1 MHz
Frequency

MPS–A18 *npn*

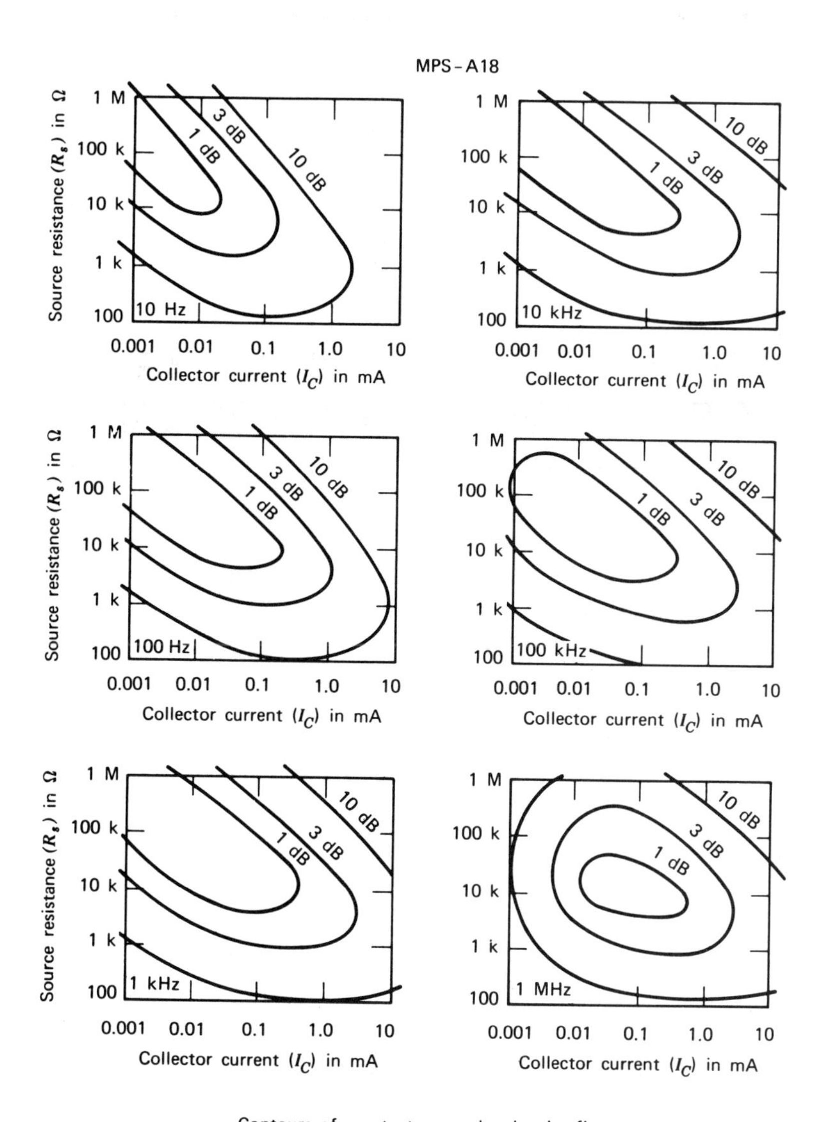

Contours of constant narrowband noise figure

APPENDIX D

RANDOM NUMBER GENERATOR PROGRAM* (WRITTEN IN PASCAL)

```
Program NOISE(input,output);

CONST
  blank= ' ';
  arraysize=6000;
  zero=0;
  index=100;
  positive='POS';
  negative='NEG';
TYPE
  Random_Number=array [1..arraysize] of real;
  Random_Set=array [1..55] of real;
VAR
  voltage_number,seed,counter,N1,N2:integer;
  numbers,output_node,neg_node,pos_node,Nexti:integer;
    Nextip,I_set:integer;
  sum,mean,variance,rms,sigma,G_set:real;
  time_step,maxtime,time,next_time,last_time:real;
  X:Random_Number;
  result:text;
  Bunch:Random_Set;
```

*This program developed by Ahmad Dowlatabadi.

```
function Random3(var seed:integer):extended;

  const
     Big=1e9;
     Local_seed=161803398;
     Factor=1e-9;

  var
     N1,N2,counter:integer;
     j,k:real;

  begin
    if (seed<zero) then
        begin
            j:=Local_seed+seed;
            if j>=zero then j:=j-Big*trunc(j/Big)
            elsej:=Big-abs(j)+Big*trunc(abs(j)/Big);
            Bunch[55]:=j;
            k:=1;
            for N1:=1 to 54 do begin
        N2:=21*N1 mod 55;
        Bunch[N2]:=k;
        k:=j-k;
        if (k<zero) then k:=k+Big;
        j:=Bunch[N2]
      end;
      for counter:=1 to 4 do begin
      for N1:=1 to 55 do begin
        Bunch[N1]:=Bunch[N1]-Bunch[1+((N1+30) mod 55)];
        if Bunch[N1]<zero then Bunch[N1]:=Bunch[N1]+Big;
        end;
      end;
      Nexti:=0;
      Nextip:=31;
      seed:=1
    end;
   Nexti:=Nexti+1;
   if Nexti=56 then Nexti:=1;
   Nextip:=Nextip+1;
```

```
    if Nextip>=56 then Nextip:=1;
    j:=Bunch[Nexti]-Bunch[Nextip];
    if j<zero then j:=j+Big;
    Bunch[Nexti]:=j;
    Random3:=j*Factor;
  end;

function Gaussian (var seed:integer):extended;


  var
    Factor,radius,v1,v2:real;


  begin


    if I_set=0 then begin
      repeat
        v1:=2*Random3(seed)-1;
        V2:=2*Random3(speed)-1;
        radius:=sqr(v1)+sqr(v2);
      until radius<1;
      Factor:=sqrt(-2*ln(radius)/radius);
      G_set:=v1*Factor;
      Gaussian:=v2*Factor;
      I_set:=1;
    end
    else begin
      Gaussian:=G_set;
      I_set:=0;
    end;
  end;


BEGIN


  for counter:=0 to arraysize do
    begin
      X[counter]:=0;
    end;
```

```
assign(result,'noise.cir');
rewrite(result);
writeln('Enter the number of random numbers needed.');
readln(numbers);
writeln('Enter an integer for seed.');
readln(seed);
writeln('Enter the duration of simulation');
writeln('(Ex. 1ms = 1e-3)');
readln(maxtime);
if seed>0 then seed:= -1*seed;
if seed = 0 then seed:= -1;
sum:= 0;
mean:= 0;
rms:= 0;
variance:= 0;
time:= 0;
counter:= 0;
I_set:= 0;

(*****************************************************************

   Calculating the mean and standard deviation (rms)
    of the set.

*)

  for counter:= 1 to numbers do
    begin
      X[counter]:= Gaussian(seed);
      sum:= X[counter] + sum;
    end;
  mean:= sum / numbers;
  for counter:= 1 to numbers do
    variance:= variance + sqr(X[counter]- mean);
  rms:= sqrt((variance) / numbers);
  write(result,'*The mean and standard deviation');
  writeln(result,' before transformation are');
  writeln(result,'*mean = ',mean,blank,'rms = ',rms);
  sigma:= 0;
```

```
(*****************************************************

    Calculating new set of numbers based on the
     obtained mean and rms values such that the new
     numbers have mean = 0 and rms = 1.

*)

  for counter:= 1 to numbers do
    X[counter]:= (X[counter]- mean) / rms;

(*****************************************************

    The .Subckt is written here.

*)

  time_step:= maxtime / numbers;
  if (numbers mod 18) = 0
      then voltage_number:= (numbers div 18)
      else voltage_number:= (numbers div 18) + 1;

  for N1:= 1 to voltage_number do begin
    N2:= 0;
    writeln(result);
    neg_node:= index*(N1- 1);
    pos_node:= index*(N1);
    if N1 = 1 then
      begin
      writeln(result,'.Subckt TDNG',blank,positive);
      writeln(blank,negative);
      write(result,'V',N1,blank,pos_node:5,blank);
      write(negative:5);
      end
      else
      if N1 = voltage_number then
      write(result,'V',N1,blank,positive:5,blank);
      write(neg_node:5)
        else
      write(result,'V',N1,blank,pos_node:5,blank,
      write(neg_node:5);
```

```
    write(result,blank,'pwl(0s  0');

   if N1 = 1
     then writeln(result)
     else begin
       writeln(result,blank,time:12,'s',blank,zero);
     end;
   repeat
     N2:= N2 + 1;
     time:= time + time_step;
     if N2 mod 3 = 1 then write(result,' + ');
     write(result,time:12,'s',blank);
     write(X[N2 + (N1- 1)*18]:10:8,blank);
     if N2 mod 3 = 0 then writeln(result);
   until ((N2 + (N1- 1)*18> = numbers) or (N2> = 18));

   if N2 mod 3 = 0 then write(result,' + ');
     next_time:= time + time_step;
     writeln(result,next_time:12,'s',blank,zero,')');

   end;
writeln(result);
write(result, 'RLOAD', blank, positive, blank);
writeln(negative, blank, '1');

writeln(result,'.ends');
close(result);
end.
```

The program is interactive and is run using Turbo Pascal. When the program is executed, the following lines will appear on the screen and prompt the user for information.

```
>Enter the number of random numbers needed.
50
>Enter an integer for seed.
1
>Enter the duration of simulation (Ex. 1 ms = 1e- 3)
5e- 6
```

Note that the numbers 50, 1, and 5e-6 were the numbers used to generate Example 4-4. The output file which is created is identified as "noise.cir" by default in this program. This file must have the additional circuit nodes and rms noise level added as shown in Fig. 4-14.

APPENDIX E

ANSWERS TO SELECTED PROBLEMS

CHAPTER 1

1. 1 kΩ: 895 $nV/Hz^{1/2}$, 4 $\mu V/Hz^{1/2}$, 17.9 $\mu V/Hz^{1/2}$
   50 kΩ: 6.32 $\mu V/Hz^{1/2}$, 28.3 $\mu V/Hz^{1/2}$, 126.5 $\mu V/Hz^{1/2}$
   1 MΩ: 28.3 $\mu V/Hz^{1/2}$, 126.5 $\mu V/Hz^{1/2}$, 565.7 $\mu V/Hz^{1/2}$
2. 8.9 $nV/Hz^{1/2}$
3. 59.5 kHz
4. 100 Hz
5. 28.3 $\mu V/Hz^{1/2}$, 56.6 $pA/Hz^{1/2}$
6. 266×10^{-12} V^2/Hz, 16.3 $\mu V/Hz^{1/2}$
7. $R = 3.5\ \Omega$, $r_e = 2.5\ \Omega$
8. $\Delta f = \pi f_2/8 = 1.15\ f_{-3\,dB}$
9. 933 Hz
10. Almost three decades
11. 375 Hz
14. 941 Hz
15. (a) $R_x = 13.5\ \Omega$, (b) $V_{o1(rms)} = 238\ mV_{rms}$, $V_{o2(rms)} = 149\ mV_{rms}$.
    (c) The S/N ratio for the diode circuit is larger.
16. 637 nV, 20 μV, 31.6 μV
17. (a) 774597, (b) 11.5×10^6

CHAPTER 2

2. $F = 1$, NF = 0 dB; $F = 2$, NF = 3 dB, $F = 3$, NF = 4.77 dB
3. NF = 3 dB
6. (a) $E_{ni} = 1.14\ \mu V$, $E_{no} = 2.34\ \mu V$ (b) $E_{ni} = 1\ \mu V$, $E_{no} = 90.9\ \mu V$
7. (a) $509 \times 10^{-18}\ V^2/Hz$; (b) $T_s = 627$ K; (c) $R_n = 21.6\ k\Omega$
8. $E_{ni}^2 = 4 \times 10^{-16}\ V^2$
9. $E_{ni} = 18.3$ nV
10. 10.32
11. (a) $E_{ni}^2 = 3.92 \times 10^{-15}\ V^2$ (b) $E_{no}^2 = 4.18 \times 10^{-13}\ V^2$
12. (a) $K_t = 25.36$ (b) $E_{no}^2 = 4.68 \times 10^{-12}\ V^2$ (c) $E_{ni}^2 = 7.27 \times 10^{-15}\ V^2$

CHAPTER 3

1. 20.8 nV
2. (a) $A = 100$, $B = 109.9$, $C = 9.9$, $D = 99$K; (b) $E_{no}^2 = 1.09 \times 10^{-11}\ V^2$ (c) $E_{ni}^2 = 1.09 \times 10^{-15}\ V^2$
3. (a) R_1: $4.025 \times 10^{-13}\ V^2$; R_2: $4.15 \times 10^{-13}\ V^2$; R_F: $5.95 \times 10^{-18}\ V^2$; R_L: $0\ V^2$; (b) E_n: $8.26 \times 10^{-13}\ V^2$; I_{n1}: $1.4 \times 10^{-15}\ V^2$; I_{n2}: $1.43 \times 10^{-15}\ V^2$; (c) $E_{no} = 161\ \mu V$
4. I_n: $5.68 \times 10^{-15}\ V^2$
5. $2.475\ k\Omega$
6. (a) $E_{no}^2(R_1) = 2.01 \times 10^{-12}\ V^2$
   $E_{no}^2(R_2) = 1.02 \times 10^{-12}\ V^2$
   $E_{no}^2(R_3) = 1.97 \times 10^{-12}\ V^2$
   $E_{no}^2(R_4) = 1.93 \times 10^{-16}\ V^2$
   $E_{no}^2(R_5) = 1.93 \times 10^{-14}\ V^2$
   (b) $E_{no}^2(E_n) = 4.12 \times 10^{-12}\ V^2$
   $E_{no}^2(I_n) = 1.5 \times 10^{-12}\ V^2$
7. (a) $E_{ni}^2(R_s) = 1.61 \times 10^{-16}\ V^2$
   $E_{ni}^2(E_n) = 1.13 \times 10^{-15}\ V^2$
   $E_{ni}^2(E_z) = 5.06 \times 10^{-14}\ V^2$
   (b) $E_{no} = 1.61$ mV
8. $E_{no}^2(R_{s1}) = 1.55 \times 10^{-12}\ V^2/Hz$, $E_{no}^2(R_{s2}) = 1.55 \times 10^{-12}\ V^2/Hz$, $E_{no}^2(R_1) = 1.74 \times 10^{-15}\ V^2/Hz$, $E_{no}^2(R_2) = 5.8 \times 10^{-17}\ V^2/Hz$, $E_{no}^2(E_n) = 9.61 \times 10^{-14}\ V^2/Hz$, $E_{no}^2(I_{n1}) = 8.65 \times 10^{-13}\ V^2/Hz$, $E_{no}^2(I_{n2}) = 8.66 \times 10^{-13}\ V^2/Hz$, $K_t = 31$

9. $E_{no}^2(R_1) = 1.45 \times 10^{-14}$ V^2/Hz, $E_{no}^2(R_2) = 1.45 \times 10^{-14}$ V^2/Hz, $E_{no}^2(R_3) = 1.45 \times 10^{-14}$ V^2/Hz, $E_{no}^2(R_4) = 1.45 \times 10^{-14}$ V^2/Hz, $E_{no}^2(R_F) = 4.83 \times 10^{-16}$ V^2/Hz, $E_{no}^2(E_{n1}) = 9.0 \times 10^{-14}$ V^2/Hz, $E_{no}^2(I_{n1}) = 3.60 \times 10^{-19}$ V^2/Hz, $E_{no}^2(E_{n2}) = 9.3 \times 10^{-14}$ V^2/Hz, $E_{no}^2(I_{n2}) = 3.6 \times 10^{-13}$ V^2/Hz, $E_{no}^2 = 6.01 \times 10^{-13}$ V^2/Hz
10. $E_{no}^2 = 2.82 \times 10^{-16}$ V^2
11. The inverting amplifier has twice the equivalent input noise as the noninverting configuration.

CHAPTER 4

1. $KF = 6.4\text{E} - 17$, $AF = 1$, $KF = 1.6\text{E} - 16$, $AF = 1$
2. $E_n = 10$ nV/Hz$^{1/2}$, $f_{nce} = 200$ Hz, $I_n = 1.2$ pA/Hz$^{1/2}$, $f_{nci} = 1$ kHz.
3. -37.3 dB
4. Same excess noise. Gain is doubled.
7. 195 Hz
8. (a) $E_{no} = 46.5\ \mu$V; (b) $E_{no} = 397\ \mu$V

CHAPTER 5

1. $E_{ni}^2(R_s) = 1.61 \times 10^{-16}$ V^2, $E_{ni}^2(E_n) - 4.43 \times 10^{-15}$ V^2, $E_{ni}^2(I_n) = 1.16 \times 10^{-17}$ V^2; $E_{ni}^2 = 4.6 \times 10^{-15}$ V^2
2. $E_{ni}^2(R_L) = 1.078 \times 10^{-18}$ V^2
3. $E_s = 11.3$ nV, $E_L = 35.8$ nV, $E_x = 2.83$ nV, $I_{nc} = 6.8$ pA, $\beta = 150$, $I_{nb} = 0.555$ pA, $I_f = 1.754$ pA
4. (a) MPS-A18; (b) 2N4403
5. (a) $E_{ni}^2 = 5.09 \times 10^{-16}$ V^2; (b) $T_s = 354$°C; (c) $R_n = 21.6$ kΩ
7. The *pnp* gives smaller equivalent input noise because the base spreading resistance is lower due to *n*-type material.
8. Amplitude, duration, and frequency of occurrence.
9. $E_n = 3.63$ nV/Hz$^{1/2}$, $I_n = 0.979$ pA/Hz$^{1/2}$, $F_{\text{opt}} = 1.08$ dB, $R_{\text{opt}} =$ 6.9 kΩ

CHAPTER 6

1. $W/L = 578.9$, $I_D = 5.34$ mA
2. (a) $g_m = 317\ \mu$A/V, $I_{nd} = 1.84$ pA/Hz$^{1/2}$, $r_{ds} = 2$ MΩ, $I_{ng} = 0.031$ fA/Hz$^{1/2}$, $I_f = 288$ pA/Hz$^{1/2}$; (b) $R_{\text{opt}} = 29.3$ GΩ; (c) $I_D = 125\ \mu$A, $V_{GS} = 3.577$ V, $V_{DS} = 5$ V.

3. (b) $F = 1.0086$, NF = 0.0371 dB
4. (a) 2N3821 (b) 2N3631
5. (a) 4.88 nV/Hz$^{1/2}$ (b) 1.6 kHz (c) $I_D = 16$ mA, $V_{GS} = 19.9$ V
6. $R_{\text{opt}} = 3.33$ MΩ, $E_{ni}^2 = 2.59 \times 10^{-15}$ V^2/Hz
7. 2.31 nV/Hz$^{1/2}$, 56.6 fA/Hz$^{1/2}$

CHAPTER 7

1. (a) $\frac{2}{3}$ (b) $\frac{1}{3}$ (c) $\frac{5}{6}$ (d) $\frac{10}{11}$
2. 4.1 μV
3. NF = 1.186 dB at resonance; NF = 19.71 dB at 3 kHz
4. (b) $E_{ni} = 2$ μV at 3 kHz, $E_{ni} = 16.7$ nV at 10 kHz, $E_{ni} = 183.3$ nV at 30 kHz
5. (a) 678.4; (b) $E_{ni}^2(R_s) = 1.61 \times 10^{-16}$ V^2, $E_{ni}^2(E_n) = 1.6 \times 10^{-17}$ V^2, $E_{ni}^2(I_n) = 1.28 \times 10^{-16}$ V^2; $E_{ni} = 17.43$ nV; (c) $\Delta f = 275$ kHz, $E_{no} = 6.24$ mV

CHAPTER 8

2. (a) $I_{sh} = 98$ fA/Hz$^{1/2}$, $I_f = 30$ nA/Hz$^{1/2}$ (b) 74.9 kHz (c) $E_n = 600$ nV (d) Area of the diode.
3. (a) 850.7 kHz (b) 241.3 μV
4. (a) $f_{os} = 159.15$ kHz, $f_{op} = 245.6$ kHz (b) $I_{ni} = 2.05$ μA

CHAPTER 9

1. $R_{\text{opt}} = 3.33$ MΩ, $E_{ni} = 131$ nV/Hz$^{1/2}$
2. $E_n = 4$ nV/Hz$^{1/2}$, $I_n = 0.9$ pA/Hz$^{1/2}$, $R_{\text{opt}} = 4.44$ kΩ, $E_{ni} = 53.3$ nV/Hz$^{1/2}$, NF = 5.48 dB
3. $T = 5$, $V_2 = 10$ mV
4. (a) Choose $I_C = 10$ μA, $E_{ni} = 7.38$ μV, NF = 0.546 dB; (b) choose $I_C = 30$ μA, $E_{ni} = 7.19$ μV, NF = 0.295 dB.

CHAPTER 10

1. (b) C_E, 5.96 Hz (c) 21.23 kHz
2. (b) C_E, 2.083 Hz (c) 54.35 kHz
3. (b) C_E, 1.27 Hz (c) 14.67 kHz

4. $E_{ni}^2 = 1.55 \times 10^{-16}$ V^2, R_s dominates at 10 Hz
   $E_{ni}^2 = 1.35 \times 10^{-16}$ V^2, R_s dominates at 10 kHz
5. (a) $E_{ni} = 37.3$ nV at 10 Hz, $E_{ni} = 6.3$ nV at 10 kHz
   (b) $E_{ni} = 30.32$ nV at 10 Hz, $E_{ni} = 6.3$ nV at 10 kHz
6. $E_{ni}^2 = 1.93 \times 10^{-14}$ V^2 at 10 Hz, R_s dominates
   $E_{ni}^2 = 1.63 \times 10^{-14}$ V^2 at 10 kHz, R_s dominates
7. $E_{ni}^2 = 6.5 \times 10^{-11}$ V^2 at 10 Hz, R_s dominates
   $E_{ni}^2 = 6.71 \times 10^{-14}$ V^2 at 10 kHz, R_s dominates
8. $E_{ni}^2 = 2.31 \times 10^{-17}$ V^2

CHAPTER 11

3. (a) $V_m = V_{\text{ref}}(m/2^N)$; (b) $V_m = V_{\text{ref}}[(2m - 1)/(2^{N+1} - 1)]$; (C) $V_m = V_{\text{ref}}[(2m - 1)/(2^{N+1} - 2)]$

CHAPTER 12

1. $E_T = 5.9\ \mu$V
2. NI $= -12.88$ dB, $E_{ex} = 598.7 \times 10^{-9}$
4. $I_{\text{DC}} = 50.24\ \mu$A

CHAPTER 13

1. For $W_n/L_n = 4$, $W_p/L_p = 12.77$
2. E_{no}(diode) $= 463.3$ pV/Hz$^{1/2}$,
   E_{no}(transistor) $= 1.374$ nV/Hz$^{1/2}$

CHAPTER 15

2. NF $= 7.96$ dB

APPENDIX F

SYMBOL DEFINITIONS

To designate electronic circuit quantities the system of symbols used in this book conforms to standard practice in the semiconductor field wherever possible, unless this adds ambiguity or creates confusion.

1. Dc values of quantities are indicated by capital letters with capital subscripts (I_B, V_{DS}). Direct supply voltages have repeated subscripts (V_{BB}, V_{CC}).
2. Rms values of quantities are indicated by capital letters with lowercase subscripts (I_n, V_s).
3. The time varying components of voltages and currents are designed by lowercase letters with lowercase subscripts (g_m, r_d).

SYMBOLS

Symbol	Definition	First Used in Equation
a	Number of bursts per second	5-38
A_b	Buffer voltage gain	13-1
A_v	Voltage gain	1-9
A_{vo}	Peak voltage gain	1-9
β	Feedback factor	3-1
BF	Bit factor	11-2
β_o	Short circuit current gain in BJT	5-1
BW	Bandwidth	5-39
C	Capacitance	1-56

Symbol	Definition	First Used in Equation
C	Correlation coefficient	1-31
C_{eq}	Reflected capacitance	6-13
C_{gd}	Gate to drain capacitance	6-6
C_L	Load capacitance	6-6
C_μ	Collector–base capacitance	5-5
C_{ox}	Gate oxide capacitance	6-16
C_π	Base–emitter capacitance	5-5
C_p	Shunt capacitance	7-10
Δf	Noise bandwidth	1-3
ϵ	Meter error	15-28
E_a	Activation energy	6-39
E_A, E_B, E_C	Noise voltage of resistors R_A, R_B, R_C	10-7
E_C	Noise voltage of coupling network	7-1
E_f	$1/f$ noise voltage	1-47
E_n	Equivalent input noise voltage	2-7
E_{n1}, E_{n2}	Noise voltage of stage 1 and 2	3-16
E_{ni}	Equivalent input noise voltage	2-7
E_{no}	Output noise voltage	1-27
E_{ns}	Noise of source resistance	10-5
E_{nT}	Total input noise voltage generator	10-33
E_π	Noise voltage drop across r_π	5-13
E_p	Noise voltage of popcorn noise	15-32
E_{ref}	Noise voltage of reference	11-17
E_{sh}	Shot noise voltage	1-55
E_t	Thermal noise voltage of a resistance	1-6
E_{tp}	Thermal noise voltage of r_π	3-16
E_u	Noise voltage of unregulated supply	13-1
E_x	Thermal noise voltage of base spreading resistance	5-7
E_Z	Noise voltage generator of Zener diode	12-12
F	Noise factor	2-9
f	Frequency	1-9
f_2	−3-dB high-frequency corner	1-15
f_a	System −3-dB cutoff frequency	1-17
f_h	Upper frequency limit	1-45
f_L	Noise corner frequency	5-11
f_l	Lower frequency limit	1-45
F_{opt}	Noise factor at R_o	2-14
f_T	Current cutoff frequency	5-5
G	Conductance in siemens	1-25
γ	Exponent in $1/f$ model	5-34
G_a	Available power gain	2-19
G_d	Channel conductance	6-35

Symbol	Definition	First Used in Equation
g_{ds}	Output conductance	6-18
g_m	Transconductance	5-1
G_o	Peak power gain	1-8
I_B	Base current	5-8
I_{bb}	Burst noise current	5-37
I_C	Collector current	5-2
I_D	Dc drain current	6-15
I_d	Noise current of a diode	4-3
I_{DC}	Direct current in amperes	1-50
I_{DQ}	Quiescent drain current	6-26
I_E	Emitter current	1-51
I_f	$1/f$ noise current	5-10
I_{GSS}	Dc reverse saturation gate current	6-33
I_n	Equivalent input noise current	2-7
I_{n1}, I_{n2}	Noise current of stage 1 and 2	3-16
I_{nb}	Shot noise of base current	5-8
I_{nc}	Shot noise of collector current	5-9
I_{nd}	Drain noise current generator	6-24
I_{ng}	Gate noise current generator	6-23
I_{no}	Output noise current	5-13
I_{np}	Thermal noise current of R_p	7-7
I_{nT}	Total input noise current generator	10-34
I_S	Reverse saturation current	1-51
I_{sh}	Shot noise current	1-50
I_t	Thermal noise current of a resistance	1-25
I_{tot}	Total noise current	15-26
k	Boltzmann's constant (1.38×10^{-23} W-s/K)	1-3
K_1	Dimensional constant	1-45
K_{cm}	Common mode gain	10-38
K_{dm}	Differential gain	10-36
K_p	Transconductance parameter	6-15
K_r, K_b	Voltage gains	13-1
K_t	System transfer gain	2-1
K_{tr}	Reflection coefficient	6-28
L	Gate length	6-15
λ	Channel length modulation parameter	6-15
L_{eff}	Effective channel length	6-26
L_p	Parallel inductance	7-13
LSB	Least significant bit	11-28
m	Excess noise constant	12-2
μ	Mean or average value	1-1
μ_o	n–channel mobility	6-16
N	Number of bits	11-1

Symbol	Definition	First Used in Equation
N	Number of stages	10-45
NEP	Noise equivalent power	8-2
NF	Noise figure in dB	2-11
N_f	Noise power	1-45
NI	Noise index	12-5
N_t	Available noise power	1-3
P	Noise coefficient	6-35
P_E	Error probability	11-28
q	Electronic charge (1.602×10^{-19} Coulombs)	1-50
R	Resistance, real part of Z	1-6
r'_x	$1/f$ base spreading resistance	5-12
R_B	Bias resistance	8-1
R_C	Collector resistance	10-1
r_d	Diode dynamic resistance	12-12
R_{DRAIN}	Equivalent drain noise resistance	6-25
r_{ds}	Output resistance	6-18
R_E	Emitter resistor	10-1
r_e	Shockley emitter resistance	1-54
R_{eq}	Reflected resistance	6-10
R_F	Feedback resistance	11-16
R_L	Load resistance	1-5
R_n	Equivalent noise resistance	2-16
R_o	Optimum noise source resistance	2-13
R_p	Parallel resistance	3-9
r_π	Base–emitter resistance	5-1
r_r	Output resistance of reference	13-1
R_S	Resistance of signal source	1-5
R_{TH}	Thevenin resistance	11-19
r_x	Base spreading resistance	5-7
σ	Standard deviation	1-1
S/N	Signal to noise ratio	7-3
S_f	Spectral density of noise voltage	1-47
T	Temperature in kelvins (K)	1-3
T	Transformer turns ratio	8-5
τ	Trapping time constant	6-38
T_p	Period	1-2
T_s	Equivalent noise temperature	2-17
V	Voltage (not noise voltage)	1-32
V_{BB}	Bias voltage	8-1
V_{BE}	Base–emitter voltage	1-51
V_C	Common mode signal	10-28
V_{CC}	Collector supply voltage	10-8
V_D	Differential difference signal	10-28

Symbol	Definition	First Used in Equation
V_{DS}	Drain to source voltage	6-15
V_{ex}	Excess noise voltage	4-6
V_{GS}	Gate to source voltage	6-15
V_{in}	Input signal voltage	2-1
V_{ref}	Reference voltage	13-4
V_{rms}	Rms voltage	1-2
V_s	Signal voltage at source	5-15
V_{so}	Output signal voltage	2-1
V_T	Threshold voltage	6-15
W	Gate width	6-15
ω	$2\pi f$	1-56
X_C	Capacitive reactance	10-1
Y_1	Input admittance	6-8
Z_C	Impedance of coupling network	7-2
Z_{in}	Input impedance	2-2
Z_s	Series impedance	8-3

PREFIXES

The following prefixes are used to indicate decimal multiples or submultiples of units:

Multiple	Prefix	Symbol
10^{12}	tera	T
10^{9}	giga	G
10^{6}	mega	M
10^{3}	kilo	k
10^{-3}	milli	m
10^{-6}	micro	μ
10^{-9}	nano	n
10^{-12}	pico	p
10^{-15}	femto	f

Thus 1.7 GΩ is an impedance equal to $1.7 \times 10^{9}\ \Omega$, and 6.0 fA is a current of 6.0×10^{-15} A.

INDEX